Agitator Design for Gas–Liquid Fermenters and Bioreactors

Agitator Design for Gas–Liquid Fermenters and Bioreactors

Gregory T. Benz

Benz Technology International, Inc.,
Clarksville, OH, USA

WILEY

AIChE
The Global Home of Chemical Engineers

A Joint Publication of the American Institute of Chemical Engineers and John Wiley & Sons, Inc.

Published by John Wiley & Sons, Inc., Hoboken, New Jersey.
Published simultaneously in Canada.

Registered Office
John Wiley & Sons, Inc., 111 River Street, Hoboken, NJ 07030, USA

Editorial Office
John Wiley & Sons, Inc., 111 River Street, Hoboken, NJ 07030, USA

For details of our global editorial offices, customer services, and more information about Wiley products visit us at www.wiley.com.

Wiley also publishes its books in a variety of electronic formats and by print-on-demand. Some content that appears in standard print versions of this book may not be available in other formats.

Library of Congress Cataloging-in-Publication Data

Names: Benz, Gregory T., author.
Title: Agitator design for gas–liquid fermenters and bioreactors / Gregory T. Benz, Benz Technology International, Inc., Ohio, US.
Description: First edition. | Hoboken, NJ, USA : Wiley, 2021. | Includes bibliographical references and index.
Identifiers: LCCN 2020051152 (print) | LCCN 2020051153 (ebook) | ISBN 9781119650492 (hardback) | ISBN 9781119650508 (adobe pdf) | ISBN 9781119650539 (epub)
Subjects: LCSH: Bioreactors–Equipment and supplies. | Fermentation–Equipment and supplies. | Mixing machinery–Design and construction. | Gas-liquid interfaces.
Classification: LCC TP248.25.B55 B46 2021 (print) | LCC TP248.25.B55 (ebook) | DDC 660/.28449–dc23
LC record available at https://lccn.loc.gov/2020051152
LC ebook record available at https://lccn.loc.gov/2020051153

Cover Design: Wiley
Cover Image: © Courtesy Gregory T. Benz

MIX
Paper from responsible sources
FSC® C013604

I dedicate this book to my late father-in-law, Richard Durchholz, for inspiring me as an engineer and a person; to Wayne Ramsey, for mentoring me and giving me the opportunity to design the largest fermenters built by Chemineer up to that point; to Ms. Jian Li, my colleague and friend, for helping me to succeed in managing the China office and understanding Chinese culture, and my wife, Kim Benz, for encouraging me and supporting me in the massive undertaking of writing this book.

Contents

Preface

This is a book about fluid agitation, as applied to gas–liquid systems such as fermenters or bioreactors (We will use those terms interchangeably in this text.). The specific focus is on mechanically agitated systems, consisting of a closed vessel with a rotating shaft and impellers, as this is the most common and versatile way to achieve process objectives in a gas–liquid system. Though airlift and bubble columns have also been used, they will not be discussed in any detail here, as that is not the focus of this book.

Many books have been written about fluid agitation. Many books have also been written about fermentation. Much, though not all, of the material in this book has been covered in such books. However, all such books cover much more than agitator design for bioreactors. For example, typical books on agitation cover topics such as solids suspension (almost never an issue in fermentation), highly viscous systems (>50 000 cP), specialized impellers such as helical ribbons, anchors, augers, and others that have no use in fermenters, mixing in high-yield stress fluids such as paper stock, etc. Likewise, books on fermenter design usually cover some topics on agitator design but also cover feeding strategies, reaction kinetics, cell metabolism, sensitivity to concentration and temperature changes, product recovery, and a whole host of other topics. Little has been published in such books about how to acquire the proper pilot data for agitator design, or how to minimize energy consumption.

The main purpose of this book is to be a single-source reference on all the major issues related to agitator design for bioreactors. It is intended to save the reader time by avoiding the need to consult multiple references or sift through many pages of text to find what is needed specifically for fermenter agitator design. This book will also cover important related topics such as heat transfer, power cost, basic agitator mechanical design, and vendor bid evaluation.

Though some introductory fundamental theory is included, the main focus is on practical application of theory to real-world agitator design. This book is more of a how-to book than an academic treatise. The relative brevity of the book is intentional. It is hoped that the brevity will encourage people to actually read the entire book, not just skim an occasional page or chapter.

This book is intended to be useful for a variety of people. Since it is primarily a technical document, most readers will have a science or engineering degree. Many will be Chemical Engineers. Some will be chemists or microbiologists tasked with operating facilities in a way that can produce scalable data. Academic degrees among readers will vary from Bachelor up through Post-Doc. Most readers will be employed by companies using bioprocessing to make valuable products as well as many making commodity products. Some will work for agitator manufacturers. If used as a course supplement, some will be college students or professors. Top-level managers may want to skim the contents to make sure their teams are properly staffed and have a high-level view of what their team should be doing. They will find the overview and flow chart described in Chapter 2 especially useful. Chapters on energy use optimization will also be of interest to business unit managers. Information on bid evaluation should be of interest to procurement professionals. Although written primarily for users of agitation equipment and operators of fermentation facilities, engineers employed by agitator manufacturers will likely find it of interest as it provides a deeper window into the details of these applications than they are accustomed to, as well as how their bids may be viewed in a competitive environment.

A note about symbols: rather than make the reader refer to a list of symbols in the appendix, each chapter has the symbols used in that chapter at the end. That should save the reader some time. Also, it lets the author use the same symbol for different purposes in different contexts, reducing the number of symbols needed. For example, C means off bottom impeller clearance in most cases, but in the context of mass transfer correlations, it is used as an exponent, and it can also mean dissolved gas concentration.

Most of the book is focused on gas–liquid agitation, as that is the controlling parameter for most bioreactors. By that I mean the agitator is primarily designed to disperse gasses into liquids. This does not mean evolving gas from solution, which is a separate case. The fundamentals presented are applicable to other processes as well, such as miscible liquid blending, but design procedures for these problem categories are not presented here.

Gregory T. Benz
Benz Technology International, Inc.

Foreword

Genetic modification, microbiome, green technology, renewable fuels and chemicals, bio-degradable plastic, pandemic recovery, prebiotics, probiotics, agricultural biologics, world food shortage, meatless meat, animal free dairy, human and animal health. What do these important concepts have in common? They all rely on the use of bioreactors to realize the ultimate benefit to current and future generations.

The most powerful of these products utilized in human and animal health can generate the world supply in quantities measured in pounds. Vaccines, antibiotics, probiotics, prebiotics, and others have a large portion of their cost included in research and development, clinical trials, and regulatory approval processes that bring challenge to this business space. In these cases, the bioreactors capital and operational cost impact to the cost of goods sold is small compared to the margins and returns of a successful product launch. These applications historically required a focus on agitation and reactor design with a focus on functionality versus a minimization of operating cost. These products are apportioned in quantities measured in microgram to gram quantities with price measured in millions of dollars per pound in some cases.

On the other end of the spectrum are commodity products utilized every day in quantities measured in tens to hundreds of millions of tons per year. Fuel, polymers, industrial chemicals, animal feed ingredients, and the like. These products' sales prices are measured in pennies to dollars per pound and operate on tight margins. Making these products in bioreactors is more challenging as a result requiring a focus on things such as reactor design, power optimization between the agitator and air compressor can be a competitive advantage or define the success or failure of a venture.

The teams I worked with directly had the pleasure of working with Greg Benz for the past 15 years on commodity products. From development to commercialization, the details of reactor design mattered significantly in these projects. The information provided in this book allowed the proper questions to be asked during process design. Bench, pilot, and demonstration trials were designed to be commercially applicable as a result. This allowed for realistic process design, rate, titer, and yield demonstrations to be applied to financial and process modeling early in the process. It also prevented mistakes that saved hundreds of thousands of dollars through effective understanding prior to spending significant development dollars.

Our team worked with the smallest start-ups to the largest most established biotech companies in the world as a contract research and manufacturing operation. Each time agitation questions are asked, Greg is the go-to expert that everyone already knows and has positive experiences with. Greg's knowledge and experience in this area is of significant importance to realizing the benefit of modern biological technology. I am happy to see that he has decided to put his knowledge and experience in a more detailed writing as I have referenced his course materials hundreds of times in the past 15 years. Thank you to Greg, the biotechnology industries favorite "Professional Agitator."

Jeremy Javers PhD
St. Joseph, MO
1 September 2020

Foreword for Greg Benz

Bioreactor agitator engineering is a broad mosaic. The image is simple and clear from a distance, but as the viewer moves closer, a multitude of distinct individual pieces come into view. Likewise, several diverse disciplines converge in this specialized field: microbiology, transport phenomena, machine design, metallurgy, and reliability engineering. During a project, this list is expanded to include manufacturing and procurement. For the practitioner, the challenge is significant. What information is important? What solutions are time-tested? What are the common pitfalls? How should all of these pieces be assembled into a unified design?

There are many books and articles available on the design of agitators and bioreactors. However, when the time comes to prepare drawings and make purchases for an actual project, it becomes apparent that those resources are missing large swaths of practical information to guide the reader's design choices. How are bioreactor agitators designed in real life? This comprehensive book addresses both the broad background and the small details needed to deliver a good project, from design through delivery.

I was excited to learn that Greg Benz was writing this book. We have worked together for many years designing equipment for bioprocessing facilities, from cellulosic ethanol to enzyme production to hydrogen-rich gas fermentation. He has been a trusted mentor and a patient teacher.

Greg is an accomplished practitioner, a true craftsman. His career has spanned the full scope of the design, manufacturing, and operation of mixing systems, with a special focus on gas–liquid systems for bioreactors. Through his years at Chemineer, and later as a well-known and respected mixing consultant, he has perhaps overseen more bioreactor agitator designs than anyone in the field. His expertise helped to establish industrial biotechnology as a mature industry.

During our years working together, Greg has offered insight on many questions not generally answered in fermentor design books, such as: What is the best way to seal a shaft? What is better: small, fast agitators or big and slow? What are the most common failure modes? Is metal surface polishing really necessary in comparison with other contamination sources? How much polish? What are the most common failure modes? How much overdesign should be included? Bubble columns versus stirred tanks? What are the latest innovations? How does fed-batch impact agitation design? What information should we gather at pilot scale to ensure commercial-scale success? How should the fermenter be controlled to maintain a dissolved oxygen level: vary the air or vary the motor speed? How do agitation performance and power draw change if the mixer is on speed control? How are baffles designed? How do we clean underneath an impeller? How can thermal expansion be handled during cleaning and steam-in-place? What heat transfer coefficient should we expect from internal coils? External jackets? What vendors are reliable? How do we install this equipment, anyway?

Until now, answers to these questions have been difficult to find, making this book a treasure trove for a practicing engineer. Additionally, this valuable information will fuel the progress of biotechnology, which provides food and energy resources to people around the world.

Few engineers possess Greg's wealth of expertise and fewer still take the time to meticulously summarize their knowledge for the benefit of future generations. That he did so makes me very glad.

Keith Flanegan, P.E.
IdeaCHEM, Inc.
September 2020

1

Purpose of Agitator Design

The purpose of using the agitator design principles in this book is to ensure, to the extent possible, that the user of agitation equipment achieves the process objectives and does so in a reliable and economical manner.

Agitators are employed in many different industries. The process results/objectives desired from the agitators vary by industry and by application within each industry. Since an agitator is ultimately nothing more than a kind of pump, and the agitated tank is essentially a deadheaded pump, it would be ideal if the objectives could be stated in purely physical terms, mostly related to flow and head. For example, some would describe agitation in terms of pumping capacity, characteristic fluid velocity [1], G-value [2], or other physical terms.

Some process results correlate well with simple physical measurements of agitation. For example, the ability to overcome density differences or viscosity ratios correlates well with characteristic fluid velocities [1]. However, many other process objectives do not correlate well with such simple measures. Examples of process results that have complex relationships to agitation and do not correlate well with pumping capacity, fluid velocity, or other simple measures would include blend time, mass transfer rate, heat transfer rate, off-bottom solids suspension, solids suspension degree of uniformity, solids suspension cloud height, rate of particle attrition or shear damage, dissolved oxygen spatial distribution, reaction rate, reaction product distribution, and many others.

Since this book is about agitator design for fermenters/bioreactors, we will focus on the attributes of agitator design most important for those applications. The most important process result is normally the mass transfer rate (MTR), often called the OTR, or oxygen transfer rate, when oxygen is the species being transferred. This is generally the dominant design requirement.

The mass transfer rate depends on more than just agitation, of course. It also depends on the airflow, the properties of the broth, the organism's ability to absorb the transferred gas (OUR, or oxygen uptake rate for aerobic systems), and a host

Agitator Design for Gas–Liquid Fermenters and Bioreactors, First Edition. Gregory T. Benz.

of other factors. The principle agitation parameter for a given system is the power invested under gassed conditions. Therefore, the principle purposes of agitator design in this book are enumerated below and expanded upon in subsequent chapters. In most chapters, we will describe results based on the gas being oxygen. Chapter 11 will delve into how to handle other gasses.

- Provide sufficient power input to facilitate the required mass transfer rate. This will vary with tank geometry, scale of operation, pressure, temperature, allowable minimum dissolved gas concentration, and gas flowrate.
- Use an impeller system designed to maximize fluid mixing and thereby minimize concentration gradients, while still dispersing gas.
- Provide sufficient overall mixing. Usually, the agitation required to disperse gas is more than ample for other mixing requirements.
- Optimization of power used. The same mass transfer rate can be achieved with different combinations of airflow and agitator power. The total power of agitator and compressor goes through a minimum. Ideally, the design should use that minimum unless other factors override this desire.
- Optimization of capital cost. Within a certain design power, there is a range of acceptable agitator designs. But there can be differences in capital cost among different designs.
- Optimization of total batch cycle energy costs. Since batch processes have different OTR requirements at different stages of the batch cycle, the power costs can be optimized at each stage, thereby minimizing total energy used per batch.
- Optimization of total system economics. Tank geometry affects capital and energy costs of both the tank itself and the agitator
- Assure the final design has the utmost in mechanical integrity. This includes the tank and the mounting arrangement. Historically, agitators for gas–liquid contacting have had higher mechanical failure rates than those used for simple liquid blending, yet the cost of downtime can be very high. We aim to remedy that by promoting design principles that lead to minimal downtime.
- Choose vendors that not only build a good product, but can support it in the field.

References

1 Hicks, R.W., Morton, J.R., and Fenic, J.G. (1976). How to design agitators for desired process response. *Chemical Engineering Magazine*: 22–30.

2 Benz, G.T. (2007). The G-value for agitator design: time to retire it? *Chemical Engineering Progress* 103: 43–47.

2

Major Steps in Successful Agitator Design

This chapter presents an overview of the main steps and logic required to achieve the best agitation system design. Subsequent chapters will provide more technical details and fundamental concepts so that each step can be undertaken. Figure 2.1 provides a graphic summary of these steps. We will describe each one in more detail in the following paragraphs. The flow chart concept used here was inspired by the procedures in Ref. [1], but is expanded upon in more detail here specifically for bioreactor design.

Define Process Results

The first step in agitator design, or, for that matter, the design of any kind of process equipment, is to define the expected process result. For agitators, that could be a number of different things, such as degree of solids suspension, blend time to some specified degree of uniformity, characteristic fluid velocity, heat transfer coefficient, etc. While some or all of these process results may be needed or applicable to bioreactor design, in general, the requirement for a certain mass transfer rate is the most important and difficult to achieve. In other words, when an agitator is designed for mass transfer, the other process requirements are normally exceeded.

There are two exceptions to this. One is when the mass transfer requirement is very low (say, less than 10 mmol/l-h). This is sometimes called micro-aeration. In such a case, there may be minimum liquid velocities or blend time requirements. However, we feel that such cases are covered well in the general literature, such as in Refs. [1,2]. Therefore, we will not describe agitator design where velocity or blend time is the required results for low viscosity liquids. By "low viscosity," we typically mean that the viscosity is less than 1000 cP. Viscosities less than 1000 cP typically have little effect on power draw or blending performance. However, heat

Agitator Design for Gas–Liquid Fermenters and Bioreactors, First Edition. Gregory T. Benz.

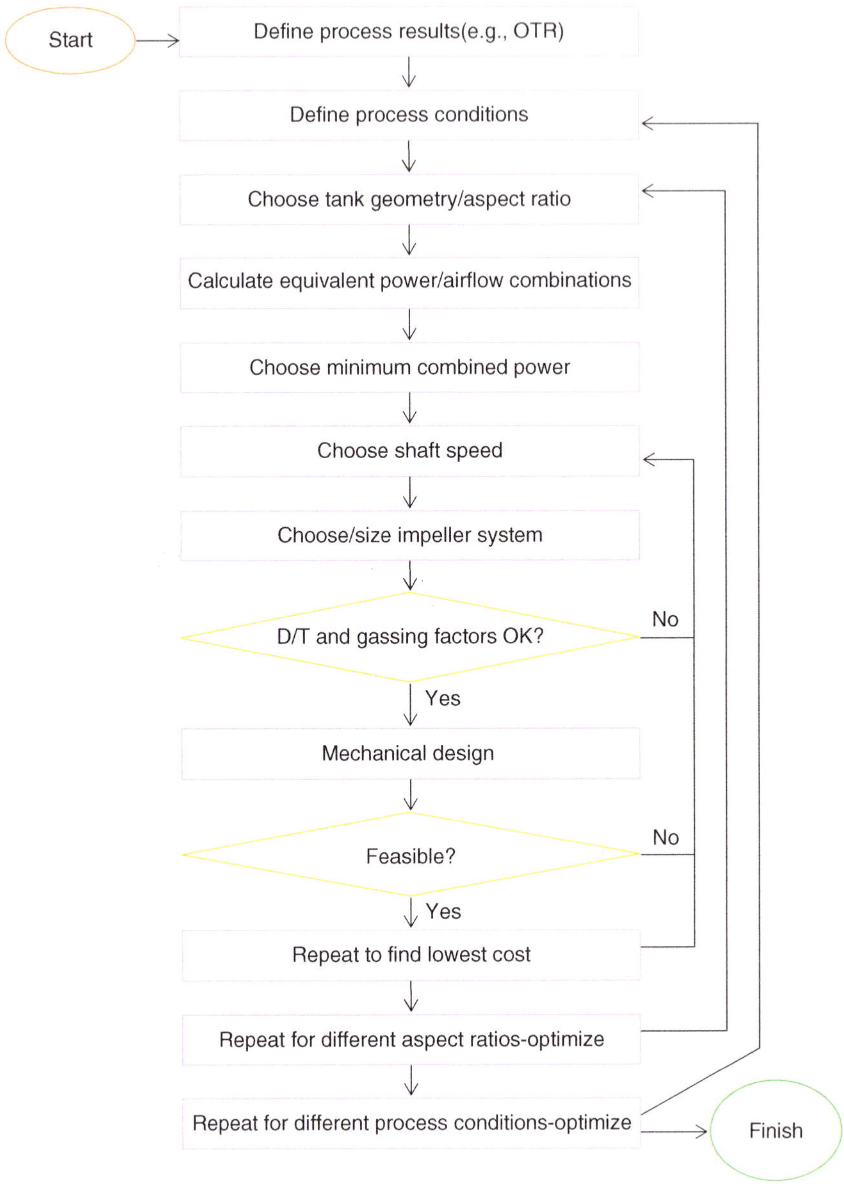

Figure 2.1 Agitator design flow chart.

transfer is affected at all viscosities, and mass transfer is affected when viscosity gets above approximately 50–80 cP.

The other exception is fermentation of highly viscous liquids, such as Xanthan gum or Gellan gum. At peak concentrations in the broth, such materials may have viscosities at a shear rate of 1 per second of 30 000 cP or even higher and apparent viscosities at the impeller of 2 000–10 000 cP. They are also quite non-Newtonian. We will describe some viscosity models and effects in Chapter 3 and specific issues with viscous fermenter design in Chapter 12.

With the foregoing in mind, the first step in our flow chart, defining process results, will focus on the required mass transfer rate, MTR. Since most fermenters consume oxygen, and the feed gas is air, most of this book will use aerobic fermentation with air feed for examples and calculations. So, we will usually refer to the mass transfer rate as the OTR, or oxygen transfer rate. Units are normally either mass per volume–time or moles per volume–time. The most common units of this type are mg/l-h or mmol/l-h. Relatively speaking, an "easy" fermentation would have an OTR of less than 100 mmol/l-h, an "average" one would have around 150–200, and a difficult one would be 300 and up. There are huge implications on equipment size and power costs at these different levels.

Because mass transfer correlations are generally no more accurate than about ±30% when developed for the actual broth and can be much greater in error if generic, published correlations are used, the design OTR should be increased over the required OTR by a suitable factor.

Chapter 11 will deal with cases where the feed gas is not air. For such cases, it may not be possible to optimize power the way we present it in this flow chart, as the cost of the feed gas is not just the power required to deliver it to the tank, and there may be other process constraints.

Note that evolving gas from solution is a separate issue from dispersing gas. Evolving gas is already dispersed, though it affects power draw, performance, and mechanical behavior in a similar way to dispersing gas.

Define Process Conditions

All conditions impacting the agitator design, the mass transfer rate, and ancillary functions such as heat transfer must be delineated. A partial list follows:

- Fluid density (e.g. specific gravity or density such as kg/m^3)
- Fluid viscosity (e.g. cP or Pa-s). If the fluid is non-Newtonian, the model and parameters describing it must be included
- Operating temperature
- Mean barometric pressure
- Back pressure in vapor space
- Heat capacity of the process fluid at constant pressure, C_P. Sample units J/g-C or BTU/lb-F
- Thermal conductivity of the process fluid, k. Sample units J/h-m-C or BTU/h-ft-F.

- For the ancillary heat transfer calculations, need the above thermal properties for the heat transfer medium as well
- For the heat transfer calculations, need the available flow and temperature of the heat transfer medium
- Concentration of oxygen (or other species to be transferred) in the feed gas (usually, we will assume air at 21% oxygen)
- Henry's law constant, or alternatively, saturation concentration at feed gas concentration at process temperature at a reference pressure (typically 1 atm)
- Minimum required dissolved oxygen concentration for organisms to thrive
- Maximum allowable CO_2 concentration in the exit gas, either as mole fraction or as actual partial pressure
- Any other process constraints affecting design

Choose Tank Geometry

In principle, many tank shapes can be used. That can include cylindrical, rectangular, and spherical tanks. However, odd-shaped tanks may be hard to baffle and agitate properly. Rectangular tanks may be harder to clean and sterilize if they have sharp corners. Most tanks used in this industry are cylindrical. Most are mounted with their axis vertical. However, the author knows of at least one installation where a multitude of horizontal cylindrical tanks were used. This is decidedly not recommended, for a host of reasons. Just a few worth mentioning: Multiple gear drives per tank are required, increasing agitator cost. Very random hydraulic forces occur, causing more frequent mechanical failure. Low absolute liquid height fails to take advantage of higher oxygen solubility at the bottom of the tank due to liquid head. Harmonic flows in the sparge system can occur.

For the purposes of this book, we will stick to vertical cylindrical tanks. With that restriction, the geometry to be decided is the ratio of liquid height to tank diameter (Z/T), often referred to as aspect ratio. Fermenters have been built with a wide variety of aspect ratios, for various reasons. The most common or popular designs normally have aspect ratios between 2 and 3, but that may be more related to tradition than because it is optimum for a particular set of circumstances.

Chapter 19 discusses aspect ratio in detail, in terms of its effect on capital and power costs, which provides an opportunity for optimization. However, we have to begin somewhere to go through the rest of this flow chart. In the absence of any restrictions on geometry, an aspect ratio between 2 and 3 seems reasonable to start with.

There can be restrictions on aspect ratio due to building constraints. If the vessel must fit within a given floor space, that may place restrictions on the diameter, forcing a certain minimum aspect ratio. Sometimes, local building codes carry height restrictions. So, the allowable aspect ratio range may be bound by such constraints, among others.

Calculate Equivalent Power/Airflow Combinations for Equal Mass Transfer Rate

It is possible to achieve the same mass transfer rate using a small amount of air and a lot of agitator power, or a lot of air with low agitator power, and an infinite number of steps between.

There are upper and lower airflow limits, however. The minimum airflow is where the OTR is stoichiometrically balanced; that is, the molar flow of oxygen in the incoming air stream exactly matches the molar rate of consumption. In other words, this would require 100% mass transfer. That would require an infinite amount of agitator power!

The upper bound would be when either the vessel or the agitator is flooded. We will define these conditions in Chapters 4 and 8. Suffice it to say for now that there is such a thing as too much airflow.

To keep calculation effort reasonable, the calculations should be performed incrementally: say, starting with 25% more than the minimum and stepping in about 5-10% increments on that, up to the maximum.

Chapter 8 describes this procedure in detail.

Choose Minimum Combined Power

From the above combinations of agitator power and airflow, calculate the agitator brake power, which is the shaft power divided by the mechanical efficiency of the agitator, which includes gear drive losses as well as seal losses. Calculate the compressor brake power, including pressure losses through lines, filters, and sparger as well as compressor efficiency. Add the agitator and compressor brake power together and choose the combination with the lowest total, unless other constraints govern, such as CO_2 in the exhaust gas.

Choose Shaft Speed; Size Impeller System to Draw Required Gassed Power

It is possible to invest the same power at different shaft speeds by using different impeller sizes. In essence, high shaft speeds use smaller impellers and low shaft speeds use larger impellers. There are process, mechanical, and cost implications to this decision, as described in Chapters 6 and 17 and Ref. [2]. So, when we choose an initial speed, we may have to go back and choose another, as in the decision diamonds in Figure 2.1.

Though not all gear drives match the American Gear Manufacturer's Association (AGMA) standard speeds, they are a good place to start, prior to engaging in detailed mechanical design. The speeds relevant for agitator design, in rpm, are 30, 45, 56, 68, 84, 100, 125, 155, 190, 230, 280, and 350 rpm. Laboratory units may have considerably higher speeds than this range.

Decision Point: D/T and Gassing Factors OK?

As described in Chapter 6, D/T has effects on power, performance, and gassing factors (gassing factor is the ratio of power draw in the gassed condition to that in the ungassed condition). For example, we have found that designs requiring a D/T of more than 1.0 are unlikely to be successful. In general, smaller D/T ratios have less impact of gas flow changes on power draw than large ones, but create a less uniform bubble size and may be difficult to design mechanically. Also, the need for internal heat transfer surfaces may limit the maximum D/T.

If the chosen shaft speed causes problems with gassing factors or mechanical interference, go back and choose a different shaft speed. If it is OK, go to the next step.

Mechanical Design

This actually involves several things. It includes how the agitator is to be mounted (Chapter 16), gear drive selection and shaft/impeller design (Chapter 17). Some designs may not be feasible due to shaft critical speed or a complex shaft design, such as one requiring multiple steady bearings. The mechanical design at the chosen shaft speed should be deemed feasible or not.

Decision Point: Is the Mechanical Design Feasible?

If the answer is no, go back and try a different shaft speed and repeat until one or more feasible designs are found.

Repeat to Find Lowest Cost

There may be several mechanically feasible designs at different shaft speeds. These different designs may have different costs. Higher speed means less torque and a less expensive gear box. However, the shaft design may be more expensive. There is no straightforward rule of thumb for this; each design must be fleshed out and a cost estimate made. In general, we would choose the least capital cost design unless there are other constraints. Once a semi-final design is selected (for the entire optimization process we go through here, not just shaft speed), equipment vendors are generally helpful in optimizing capital cost.

Repeat for Different Aspect Ratios

All of the previous steps were within the confines of the starting aspect ratio. So, up to this point, hopefully we have an optimum design for that ratio. However, that ratio may not be optimum overall. So, ideally the entire process should be repeated over the range of aspect ratios that are not constrained by other factors, such as site restrictions and shop-fabricated vs field-fabricated issues. Only by doing this will we find the economically optimum design. The capex and opex of the agitator, vessel, and compressor should ideally be included.

Repeat for Different Process Conditions

All of the above was for the process conditions chosen at the start. But for some processes, these conditions can also be varied within limits. For example, the back pressure on the vessel can be varied, though there may be an upper limit, such as that required to allow exit of CO_2. But, for example, raising the pressure from 0 to 0.5 bar-g may reduce agitator power requirements by 15–20%. Operating at a lower temperature increases oxygen solubility, reducing power but also reducing the metabolic processes within the organism. Lowering the peak cell population density can lower OUR but because the production rate will also be lowered, more total volumetric capacity will be required, albeit with a lower total power input. This is a classic case of capex vs opex. So, there are many potential options here.

Finish

When all of the steps are completed as many times as it takes to get the final optimum, the capex and opex per unit of capacity will be optimized. As you may have surmised, that is a lot of work. However, the savings could be quite significant. Moreover, a very experienced agitator designer can quickly go through the optimization for a given set of conditions and aspect ratio by instinctively avoiding designs that his experience indicates are poor or infeasible.

The balance of the book provides background information and details needed to complete these steps to the degree possible. Outside resources will be needed for cost data. The individual chapters are not organized as extensions of the step-by-step procedure, but, rather, as sources of information and calculation methods, as well as providing enough fundamental understanding to use the procedures described herein.

Summary of Chapter

This chapter has presented a series of steps to arrive at optimum fermenter design and operation. All of these steps will be covered in this book, in varying degrees of detail. The book will not follow this logic chapter by chapter, as a lot of background information and principles must be established before optimization can begin. The next several chapters will do that. The optimization steps begin in Chapter 8. There will also be several chapters after those covering optimization that will deal with special issues such as heat transfer, aspect ratio, and viscous fermentation. Sorry if this summary seems a bit repetitively redundant after the section "Finish."

List of Symbols

C_P Heat capacity at constant pressure
D Impeller diameter
k Thermal conductivity
T Tank diameter
Z Liquid height

References

1 Hicks, R.W., Morton, J.R., and Fenic, J.G. (1976). How to design agitators for desired process response. *Chemical Engineering Magazine*: 22–30.
2 Fasano, J.B., Bakker, A., and Penney, W.R. (1994). Advanced impeller geometry boosts liquid agitation. *Chemical Engineering* 7 pages.

3

Agitator Fundamentals

Before delving into details of agitation specific to bioreactors, we must establish a common framework of terminology and principles common to all agitation systems. This chapter will cover basic terminology, how experimental data are usually correlated, and some basic viscosity models used in fermentation broths.

Agitated Tank Terminology

A very simplified view of an agitated tank may be found in Figure 3.1. Though simplified, all of the main elements of an agitator and tank may be seen there. A schematic view with components labeled and a few major nomenclature symbols may be found in Figure 3.2.

An agitated tank consists of a number of elements and is dimensionally described by a number of symbols. We will go through these more or less in the order of power flow, referring to the nomenclature of Figure 3.2.

Prime Mover

Motive energy is provided to an agitator by means of a prime mover, which provides power in a rotary fashion. Usually, it is an electric motor, as shown in Figure 3.2. For fermenters, a variable speed drive is often provided, usually by means of a variable frequency drive (VFD), though other technologies are possible. In principle, many other rotary power sources could be used. Some that the author has seen used include air motors, DC motors, hydraulic motors, and even diesel engines. But, probably more than 99% of the time, the prime mover will be an electric motor.

Agitator Design for Gas-Liquid Fermenters and Bioreactors, First Edition. Gregory T. Benz.

Figure 3.1 Agitated tank. *Source:* Photo courtesy Chemineer, a brand of NOV. Permission granted by NOV.

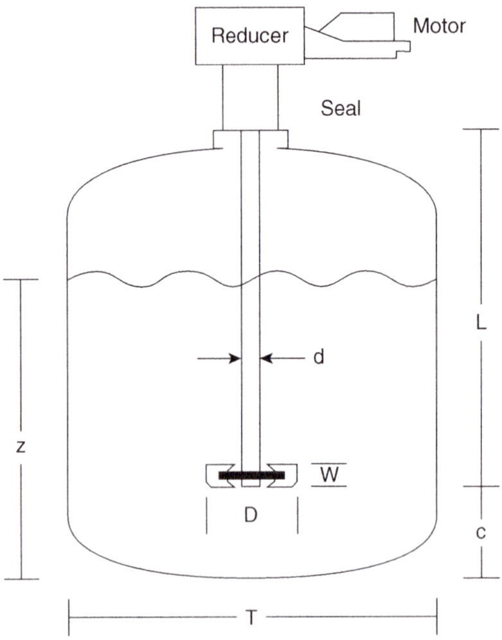

Figure 3.2 Agitated tank sketch.

Reducer

Most agitator designs do not operate at direct motor speed, except in very small tanks. The reducer decreases the shaft speed below motor speed and increases torque. In most agitator designs, the reducer must also support the weight of the shaft and impellers, the thrust due to tank pressure or vacuum, and the bending moment created by random fluid forces acting on the impellers. In some cases, those forces are supported by a separate set of bearings, and the shaft is flexibly coupled to the reducer.

The two most common reducer designs in industry are belt drive and gear drive. Most fermenter agitators use gear drives. More discussion of drive types will be found in Chapter 17.

Shaft Seal

Although not all agitators have shaft seals (some are mounted on open-top tanks or basins), those used in fermenters almost always do. The purpose of the seal, in addition to maintaining tank pressure or vacuum, is to isolate tank contents from the outside environment. This may be done to keep foreign matter from contaminating the broth or to protect plant personnel from exposure to potentially harmful organisms or gases. Often, the shaft seal area is heated to create a sterile barrier. More information on shaft seals will be found in Chapter 15.

Wetted Parts

The power and torque from the reducer are transmitted to the tank contents by means of a shaft with diameter d, extending a distance L from the mounting flange. On the shaft are mounted one or more impellers of diameter D and actual blade width or height, W, located off bottom at a distance C. For this book, we measure C from the bottom edge of the impeller. Some other sources use the centerline of the impeller. We also define D as the flat-to-flat dimension of the blades in plan view, rather than the swept circle, called D_S. This makes a difference of about 1% in the calculated diameter for a pitched blade turbine of standard design, for example. This is illustrated in Figure 3.3, along with a few other relevant dimensions, such as the blade thickness, t_b.

For multiple impellers, we would use subscripts such as D_1, D_2, C_1, and C_2.

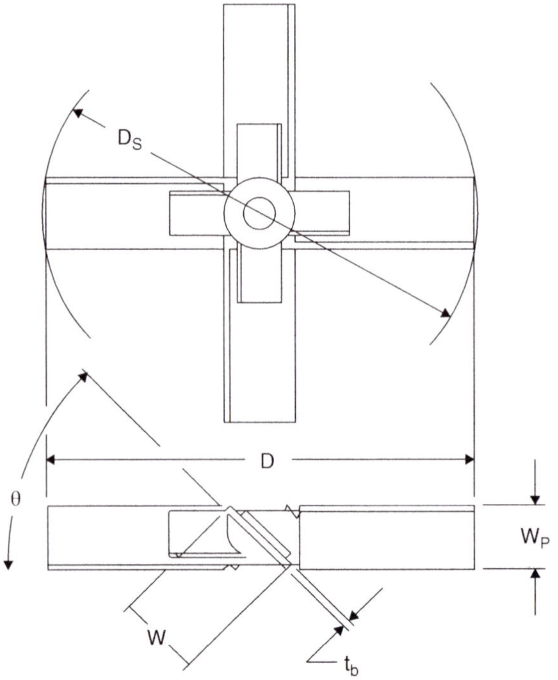

Figure 3.3 Swept diameter.

Tank Dimensions

The tank diameter is designated as T. The liquid level is designated as Z. Other tank dimensions, not shown on the sketch, could include head depths, straight side, nozzle projections, baffles (width, length, and offset from wall), and any relevant internals.

How Agitation Parameters Are Calculated

Agitation systems, just as any other system producing or modifying fluid flow, must obey the laws of physics. In terms of mathematical models, they obey the equations of continuity and the Navier–Stokes equations. Unfortunately, those equations can usually only solve problems analytically in relatively simple geometries, such as flow in a pipe, and, often, only in laminar flow. Such equations can be supplemented by various turbulence models.

An agitated tank, however, is a very complex geometry. Most would agree that it is all but impossible to solve the equations of motion for an agitated tank by analytical methods. In modern times, there have been many successful attempts to model agitated tanks by using numerical methods, which in essence convert differential equations into a series of algebraic approximations. Those approximations can be very good, depending on the skill of the modeler and the computational power used. These methods are often called CFD (Computational Fluid Dynamics) and sometimes called CFM (Computational Fluid Mixing.) Chapter 14 describes some of the uses of CFD as applied to fermenter design.

The traditional way of solving agitation problems is quite different. The approach that has been used in most studies, and which is still the staple of agitator design, is to use the equations of motion to derive dimensionless number groups and then correlate experimental data in terms of those dimensionless numbers. That is the approach we will take for the majority of this book.

We will not show the derivation of the dimensionless numbers, but will describe the ones important for our use in designing agitators, and how they are used, especially for fermenter design.

Some readers may be unfamiliar with the concept of dimensionless numbers, so we will give a brief description here, prior to getting into the commonly used dimensionless numbers.

A dimensionless number is a ratio of quantities such that the dimensions and units in the numerator exactly match the dimensions and units in the denominator, thereby canceling all dimensions and units. The resulting dimensionless number has no units or dimensions; it is just a scalar number. It also does not depend on what units are used, though converting dissimilar units to a consistent set of units will assist with the math.

A rather trivial example is the concept of aspect ratio of a cylinder, which equals its height or length divided by its diameter. A 5-ft. tall cylinder with a 12 in. diameter has an aspect ratio of 5. That is because 5 ft. is 5 times as much, in terms of its dimension (length), as 12 in. But the math would be more obvious and less prone to error if we first converted the diameter to feet by dividing by 12, or, alternatively, converting the height of the cylinder to inches by multiplying by 12. But the important point is that it is the ratio of the actual physical dimensions and is not unit dependent. We could have stated the dimensions as meters, microns, or cubits; the dimensionless number we are calling aspect ratio would still be 5.

In the next several sections, we will cover the major dimensionless numbers used in fermenter agitator design, and then we will show how some of them are used.

Reynolds Number

The Reynolds Number is the most widely used dimensionless number in fluid agitation. Many other dimensionless numbers are functions of it, as we will see in many subsequent chapters. Conceptually, the Reynolds number represents a ratio of inertial forces to viscous forces. When the Reynolds number is high, inertial forces dominate. This is the turbulent flow regime. When the Reynolds number is low, viscous forces dominate. This is the laminar regime. Intermediate Reynolds numbers constitute the transition flow regime and may exhibit attributes of both turbulent and laminar flow. In fact, in an agitated tank, there can be regions of laminar flow and regions of turbulent flow in the same tank when operating in the transition flow range of Reynolds numbers.

Mathematically, the Reynolds number is the product of a reference dimension times a reference velocity times the fluid density, divided by the fluid viscosity. The chosen reference dimension and velocity depend on the system being studied. For example, for a pipe, the typical reference dimension is the pipe inside diameter, and the reference velocity is the bulk velocity in the pipe.

For an agitated tank, it is customary to use the impeller diameter, D, as the reference length dimension. Likewise, it is customary to use a form of the impeller tip speed, πND, as the reference velocity. However, to avoid building a constant in a dimensionless number, π is dropped, so ND is used for velocity. The resulting expression is as follows:

$$N_{Re} = D^2 N\rho/\mu \qquad (3.1)$$

Because all units must cancel, it is best to use a consistent set of units. SI (Systeme Internationale) units work well. Normally, this is no problem, except for viscosity. Most of the time, viscosity is stated as cP. However, to use SI units, viscosity should be in units of kg/m-s, also known as Pa-s. Fortunately, the conversion is simple. 1 cP = 1 mPa-s = 0.001 Pa-s = 0.001 kg/m-s. English units become very peculiar. If the lengths are in feet and the density is in lb/ft^3, with time in seconds, the viscosity unit to use is pound mass/foot-second. The conversion from cP is 1 cP = 6.72 E-4 pound mass/foot-second. (I have yet to see a viscometer that reads in units of pound mass/foot-second.)

For those who prefer to use inches, rpm, specific gravity, and cP, we can build in a conversion factor using those units:

$$N_{Re} = 10.7 * D^{2*} N * sg/\mu \qquad (3.2)$$

Caution: do not use the above expression without using the prescribed units!

Power Number

Power number is conceptually the ratio of power draw to impeller parameters, speed, and density. It is defined as:

$$N_P = P/\rho N^3 D^5 \qquad (3.3)$$

The main use of power number is to calculate power draw. It is a function of impeller type, Reynolds number, and various geometric factors. When using SI units, the power will be expressed in watts. If one chooses to use inches, rpm, Hp, and specific gravity, we can insert the conversion factor and rearrange for power:

$$P(Hp) = 6.555E - 14 * N_P * SG * N^3 * D^5 \quad (3.4)$$

Same caution as for Eq. (3.2)

Note that power draw is quite sensitive to both impeller diameter and shaft speed.

Pumping Number

Pumping number is the ratio of impeller discharge rate (aka primary pumping capacity) to the cube of its diameter and the shaft speed:

$$N_Q = Q / ND^3 \quad (3.5)$$

It is used to calculate the flow created by the impeller, which can be used to determine a characteristic velocity in the tank. It is a function of impeller type, Reynolds number, and geometric parameters.

The units used for N and D determine the resultant units for Q. For example, if D is in feet, and N is in rpm, Q will be expressed as cubic feet per minute. Likewise, if D is in m and N is in revolutions per second, Q will be in cubic meters per second.

There are many different ways to measure impeller pumping, and they do not all give the same results. The most widely accepted methods define a discharge area around the impeller and measure flow through it, usually by use of either laser Doppler or particle image velocimeters. It is not within the scope of this book to discuss such methods. The above methods measure what is commonly called primary impeller flow or discharge. There can also be entrained flow, which can be several times as high as the primary flow. Any claimed impeller pumping capacity should state whether it is primary flow or total flow.

Dimensionless Blend Time

Dimensionless blend time is the product of blend time (defined as the time to reach some degree of concentration variance reduction after an assigned starting time) times the shaft speed. In other words:

$$N_B = \tau * N \quad (3.6)$$

This group is used to determine the blend time to some degree of attenuation of concentration differences. It is the product of blend time and shaft speed. Essentially, it is how many revolutions of the impeller are required to achieve a certain degree of blending. It is a function of Reynolds number, impeller type, and geometric factors. Rarely is blend time a limiting factor in fermenters.

Sometimes, people will add a factor of $(D/T)^{\alpha}$ to the right-hand side of the equation to correct for geometric effects. The value of α depends on the impeller type, but is usually about 2.3 for pitched blade and straight blade turbines, and about 1.73 for propellers and hydrofoils.

Aeration Number

Aeration number, also called gas flow number, is the ratio of actual gas flow rate at the impeller (corrected for absolute pressure and temperature) divided by the impeller diameter cubed and the shaft speed:

$$N_A = Q_g / ND^3 \qquad (3.7)$$

This group is used for power draw calculations in the gassed condition, along with other dimensionless groups. It can be thought of, in a way, as being proportional to the ratio of gas flow rate to the impeller pumping capacity.

Gassing Factor

When gas is introduced into an impeller or is present in the tank, it affects the impeller power draw. Usually, it reduces the power draw, but under some circumstances, it may increase it. The ratio of gassed power to ungassed power is called the gassing factor, and it does not usually have a special symbol for it. Instead, the ratio is simply expressed as a ratio: P_g/P_u.

It is a function of impeller type, Reynolds number, Aeration number, Froude number, and geometric factors.

Nusselt Number

The Nusselt number is the ratio of the convective heat transfer coefficient times a reference length divided by the thermal conductivity of the fluid. The reference length dimension depends on the heat exchange surface. For a vessel jacket, that will normally be the tank diameter. For tubes inside the tank, it would normally be the tube OD. Expressed in terms of tank diameter, we have:

$$N_{Nu} = hT / k \tag{3.8}$$

It is used in heat transfer calculations, to determine the convective coefficient, h. It is a function of impeller type, Reynolds number, Prandtl number, and various geometric factors, as well as the ratio of local viscosity to bulk viscosity.

Froude Number

Froude number is the ratio of a reference rotational speed times a reference velocity divided by the local gravitational acceleration. For similar reasons that we used for Reynolds number, we use N for rotational speed and ND for velocity, resulting in:

$$N_{Fr} = N^2D / g \tag{3.9}$$

Note that g is not a constant; it is the local gravitational acceleration. While it is almost uniform on the Earth's surface, it will be quite different on other planets. So the same impeller operating at the same speed would have different Froude numbers on Earth, the moon, Mars, and Jupiter. This will be important if we ever build fermenters on another planet.

Conceptually, the Froude number is the ratio of inertial forces to gravitational forces. High Froude numbers mean inertia dominates. This is associated with a choppy surface and vortex formation. Low Froude numbers mean gravity dominates, which is associated with a quiet, flat surface. Froude number, in combination with Reynolds number, impeller type, baffle number, and various geometric factors, can be used to predict mean vortex depth.

In gas–liquid contacting, the impeller gassing factor is a function of Froude number, impeller type, Reynolds number, Aeration number, and geometry factors. That is its principal use for bioreactor design. It is also used in most impeller flooding correlations.

Prandtl Number

Prandtl number is simply a physical property group used in heat transfer correlations:

$$N_{Pr} = C_P\mu/k \tag{3.10}$$

The Nusselt number is a function of the Prandtl number. Care must be taken to make sure all units cancel. The most common units for the fluid properties, especially viscosity, do not cancel. SI units work well. If common English units are used, the viscosity must be converted to pounds mass/foot-hour. The conversion is 1 cP = 2.419 lb/ft-h.

Geometric Ratios

Relative size and placement of impellers in a tank affects power draw, blend time, pumping capacity, solids suspension capability, vortex formation and maybe a few other things. Some parameters are more sensitive to geometry than others. The dimensionless geometric ratios used in calculations include D/T, Z/T, C/T, S/D, O/T, and W/D. As long as the same units are used for numerator and denominator, it does not matter what system of units is used.

Baffle Number

Although this could be lumped in under geometric ratios, it affects things in different and important ways, so I decided to mention it separately. Essentially, it represents total baffle width (calculated normal to the tank wall) divided by tank diameter:

$$n_b = \#\text{of baffles} * w_b / T \tag{3.11}$$

Power draw and vortex formation depend on baffle number. At a baffle number of zero, the tank is called unbaffled and power draw is at a minimum. Under such a condition, in turbulent flow, there is a lot of swirl, with mostly tangential motion rather than axial or radial. Gas dispersion is essentially impossible to do under such a condition. As baffle number increases, power draw increases to a maximum and then falls off. In the USA, "standard" baffling is 4 baffles at 90° to the tank wall, each one being 1/12 of the tank diameter, resulting in a baffle number of 0.33. In Europe, it is more common to use a baffle width of 1/10 of the tank diameter, giving a baffle number of 0.4. The power draw is pretty constant over this range and is basically at the maximum. D/T and impeller type also interact with baffle number.

Dimensionless Hydraulic Force

When an impeller operates in turbulent flow, the loads on each blade are fluctuating about a mean. To better visualize this, imagine riding along in a car with your hand outside the window. You will note that your hand is buffeted about, with a highly variable force. This is due to turbulence, involving various shedding of vortices, etc. The same thing happens to an impeller in turbulent flow. The load on each blade varies with time, and the loads on each blade are not synchronous with each other. The result is that there will be a fluctuating net side load on the impeller, creating a bending load on the shaft. This load, sometimes called an imbalanced hydraulic force, is not due to mechanical imbalance or any lack of manufacturing precision. It is entirely due to the nature of turbulence.

The dimensionless hydraulic force equals the product of hydraulic force times impeller diameter divided by impeller torque:

$$F_d = F_h * D / T_q \tag{3.12}$$

Normally, the peak value of this ratio is correlated, so that a peak value of hydraulic force may be predicted and used in shaft and impeller design. This group is a function of Reynolds number and impeller type, as well as geometric ratios, aeration number, and Froude number, and direction of pumping for axial flow impellers.

Thrust Number

This relates impeller thrust to impeller parameters and density:

$$N_T = F_T / \rho N^2 D^4 \tag{3.13}$$

It is primarily used to determine impeller thrust for mechanical purposes. It also is used to predict cavern size for shear-thinning fluids, as will be discussed in Chapter 12. Qualitatively, it is a constant in laminar and turbulent flow, but goes through a minimum in transition flow.

Typical Dimensionless Number Curves

Figure 3.4 represents a typical log/log plot of Power number as a function of Reynolds number, with D/T as a parameter. (The D/T effect is primarily noticeable on axial impellers. Radial impellers show much less dependency on D/T, as we will see in Chapter 5.)

Examining this curve, we can observe several things. One is that the power number becomes constant for a given D/T under turbulent conditions, i.e. at high Reynolds numbers (typically, above 20 000). When a manufacturer of an impeller states it power number, it is normally the turbulent power number at a D/T of 1/3 and a C/T of 1/3. Axial impellers will tend to have a decreasing power number at increasing D/T.

Also note that the curve becomes a 45° angle at laminar flow conditions (typically, $N_{Re} < 10$). That means the product of Reynolds number and Power number is constant in that range. A simple derivation reveals that in the laminar range,

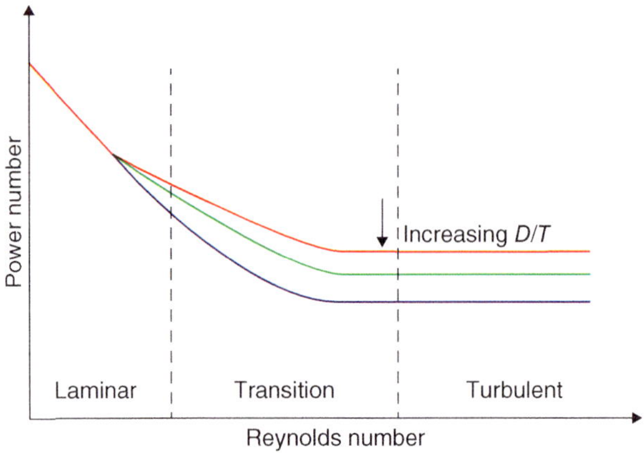

Figure 3.4 Typical power number curve.

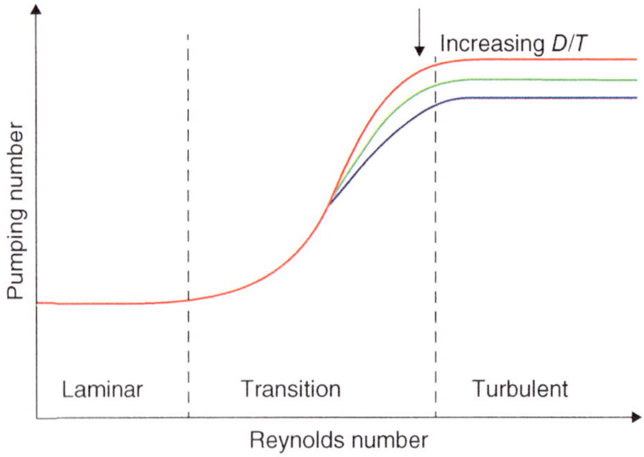

Figure 3.5 Typical pumping number curve.

$$P \propto \mu N^2 D^3 \tag{3.14}$$

The range between laminar and turbulent is called the transition range. It is a very broad range. That is because turbulence commences first near the impeller blade tips somewhere in the region of $N_{Re} = 10$, but does not reach all of the far corners of the tank until $N_{Re} > 20\,000$. The power number gradually levels out in this range. For some impellers, it goes through a minimum in the transition range.

Figure 3.5 represents a typical pumping number curve, as a function of Reynolds number and having D/T as a parameter.

In laminar flow, the pumping number becomes a constant. This means that under such conditions, the impeller pumping is independent of viscosity, though the power draw is directly proportional to viscosity.

In turbulent flow, the pumping number again becomes constant at a given D/T, though it is much higher than the laminar pumping number. As D/T increases, the pumping number generally decreases. This is because recirculating flow within the tank creates a velocity opposite the impeller discharge, impeding its flow. Nonetheless, larger impellers pump more for a given power input than smaller ones.

When vendors quote a pumping number, it is usually in turbulent flow and at a D/T of 1/3 and a C/T of 1/3. Many consider this geometry to be "standard", but there is no fundamental reason to adhere to this geometry when designing agitation equipment.

Figure 3.6 represents a typical generic blend time curve.

We can observe that blend time becomes constant in both laminar and turbulent flow. However, the laminar dimensionless blend time is often several orders of magnitude greater than the turbulent blend time. Specialized impellers have been developed for laminar flow mixing. We need not delve into these in this book, as fermenters never operate in the laminar range, because it is basically impossible to both incorporate gas into highly viscous liquids and have the gas exit in a reasonable amount of time after the gas is depleted.

Some versions of the dimensionless blend time curve incorporate the D/T effect into the Y-axis expression, as stated under the dimensionless blend time definition given earlier.

Figure 3.7 depicts gassing factor as a function of Aeration number for three different impeller types.

Gassing factor depends on Aeration number, Froude number, D/T, Reynolds number, and impeller type. Therefore, it is impossible to show the entire relationship on a simple two dimensional graph. Figure 3.7 is based on turbulent flow, a D/T of 1/3, and a Froude number of 0.5.

We can readily see that the traditional impeller for gas dispersion, the Rushton (or disk) turbine, has a rather severe drop in power at high gas flow rates. As we

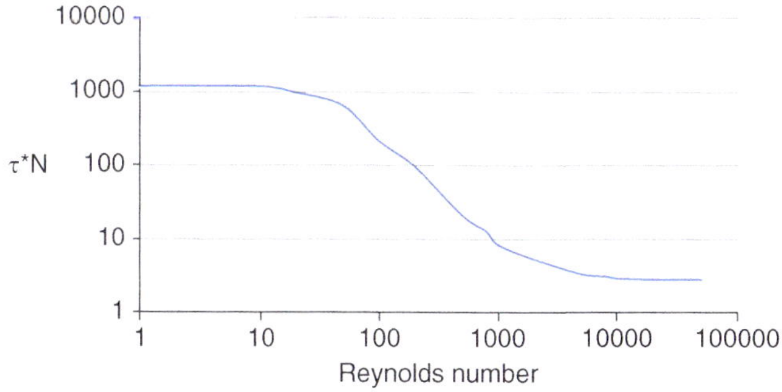

Figure 3.6 Dimensionless blend time.

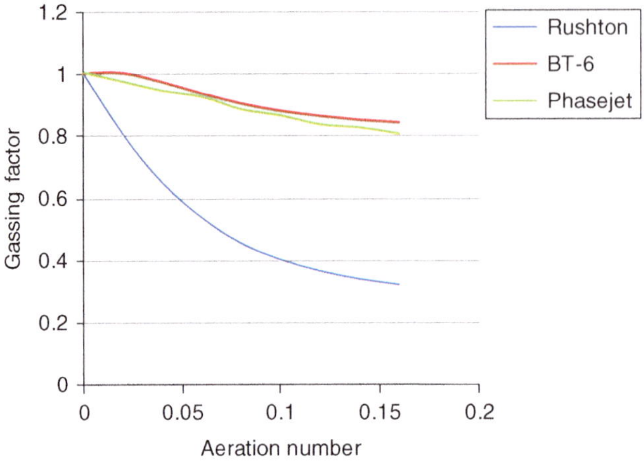

Figure 3.7 Gassing factors.

will see later, this is a problem for maintaining maximum mass transfer. That is why it has been pretty much replaced by modern, deeply concave impellers such as the BT-6 by Chemineer, the Phasejet by Ekato, and several others that have less effect of gas flow on power draw.

We will cover other dimensionless parameters, such as for heat transfer, as needed in subsequent chapters. For now, we will show some example calculations.

Example 1: Power Draw Calculation

A tank has an impeller of 1000 mm diameter, rotating at a shaft speed of 125 rpm, in a fluid with a specific gravity of 1.2. It has a known power number of 0.75. How much power will it draw?

Answer

It is best to first convert all parameters to SI units. So, we have a 1m impeller diameter, turning at a shaft speed of 125 rpm/60, or 2.0833/s, in a fluid with a density of 1200 kg/m^3. Since the power number is dimensionless, it remains at 0.75. The definition of power number, from Eq. 3.3, is:

$$N_P = P/\rho N^3 D^5 \tag{3.15}$$

Rearranging to calculate power:

$$P = N_P \rho N^3 D^5 = 0.75 * \left(1200\,kg/m^3\right)\left(2.0833/s\right)^3 \left(1m\right)^5 = 8138\,kg-m^2/s^3 \\ = 8138\,watts \tag{3.16}$$

Since it is more common to think of power in kW, this would become 8.13 kW. An 11 kW motor (a standard motor rating) would be ample to power this impeller under these conditions.

Example 2: Pumping Calculation

A tank has a hydrofoil impeller of 6 ft. diameter rotating at 30 rpm. It has a known pumping number of 0.5. How much fluid will it pump?

Answer

We do not have to do a unit conversion for this one. The units used merely determine the resultant pumping units. Equation (3.5) defines pumping number and is reproduced below:

$$N_Q = Q / ND^3 \tag{3.17}$$

Rearranging to calculate Q (pumping rate), we get:

$$Q = N_Q ND^3 = 0.5 * 30 / \text{min} * \left(6\,\text{ft.}\right)^3 = 3240\,\text{cubic feet per minute} \tag{3.18}$$

Had we used SI units, the calculated pumping capacity would have been expressed as m^3/s, but the physical quantity represented would have been the same.

Example 3: Blend Time Calculation

This really only gets interesting in transition flow, where dimensionless blend time varies with Reynolds number. But let's suppose we have an impeller with a dimensionless blend time of 10. How quickly would it blend the tank operating at 30 rpm?

Answer

If $\tau^*N = 10$, $\tau = 10/N = 10/30(1/\text{min}) = 1/3\ \text{min} = 20\ \text{s}$. Note that one can use any units for shaft speed; the resultant blend time units will be determined by this. Note also that as long as the dimensionless blend time is fixed, the result is independent of tank size. This means, for a given D/T and Reynolds number, the blend time at 30 rpm will be the same in any size tank. However, the P/V goes up exponentially when trying to keep the same D/T and the same shaft speed. So, designing for the same blend time in a large tank as a small one can be problematic.

A Primer on Rheology

It is not the purpose of this book to be a treatise on rheology. Indeed, many books have been written on the subject. However, due to the extreme difficulty of dispersing gases into viscous liquids, highly viscous liquids are not used or produced in industrially significant fermenters. (I know, some will object, saying they have viscous fermentations. But it is a matter of perspective. While a Gellan gum fermentation may seem viscous to someone operating a fermenter, it doesn't seem so viscous to those in the polymer industry. But they do not disperse gas into plastic or synthetic rubber melts. The difference is that a very high fermentation viscosity may be 5000 cP at the impeller, whereas in the polymer industry, they may see hundreds of thousands or even several million cP.)

Therefore, we will focus on a few simple models that do an adequate job of describing the behavior of fermentation broths. We will look at shear stress versus shear rate and assume that normal (i.e. at right angles to the shear plane) stresses, such as found in viscoelastic fluids, do not apply. We will also confine ourselves to time-independent models, because the fluids in fermenters are always under shear and in a quasi-steady state.

Newtonian Model

Newtonian fluids are those which have a shear stress that is directly proportional to shear rate (also called velocity gradient). Mathematically, this can be represented as:

$$\sigma_s = \mu * \gamma \tag{3.19}$$

where μ, the coefficient on shear rate, is also known as viscosity. This can be arranged to:

$$\mu = \sigma_s / \gamma \tag{3.20}$$

Most low viscosity fluids (<50 cP) are Newtonian. Intermediate and even some high viscosity fluids can be Newtonian if they are true solutions, especially if they are solutions of small molecules. For example, honey and corn syrups are often Newtonian even at 30 000 cP. Some silicone solutions are Newtonian even at 20 000 000 cP, but this is quite rare. Most suspensions of more than 5–10% solids exhibit at least some non-Newtonian behavior. High solids slurries can be very non-Newtonian.

Low-solids fermenter broths operating in a water-like fluid are usually Newtonian. High solids broths may not be, and those containing both an aqueous and an oil phase usually are not. But the overall viscosity for most fermenter broths is less than 1000 cP. The exception is some filamentous broths and those containing soluble gums, such as Xanthan. Chapter 12 will cover these in more detail.

Pseudoplastic or Shear Thinning, Model (Aka Power Law Fluid)

These fluids are called shear thinning because the slope of the shear stress vs. shear rate curve decreases as the shear rate increases. The behavior may be described by a power law model:

$$\sigma_s = M * \gamma^n \tag{3.21}$$

The exponent n is called the power law exponent. When it has a value of 1, the equation reverts to Newtonian. When it is less than 1, it is shear thinning. While it can be greater than 1, which would mean a shear thickening behavior, I have never seen a shear thickening fermenter broth. The apparent viscosity can be expressed by dividing by the shear rate on both sides of the equation:

$$\mu_a = M * \gamma^{(n-1)} \tag{3.22}$$

where M is the viscosity coefficient, or the viscosity at a shear rate of 1 (usually in units of 1/s).

Many fermenter broths exhibit this behavior at high cell density loading, or if there is a significant solids content in the broth.

Bingham Plastic

These fluids exhibit a yield stress. That is, until the fluid is exposed to a certain minimum shear stress, it will not move. Examples include some high solids slurries, toothpaste, and any fluid that can hold some of its shape after being disturbed. Above the yield stress, Bingham fluids exhibit a constant slope. Mathematically:

$$\sigma_s = \sigma_y + M * \gamma \tag{3.23}$$

and

$$\mu_a = \sigma_y / \gamma + M \tag{3.24}$$

Herschel–Bulkley

These fluids combine a yield stress with shear-thinning behavior above the yield stress:

$$\sigma_s = \sigma_y + M * \gamma^n \tag{3.25}$$

and

$$\mu_a = \sigma_y / \gamma + M * \gamma^{(n-1)} \tag{3.26}$$

One may recognize that all three previous models are special cases of the Herschel–Bulkley model, making this the most versatile model commonly in use.

These four models are described visually in Figures 3.8 and 3.9

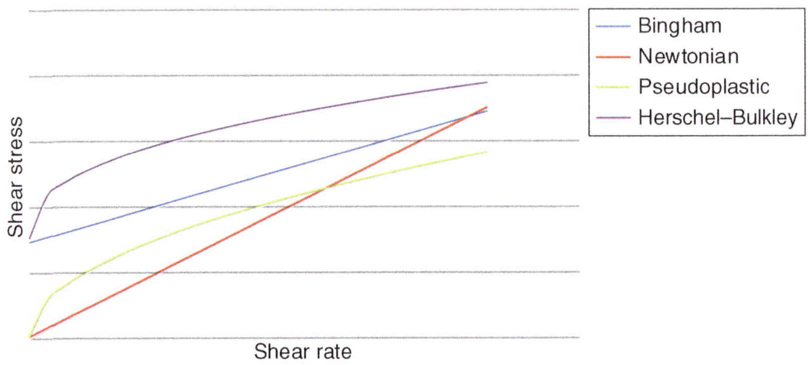

Figure 3.8 shear stress curve.

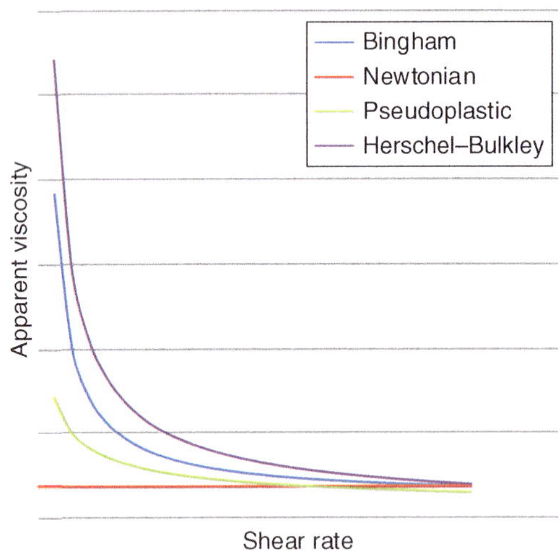

Figure 3.9 viscosity curve.

Impeller Apparent Viscosity

In laminar and transition flow, the apparent viscosity at the impeller can be evaluated using the average shear rate:

$$\gamma = dv / dx = K_M * N \tag{3.27}$$

where K_M is the Metzner–Otto constant. There are some disagreements about what that constant is for various impellers. Table 3.1 is an average of some of the published numbers.

In turbulent flow, the shear rates may be higher, so the apparent viscosity may be less for most fluids.

A Bit of Impeller Physics

Though we will get into impeller details more in Chapter 5, since this chapter is on the fundamentals, we will describe the basic physics of impeller behavior.

Some have said that some impellers invest their power more as flow, and others invest it as shear. As we will see below, this is an erroneous description.

Basically, a rotating impeller moves fluid. Agitation and gas dispersion are side effects of how much fluid is moved and how it moves it. But the macro relationship is very simple: power = flow times head, same as for pumps. Theoretical velocity head is proportional to tip speed squared or discharge velocity squared, though actual head produced depends on many other factors. Head is related to shear as well, as maximum shear depends on tip speed. So, power is more like a product of flow times shear squared, rather than a sum of flow and shear.

Some talk about the ratio of flow to power as an impeller characteristic. This, too, is erroneous, as, for a given impeller size in turbulent flow, flow is directly

Table 3.1 Metzner–Otto constants.

Impeller type	K_M
Propeller	10
Low solidity hydrofoil	10
High solidity hydrofoil	11
Pitched blade turbine	11
Radial turbines	11.5
Chaos (reversing pitch) turbines	40

proportional to shaft speed but power is proportional to shaft speed cubed. Therefore, flow/power diminishes as shaft speed increases. It approaches infinity as shaft speed approaches zero. Likewise, in laminar flow, power is proportional to shaft speed squared. So, there is no attribute of a flow/power ratio for any particular impeller. It depends on the specific application.

So, if we were to want to primarily create flow and minimize power, how would we do it? Let's start with impeller design, based on the flow vs. head relationship. We want to minimize the head term. We will look at several design attributes and the limitations imposed on them.

One attribute is the number of blades. Increasing the number of blades increases the flow, but it also increases the head, as the wake of each blade creates some degree of additivity for the net pressure, as the discharge flow will be moving at a higher velocity as we add blades. So, from a flow-for-power maximization standpoint, the optimum impeller would have only one blade. But there is a reason why we seldom see one-bladed impellers: we prefer not to break the shaft. The hydraulic imbalance would be huge! Hydraulic forces are still quite large for two-bladed impellers, so they are quite uncommon except in transition flow regimes, and even then, there will usually be several impellers, with the orientation staggered to minimize the bending moment on the shaft. Most "high flow" or hydrofoil impellers have three blades, because this is the minimum that has reasonable hydraulic stability. Using more than three blades improves hydraulic stability somewhat, but at the price of requiring more power for the flow produced.

Another attribute is blade chord angle from the horizontal plane. If the blades were infinitely thin, a zero degree angle would produce no flow and draw no power, save that created by skin drag on the blade surface. As the angle increases, flow increases and so does power, so a very shallow angle will create the most flow for the power. However, the absolute flow produced may be too small for a given diameter and speed. So, either the diameter would have to be increased or the shaft speed would have to be increased. But we are confined in diameter by the tank, so we cannot always increase the diameter; a D/T value of greater than 1 does not work very well. So, we would have to increase shaft speed in some cases in order to use a very shallow angle. But increasing shaft speed also increases head and power, negating the desired power savings from a low angle. The end result is that hydrofoil impellers available in the marketplace generally have chord angles between 15 and 30°, with 20–25° being most common. The power number range is about 0.2–0.4, with 0.3 being most common. The chord angle must never be greater than the stall angle, which is about 37° in water, or boundary layer separation will occur, greatly increasing power draw. For example, a 45° pitched blade turbine will require almost three times the power as a hydrofoil to create the same flow at the same shaft speed.

A third attribute is blade width or a related measure: impeller solidity (will be defined more in Chapter 5). As blade width increases, so does pumping, but power also increases due to higher discharge pressure. So, most high solidity impellers are used for applications that are not just about creating flow.

One of the most important attributes is shaft speed, and this applies to any impeller type. A small impeller rotating at a high speed can create the same flow as a larger impeller rotating at a slower shaft speed. The smaller impeller has a smaller discharge area; hence, its mean discharge velocity will be higher, and thus, the discharge head will be higher. Therefore, it requires more power to produce the same flow. For flow controlled applications, significant power savings can be achieved by using larger impellers at slower shaft speed. Sometimes, the capital cost is higher, but the payback time is usually a matter of months.

Of course, not all agitator design problems are flow controlled. The main application in this book, gas–liquid dispersion, is closer to being turbulent eddy dissipation controlled than anything else. So, in most cases, the power requirements will be governing, and impeller design will focus on other things besides flow.

Summary of Chapter

This chapter has presented basic material that all must know and understand in order to continue into the details required of fermenter design. It can be referred to as needed while reading the rest of the book.

List of Symbols

C	Off bottom clearance
C_P	Heat capacity at constant pressure
D	Impeller diameter, edge to edge
D_S	Swept diameter, or point-to-point diagonal diameter
dv/dx	Shear rate
F_d	Dimensionless hydraulic force
F_h	Hydraulic force
F_T	Impeller axial thrust
g	Gravitational acceleration
h	Convective heat transfer coefficient
k	Thermal conductivity
K_M	Metzner–Otto constant
M	Viscosity coefficient
N	Shaft speed

N_A Aeration number; also called gas flow number
N_B Dimensionless blend time
N_P Power number
N_Q Flow or pumping number
N_{Re} Reynolds number
N_T Thrust Number
n Power law exponent
n_b baffle number
O Impeller offset from tank center line
P Power draw
P_g Gassed power draw
P_u Ungassed power draw
Q Impeller discharge rate; also called pumping capacity
Q_g Actual gas flow at the impeller
S Impeller spacing
T Tank Diameter
T_q Impeller torque
t_b Blade thickness
W Actual blade width
W_P Projected blade width
w_b Baffle width, calculated normal to tank wall
v Liquid velocity
Z Liquid height

Greek letters

θ Blade angle from a horizontal plane
τ Blend time
μ Dynamic viscosity
μ_a Apparent viscosity
ρ Fluid density
σ_s Shear stress
σ_y Yield stress
γ Shear rate; dv/dx

References

Because the descriptions and use of dimensionless numbers in this chapter are found in many references, we chose not to do a line by line reference. Instead, one can find an introduction to these concepts in the references below, among others.

1 Dickey, D.D. and Fenic, J.G. (1976). Dimensional analysis for fluid agitation systems. *Chemical Engineering Magazine*: 7–13.

2 Paul, E. et al. (eds.) (2004). *Handbook of Industrial Mixing*. Various authors. Wiley Interscience (Look up by the name of the dimensionless number).

4

Agitator Behavior under Gassed Conditions

When an agitator is operated under gassed conditions (whether the gas is added or evolving from the process), it behaves differently from operating in liquid only. The presence of gas, in addition to possibly influencing mass transfer when the gas and liquid are not in equilibrium, can create several important effects. Among those are flooding of the impeller versus dispersing the gas, power draw effects, entrained gas, or holdup in the liquid and mechanical effects. We will go through each of these in turn, as well as the issue of what happens when there is a variable gas flowrate.

The above behaviors are mainly correlated by three dimensionless number groups: Reynolds number, Froude number, and Gassing factor. A fourth group, known as the dimensionless hydraulic force, is used for mechanical design and will be used in examples in Chapter 16.

Flooding

Flooding is a condition in which there is too much gas flowing into an impeller for it to effectively disperse it. If the main purpose of the agitator is to disperse the gas, this is obviously not a desirable condition. But how do we define it quantitatively and how do we measure it? There are actually several ways. We will describe three of them here.

$k_l a$ Method

By $k_l a$, we mean the overall mass transfer coefficient, as defined by:

$$\text{Mass transfer rate} = k_l a * (\text{driving force}) \tag{4.1}$$

Agitator Design for Gas–Liquid Fermenters and Bioreactors, First Edition. Gregory T. Benz.

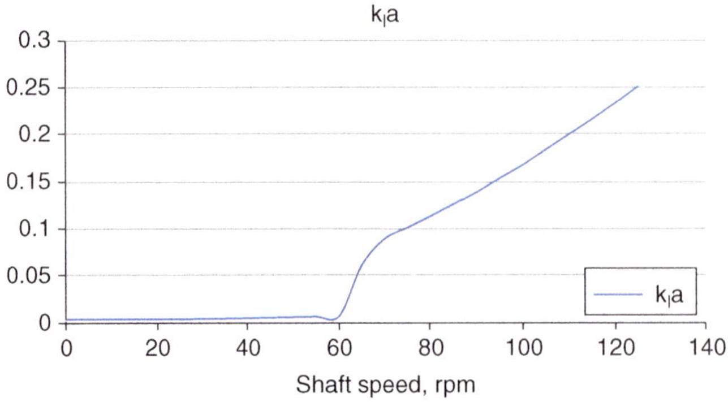

Figure 4.1 Flooding determination by using k_La.

Imagine an agitator operating at zero rpm, with gas being bubbled below the impeller. Although the impeller is not dispersing gas, there will still be some mass transfer possible, and therefore, the value of k_la will be non-zero. As the shaft speed increases, the k_la value will remain essentially constant, until a certain threshold speed is reached, at which point the k_la value will start increasing rapidly, as shown in Figure 4.1. We would call that threshold speed the flooding emergence point, by one definition. By inspection of Figure 4.1, the flooding emergence point is about 58 rpm.

One could also use mass transfer rate as the y-axis if the driving force remained constant.

While this definition of flooding has the advantage of clearly defining the point at which the agitator actually begins to influence mass transfer and thus relates to the main purpose for the use of the agitator, it is experimentally difficult to carry out. Most published definitions and correlations of flooding do not use this method, as we will see. This method does, however, have the advantage of being possible to carry out in an opaque vessel.

Power Draw Method

When gas is introduced into an impeller, it lowers power consumption, as will be described later in more detail. However, in many cases, the power draw increases when flooding, especially for Rushton or other impellers with flat vertical blades. This is due to the loss of gas pockets which streamline flow, illustrated in Chapter 5. This can be used as an indication of flooding, as illustrated in Figure 4.2

However, this phenomenon does not occur at all Froude numbers and can be difficult to detect for deeply concave impellers, as they have little or no gas pockets

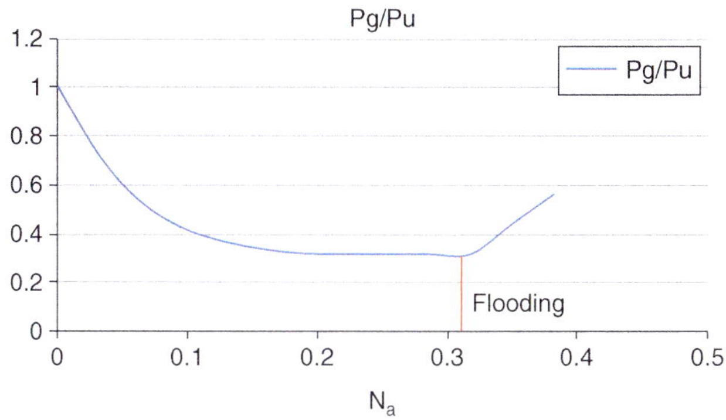

Figure 4.2 Impeller flooding by power draw.

to streamline flow. For such impellers, the gassing factor is mostly a density reduction factor, which might not change much due to flooding. An opaque vessel will work for this method also.

Visual Flow Pattern Method

The most common method of defining and determining flooding is the visual method. It is the method upon which all flooding equations in this book are based. Referring to Figure 4.3, flooding emergence occurs when the bubble flow and dispersion pattern changes from condition (a) to condition (b). This is easily observed in a transparent vessel and is a very repeatable condition. In condition (b), the bubbles are driven to the vessel wall, more or less in the plane of the impeller. This a somewhat conservative definition of flooding, as mass transfer enhancement actually begins before the bubbles are fully driven to the wall. It actually begins when the bubbles are driven even a short distance off a vertical path, as long as the impeller Froude number is at least 0.03 for Rushton turbines (the minimum Froude number may be somewhat lower for concave and deeply concave impellers, but the author is unaware of studies conducted below a Froude number of 0.03.)

Condition (c) in Figure 4.3 is called complete dispersion. It occurs when gas is driven to the bottom of the tank and some of it is recirculated into the impeller. It is not a design condition per se, but usually occurs when agitation is sufficient to support medium and high mass transfer rates (for example, above 80 mmol/l-h for an air-based fermentation at a back pressure of 1 bar-g or less.)

The transitions from condition (a) through condition (c) can be thought of as either increasing shaft speed at constant airflow or decreasing airflow at constant shaft speed.

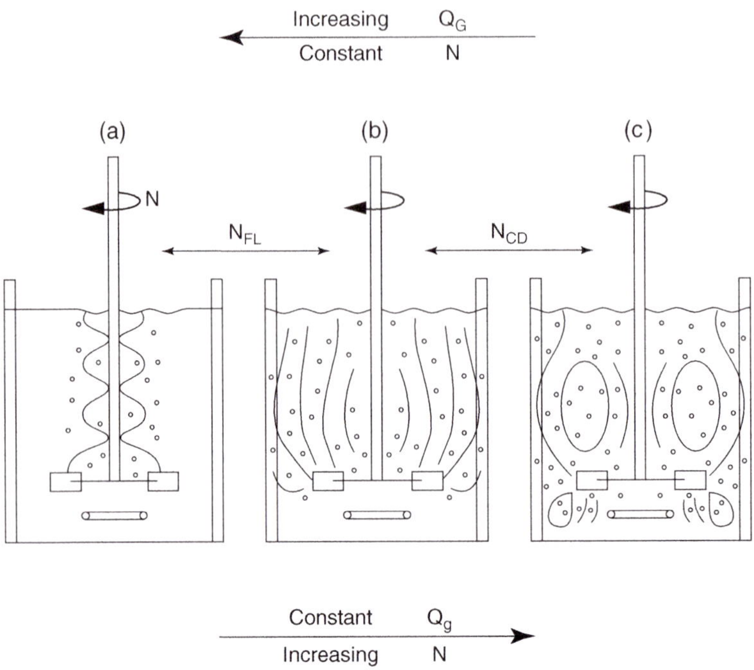

Figure 4.3 Visual flooding.

The above definitions work well for radial flow impellers, which are most commonly used as the primary disperser. Some systems use only axial flow impellers. There is no standard definition of flooding for these. The k_la method could be used, whether the impellers are up or down pumping. It would be difficult to use the visual method for an up-pumping system. For down pumping, there is a transition from direct loading (the gas goes through the impeller on the way up) to indirect loading (the gas is driven out of the impeller's path by its discharge velocity, and some gas is recirculated into the impeller from above). We do not have a figure to illustrate this. But this transition could be a sort of flooding definition.

Effect on Power Draw

As mentioned in the first section on flooding, when gas is introduced to a liquid, it affects power draw. At very low gas rates, some impellers actually have an increased power draw. Likely, this is due to increasing the discharge head imposed on the impeller. But at most gas flowrates used in fermenters and bioreactors, gas causes the power to drop.

There are two causes for this drop. One is that some of the gas is entrained, or held up, in the liquid, causing its apparent density to decrease. Remember, from the power number definition, power draw is directly proportional to fluid density. The other cause is that gas pockets may form along the backside of impellers, streamlining the flow, and reducing drag. For flat-blade turbines, such as the Rushton, this is the major cause of power draw reduction. For deeply concave impellers, the gas pockets are greatly reduced or even eliminated. Pockets also tend to be smaller on axial flow impellers. We will illustrate these pockets in Chapter 5 under each impeller. For now, note that the gassing factor for deeply concave impellers is usually between 0.85 and 1.0, whereas for Rushton impellers, it can be as low as 0.3. Axial flow impellers generally have gassing factors between these ranges.

Holdup

When gas is dispersed into liquids, it does not rise to the surface instantly. Smaller bubbles take longer to discharge at the surface than larger ones. High power/volume produces smaller bubbles, on average. And the more gas is introduced, the more that will be present in the broth at any given moment. Moreover, as the agitation gets stronger, there is more recirculation of bubbles. All of these things contribute to gas holdup.

From a fluid property point of view, surface tension, ionic strength, presence of solids, presence of a second liquid phase, and viscosity all influence holdup. Holdup may be defined as:

$$H = V_g / \left(V_g + V_l\right) \tag{4.2}$$

For a water-like fluid, the holdup may be estimated as follows, using the MKS system of units:

$$H = 1.56\left(P_g / \rho V_l\right)^{0.33} * U_s^{0.67}\left(\text{Ref.}\left[1\right]\right) \tag{4.3}$$

The effect of viscosity can be very complex. Sensel et al. [2] suggested the following:

$$H = 10.5 * \left(N_a * N_{Fr}\right)^{0.5} * \left(N_{Re}\right)^{0.1} \tag{4.4}$$

These equations are approximately valid for holdup fractions of up to 20%, yet some fermenters run at over 40% holdup. Viscosity was tested through 350 cP, but many fermenters operate at a higher viscosity than that. Also, the above equation cannot be used for mixed impeller types and diameters. For best results, an empirically derived holdup expression should be developed for the specific broth in question.

Example of Holdup Calculation

We will use equation 4.3. A fermenter with a working volume of 159 m^3 and a tank diameter of 4.11 m has an actual airflow of 0.31 m^3/s, giving us a superficial gas velocity of 3.01E-2 m/s. The fluid density is 1050 kg/m^3, and the agitator gassed power draw is 158 kW or 158 000 W. Substituting into Eq. (4.3), we get: $H = 1.56*(158000/1050*158)^{0.33} * (3.01e\text{-}2)^{0.67} = 0.147$.

Now, using equation 4.2, we can estimate the gassed volume. Re-arranging equation 4.2, we can get:

$$(1-H)V_g = HV_l \tag{4.5}$$

Substituting H: $(1\text{-}0.147)V_g = 0.147*159\ m^3$ or $V_g = 27.4\ m^3$. The total gassed volume, $V_g + V_l$, is therefore 186.4 m^3, an increase of 17% over the ungassed volume, or a ratio of 1.17 times the ungassed volume. This may also be simply calculated by saying:

$$(V_g + V_l)/V_l = 1/(1-H) \tag{4.6}$$

Holdup "War Story"

I am adding this in the hopes it may amuse readers as well as to forewarn them about a potential hazard involving holdup. A project team I led designed a fermentation system for a 250 m^3 working volume fermenter (full volume about 350 m^3) that was designed to have up to 1.5 VVM of airflow and was equipped with a 750 kW agitator. The projected holdup based on the working volume was close to 40%, or 100 m^3. During mechanical commissioning, the customer test crew decided to fill the tank with water to the brim and then start it up with 100% airflow and agitation. Luckily, they left the manway open! Within seconds of start-up, we estimate that about 100 m^3 of water, with a whole lot of entrained air, came out of the manway and hit the ceiling, then rained down on structures and people below. Fortunately, there were no injuries, though a few people got very wet. Had there been anyone looking into the manway or even on the catwalk close to the manway, serious injuries, or even a fatality could have occurred. So, test your fermenters at the operating volume, not with a full tank!

Variable Gas Flow Operation

Though some fermenters operate in a continuous mode, most operate in batch or semi-batch mode, so that it is possible to create lot numbers (in the case of pharmaceutical products) or periodically sterilize and clean to approach aseptic operation. Batch and semi-batch fermenters typically start out with a limited cell concentration and therefore a relatively low OUR. Although it is usually possible to run constant airflow, this is a wasteful practice, as less air is needed when the biomass concentration is low. Therefore, most fermenters operate with a variable airflow, starting low and increasing as needed, sometimes reducing near the end of the batch.

Airflow affects agitator power draw, as stated earlier. For some impellers, the gassing factor can be as low as 0.3 under peak airflow conditions. For others, it can be much higher. Some are as high as the density reduction factor, which is typically between 0.85 and 1.0. So, what airflow should the impeller system be designed for?

If the impeller system is designed to run ungassed at full shaft speed, it will draw less power at peak airflow conditions, where the most power is needed. Impellers that maintain a high gassing factor are preferred for this reason.

If the impeller system is designed to invest as much motor power as possible at maximum airflow and shaft speed, this is the best design for mass transfer. However, if the gas flow is reduced or eliminated, the impeller system will overload the motor, perhaps severely, unless the shaft speed is reduced. So, again, impellers with a high gassing factor are preferred, as it may even be possible to only slightly oversize the motor to allow ungassed operation.

Most fermenters have some sort of variable speed drive, the most common being variable frequency drives (VFDs). With such drives, it would be possible to design for maximum power at peak airflow and reduce speed as needed to operate at reduced or zero airflow. However, impellers with low gassing factors would have to be reduced further in speed than those with a high gassing factor, particularly considering the fact that VFDs and similar drives operate on a constant torque curve over about a 3 or 4 to 1 speed range, but not at constant power capability. That means the available power is equal to the rated motor power times the ratio of operating speed (or frequency) to full speed at the nominal power supply frequency.

The above points are illustrated in Figure 4.4.

Figure 4.4 shows a typical motor curve when operated on a VFD control. It is constant torque (% Hp = % speed) for a 4:1 ratio and variable torque (% Hp proportional to speed squared) below that. Superimposed on that curve are two operating points. Being compared are a high gassing factor impeller and a low gassing factor impeller. The high gassing factor impeller has a gassing factor of 0.85 at 1 VVM airflow and can run at 90% of full speed ungassed. The low gassing factor impeller has a gassing factor of 0.31 at 1 VVM and must have its speed reduced to 55% of full speed in order to avoid overloading the motor ungassed. This means it is impossible for the low gassing factor impeller to invest as much power at reduced airflow as would be possible with the high gassing factor impeller.

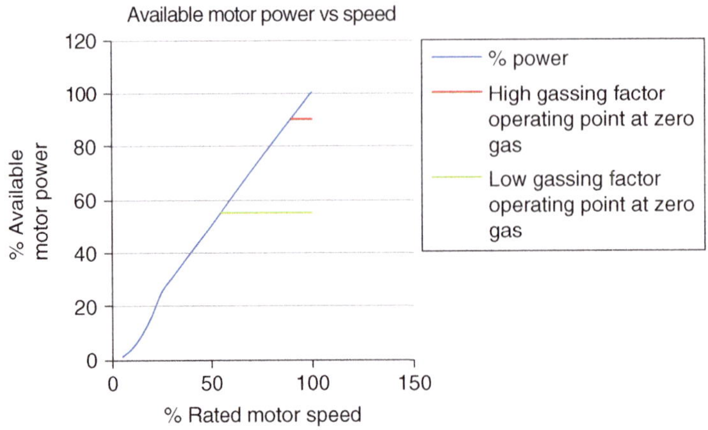

Figure 4.4 Available motor power curve using VFD.

Mechanical Effects

Impellers operating in turbulent flow have variable loads created on each blade, and a variable net side load, which is transmitted to the shaft and creates a bending moment. This was described in Chapter 3, under the heading of Dimensionless Force. In general, fermenters have very high agitation intensity, raising the value of the Dimensionless Force, even when ungassed.

When gas is introduced, the forces for radial impellers typically double. For axial impellers in up-pumping mode, the forces can more than double. In down-pumping mode, or when liquid level passes through an impeller, the loads can be triple; sometimes even more. We will discuss these issues in more detail in Chapters 5 and 17.

Summary of Chapter

This chapter has presented the basics of power draw, flooding, and mechanical effects exhibited by impellers under gassed conditions. More specific information will be presented in Chapter 5, which is about impellers used in gas dispersion.

List of Symbols

a Interfacial area/volume
k_l Liquid film coefficient

$k_l a$	Overall mass transfer coefficient (usually treated as one variable)
D	Impeller diameter
H	Fractional gas holdup
N	Shaft speed
N_a	Aeration Number
N_{CD}	Shaft speed for complete dispersion
N_{Fl}	Shaft speed for flooding emergence
N_{Fr}	Froude number
N_{Re}	Reynolds number
OUR	Oxygen uptake rate.
P	Power draw, general
P_g	Gassed power draw
P_u	Ungassed power draw
T	Tank diameter
U_s	Superficial gas velocity
V_g	Gas volume
V_l	Liquid Volume
Z	Liquid height
ρ	Liquid density

References

1 Private communication.

2 Sensel, M.E., Myers, K.J., and Fasano, J.B. (1992). "Gas dispersion at high aeration rates in low to moderately viscous Newtonian liquids." Paper 1071, AIChE 1992 Annual Meeting.

5

Impeller Types Used in Fermenters

There are many kinds of impellers used on agitators. Some have very specific purposes; some are more general purpose. In this chapter, we will focus only on those classes of impellers that have had some utility in gas–liquid dispersion applications.

Some impellers are generic, and some are proprietary. There are always new impellers being developed, so those covered in this chapter will not necessarily include all of them. Our choice of impellers to include does not necessarily denote an endorsement of a particular manufacturer or product. Our goal, rather, is to explore the different classes of impellers used, as each manufacturer normally has a variety available, and generally have offerings that are similar to those of their competitors. Manufacturers of proprietary impellers often do not publish their performance characteristics, and even when they do, there is no guarantee that the information is accurate for any particular application. Though we will provide estimated performance attributes where possible for some proprietary impellers in this chapter, it is to be understood that the numbers are based on the best information available to the author, but are not guaranteed. It is best to confirm any quantitative measures of agitator performance with the equipment manufacturer.

We will start by classifying impellers by flow pattern, and then provide examples of each with commentary and data. We will provide a summary chart of performance characteristics by type. Finally, we will briefly discuss impeller shear.

Impeller Flow Patterns

Generally speaking, turbine impeller flow patterns can be divided into three types: axial, radial, and mixed flow. A fourth flow pattern, tangential, does not exist to any significant degree in a fully baffled tank (which is necessary for good gas dispersion), so we will not discuss it further.

Agitator Design for Gas–Liquid Fermenters and Bioreactors, First Edition. Gregory T. Benz.

Axial Flow

An axial flow pattern is one in which the predominant flow direction is parallel to the shaft, at least in the region of the impeller. Obviously, it must turn around and change directions at the tank bottom and liquid surface. The flow direction can be either up or down. A stylized depiction of an axial flow pattern is found in Figure 5.1.

An axial flow pattern can be produced by impellers with varying degrees of solidity. By solidity, we mean the fraction of a circle that just envelops the impeller in plan view that is covered or occluded by the impeller. This concept is illustrated in Figure 5.2, showing what are typically described as low solidity and high solidity impellers in plan view, with a dotted circle drawn around them.

Although in principle impellers can be built with a solidity of nearly zero all the way up to 100% (with blades overlapping each other enough to cover the circle), and anywhere in between, in practice most impellers on the market are either low solidity, with a solidity value of 10–20%, or high solidity, with a solidity value of 70–100%. This is probably because there are few applications that would benefit from a medium solidity impeller. Flow velocity controlled applications will usually use low solidity, as that produces less head and therefore lower power draw for a given amount of flow. High solidity impellers are used where some pressure or a gradual

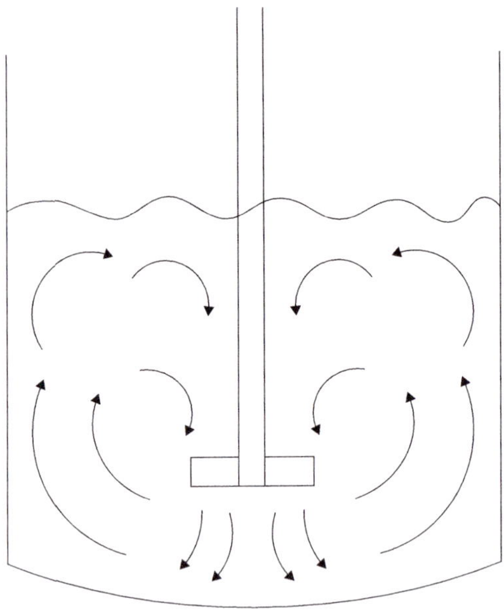

Figure 5.1 Axial flow pattern.

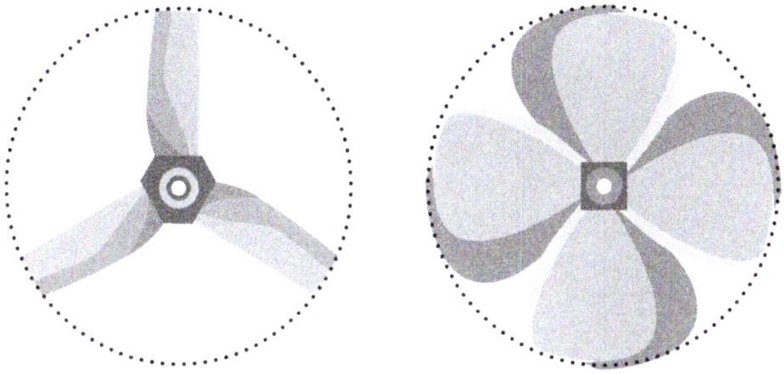

Figure 5.2 Solidity.

transition of flow and pressure is needed, such as in liquids near the boiling point, some solids suspension applications, and reasonably often for gas dispersion.

Both high solidity and low solidity impellers have been successfully used as upper impellers in fermenters with a lower radial impeller, in both up-pumping and down-pumping mode. They tend not to promote gas bubble coalescence. If there is no lower radial impeller, it is more common to use high solidity impellers. Up-pumping is more common than down pumping in such a case, because it gives better dispersion (due to gas passing through the impeller) and much better mechanical stability.

Generally, axial flow impellers will have a chord angle of 37° or less.

We will discuss impellers as a system in Chapter 6.

Radial Flow

A radial flow pattern is one in which the predominant flow pattern is radially outward from the impeller, i.e., normal to the shaft axis. It is stylistically represented by Figure 5.3.

Such a flow pattern can be beneficial when trying to agitate low liquid levels. It is also useful for dispersing gasses into liquids, as impellers producing a radial flow pattern also produce a strong system of turbulent eddies or vortices which stretch and collapse gas bubbles, creating strings of smaller bubbles and thereby increasing interfacial area.

Generally, radial flow impellers will have a blade angle of 90° from the horizontal.

Mixed Flow

For lack of a better description, this flow pattern is between axial and radial flow. Prior to the widespread use of hydrofoils, it was often called axial flow. However,

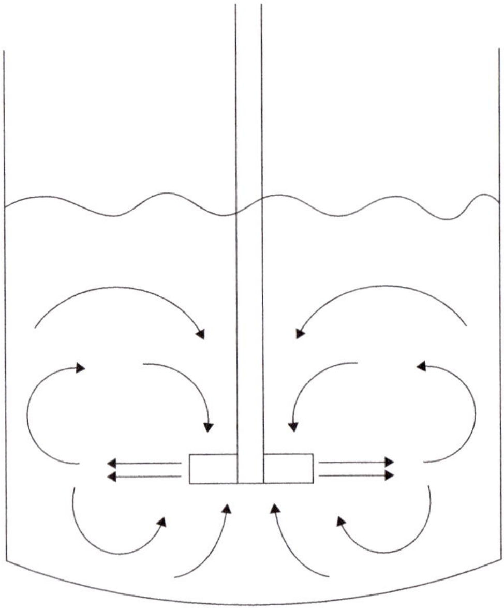

Figure 5.3 Radial flow pattern.

since hydrofoils more nearly approximate a true axial flow pattern, it is appropriate to consider mixed flow a separate category. The mixed flow pattern is stylistically depicted in Figure 5.4.

Mixed flow impellers can be useful at incorporating solids or gas from the liquid surface, as they tend to create deep vortices when operated near the surface, but only in down-pumping mode. They are not as good as either axial or radial flow impellers at dispersing gas when well submerged, as they tend to coalesce gas and then discharge larger bubbles than they received from the bulk fluid.

Mixed flow impellers generally have a chord angle of greater than 37° up to about 75°. Below 37°, the flow pattern is closer to axial, and above 75°, it is closer to radial. The most common angle is 45°.

Impellers with an inner section pitched to pump one way and an outer section pitched to pump the opposite direction actually produce a flow pattern resembling mixed flow when the pattern is time-averaged. However, the moment-to-moment flow pattern is extremely variable, even chaotic, especially when looked at in 3D. So, we are including a separate category here. It may be the first time such a category has been used in the literature.

Chaos Flow

This would describe a flow pattern in which the flow vectors change dramatically with time, and in a 3-dimensional manner. Mostly this is associated with impellers

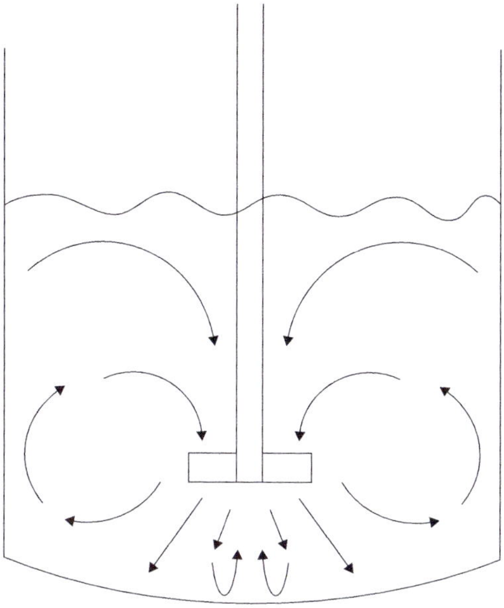

Figure 5.4 Mixed flow pattern.

that have a pitch direction change while traversing the blade from root to tip. Their time-averaged flow pattern may resemble mixed flow, axial flow or radial flow, depending on the particular impeller design and how it is applied. The flow pattern is very sensitive to D/T. Unfortunately, there is no way to give a good representation of the flow pattern using a still 2D image.

Examples of Axial Flow Impellers

Low Solidity

Most axial flow impellers used for blending and solids suspension are low solidity designs. Most have a turbulent power number of about 0.3, though some are higher and some are quite a bit lower. The low power number designs have utility for saving power usage in large tanks with gentle agitation. They are rarely used for gas–liquid contacting, as it is difficult to invest the required power with a low power number design. They must be larger than high power number impellers to invest the same power. The higher power number designs are often used in tall tanks to save weight and thereby help the shaft design.

Sometimes a vendor will create completely different geometries for their axial flow impellers, resulting in different power numbers. Sometimes they will just change the angle of attachment at the impeller hub, resulting in a different chord angle at the tip, and also affecting the power number. We will give several specific examples of impellers currently on the market in each category. Note that many manufacturers, if they publish power numbers, do not give information on how those numbers vary with D/T. When only one power number is given, it is probably based on a D/T ratio of about 1/3.

There are many low solidity hydrofoils on the market. We can only describe a few here. We will divide them up in terms of low power number ($N_P < 0.25$), intermediate ($0.25 \leq N_P < 0.4$) and high ($N_P \geq 0.4$). It is the author's opinion that impellers in each category will have similar performance at the same invested power and shaft speed.

Examples of low power number impellers include the ProQuip HFI (Figure 5.5a), the Lightnin A-510 E12 (not shown; essentially the same as the A-510 E22 and the A-310 except for blade angle) and the Chemineer XE-3 (Figure 5.5b).

Examples of intermediate power number impellers are more numerous, and include the Fusion Fluid PF3 (Figure 5.6a), the Lightnin A-510 E22 (Figure 5.6b), the Chemineer HE-3 (Figure 5.6c), and the generic marine propeller (Figure 5.6d). Marine propellers are rarely used in normal fermenters and bioreactors, but are sometimes used in mammalian cell culture reactors, due to the mistaken belief that they create less shear. (More comments on shear will be presented later in this chapter.)

High power number examples include the Chemineer SC-3 (Figure 5.7) and the Lightnin A-510 E32 (image not shown; essentially the same as the A-510 E22 but with a steeper blade angle).

Under highly gassed conditions, most low-solidity axial flow impellers will have a gassing factor of about 0.75, though Figure 5.9 gives more detail. (Gassing factors are similar for low and high solidity impellers. They may be slightly higher for high solidity in some cases).

(a) (b)

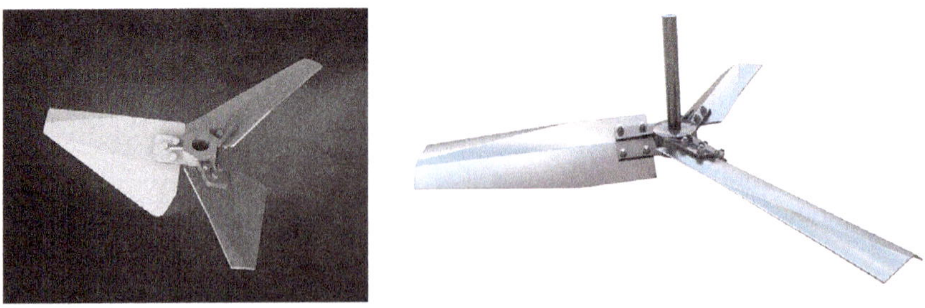

Figure 5.5 (a) Proquip HFI. *Source:* ProQuip Inc.; permission granted (b) Chemineer XE-3. Source: Photo courtesy Chemineer, a brand of NOV. Permission granted by NOV.

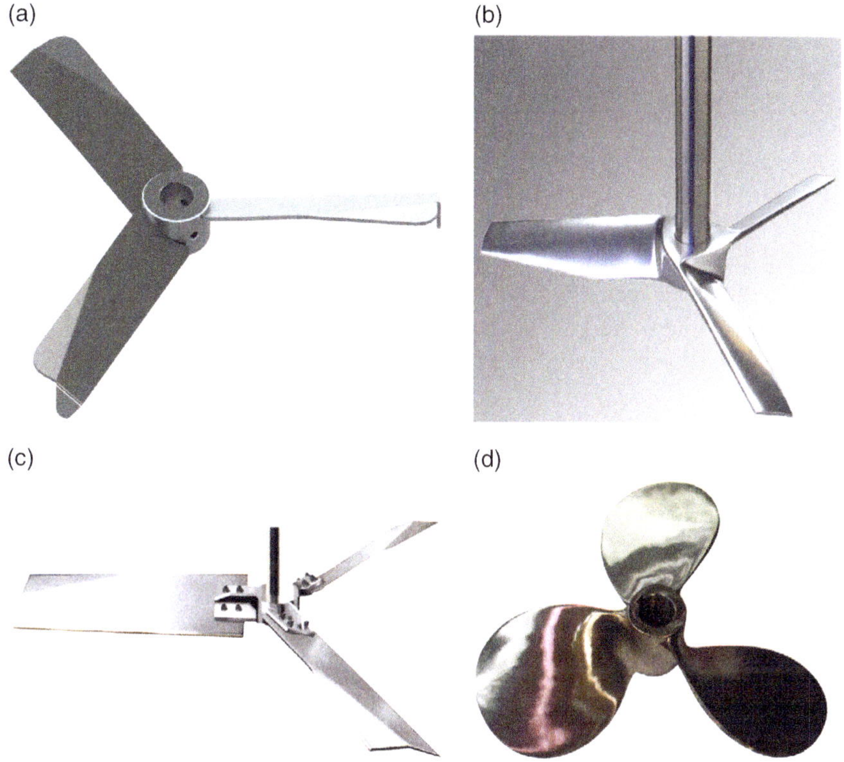

Figure 5.6 (a) Fusion Fluid PF3. *Source:* Courtesy of Fusion Fluid Equipment; permission granted. (b) Lightnin A510E22/A-310. *Source:* Courtesy of SPX FLOW, Inc. © 2019 SPX FLOW, Inc. (c) Chemineer HE-3. *Source:* Photo courtesy Chemineer, a brand of NOV. Permission granted by NOV. (d) Marine propeller. *Source:* Courtesy of Fusion Fluid Equipment.

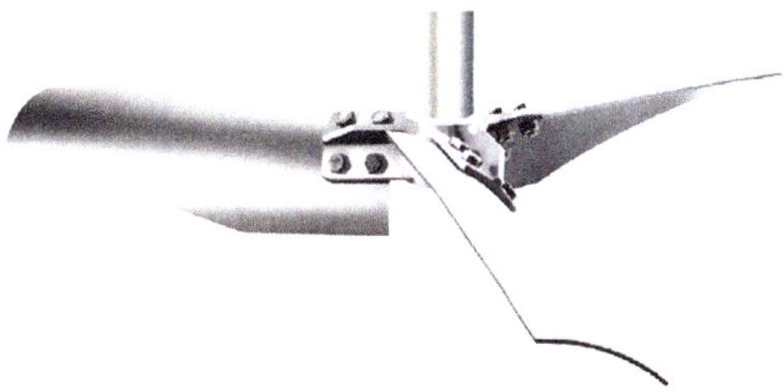

Figure 5.7 Chemineer SC-3. *Source:* Photo courtesy Chemineer, a brand of NOV. Permission granted by NOV.

Low solidity impellers have the following advantages:

- Better blending performance for the torque input. This advantage increases as the power number decreases.
- Easier to install in a vessel, due to the narrower blade width.
- Lower hydraulic force, sometimes resulting in a smaller required shaft diameter.

But they have the following disadvantages:

- Larger diameter required for a given power input. This may cause clearance problems with internals such as heat exchange surfaces
- Not effective for gas dispersion as a primary impeller

High Solidity

These are used both as secondary impellers in conjunction with a lower dispersing impeller, and as the primary dispersing device. They have been used in both up-pumping and down pumping mode. When directly loaded (gas sparged into the impeller from below) in down-pumping mode, they can create very large imbalanced hydraulic loads, corresponding to a dimensionless hydraulic force of 1 or sometimes even more. Generally this condition should be avoided.

Under heavily gassed conditions, most high solidity impellers will have a gassing factor of about 0.75, though Figure 5.9 gives more detail.

The power number range is from about 0.5 to a little more than 1.0. Lower power number versions may create more flow at a given power and speed, but must be larger than high power number impellers. The high power number versions generally will be about the same diameter as a lower radial dispersing impeller, when they are used as upper impellers.

Most have four blades as a means to achieve high solidity, but some have three blades.

Examples of high solidity impellers include the Chemineer Maxflo-W (Figure 5.8a), the Chemineer Maxflo-Y shown in up-pumping mode (Figure 5.8b), the ProQuip SHF High Solidity (Figure 5.8c), and the Lightnin A-315 (Figure 5.8d).

Advantages of high solidity impellers include:

- High power number means smaller impellers are needed to invest the required power
- Can be used to directly disperse gas, especially in up-pumping mode

Disadvantages include:

- Lower flow and blending performance for the torque invested
- Difficulty in fitting wide blades through access openings (there are special split-blade versions to help with this, but they are less sanitary)

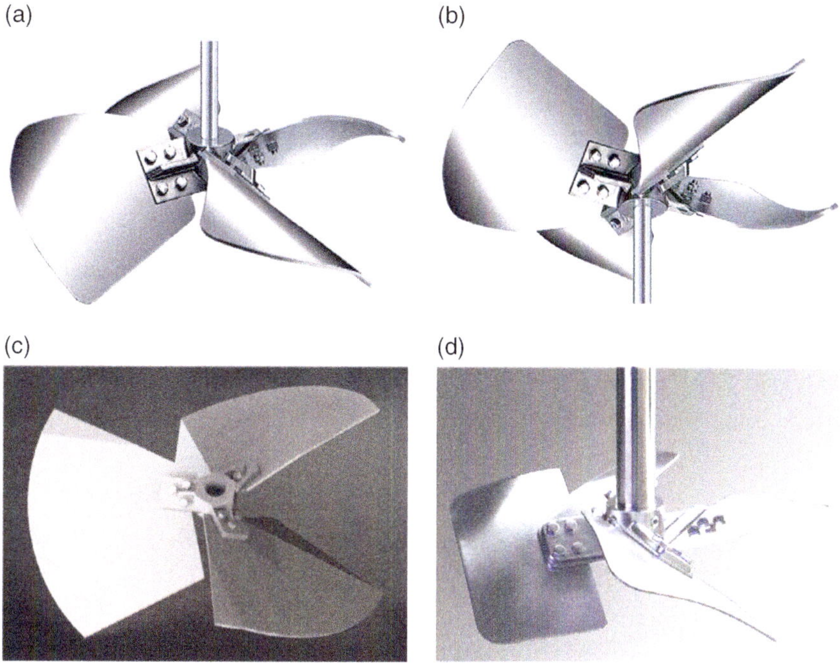

Figure 5.8 (a) Chemineer Maxflo W. Source: Photo courtesy Chemineer, a brand of NOV. Permission granted by NOV. (b) Chemineer Maxflo Y up-pumping. Source: Photo courtesy Chemineer, a brand of NOV. Permission granted by NOV. (c) ProQuip SHF High solidity. Source: ProQuip Inc.; permission granted. (d) Lightnin A-315. Source: Courtesy of SPX FLOW, Inc. © 2019 SPX FLOW, Inc.

- Greater weight
- High hydraulic forces, especially when directly gassed, down-pumping

To summarize this section, we have Table 5.1 and Figure 5.9 below, using the author's best estimates. Note that at low values of Froude and Aeration numbers, the gassing factor can exceed 1. That is, the introduction of air can cause the power draw to increase. However, this is an uncommon operating regime. A typical fermenter operates at a Froude number of about 0.4–0.6.

Up-pumping can increase the gassing factor slightly for high Froude and Aeration numbers, and decrease it for low values. Consult the equipment vendor for more specific predictions.

Example

What is the approximate gassed power draw of a 1.5 m diameter A-315 impeller, operating at 125 rpm in a 3.75 m diameter vessel, in a fluid with a specific gravity of 1.05, with an actual air flow rate of 1 m^3/s at the impeller?

First, let us calculate the ungassed power draw. From Table 5.1, we can look up a power number of 0.75. Using SI units, we can calculate the ungassed power:

Table 5.1 Axial flow power and pumping numbers (D/T~0.33) turbulent flow only.

Brand or manufacturer	Impeller designation	Solidity class	Power number	Pumping number
Chemineer	XE-3	Low	0.24	0.5
Chemineer	HE-3	Low	0.29	0.53
Chemineer	SC-3	Low	0.5	0.71
Chemineer	Maxflo W-4	High	1	0.77
Chemineer	Maxflo Y-4	High	0.5–0.9	0.6–0.75
ProQuip	HFI	Low	0.22	0.48
ProQuip	SHF	High	0.5	0.65
Fusion Fluid	PF3	Low	0.3	0.5
Lightnin	A-310	Low	0.3	0.54
Lightnin	A-510E12	Low	0.16	0.39
Lightnin	A-510E22	Low	0.3	0.51
Lightnin	A-510E32	Low	0.45	0.63
Lightnin	A-315	High	0.75	0.75
Lightnin	A-320	High	0.65	0.62
Generic	marine prop	Low	0.35	0.53

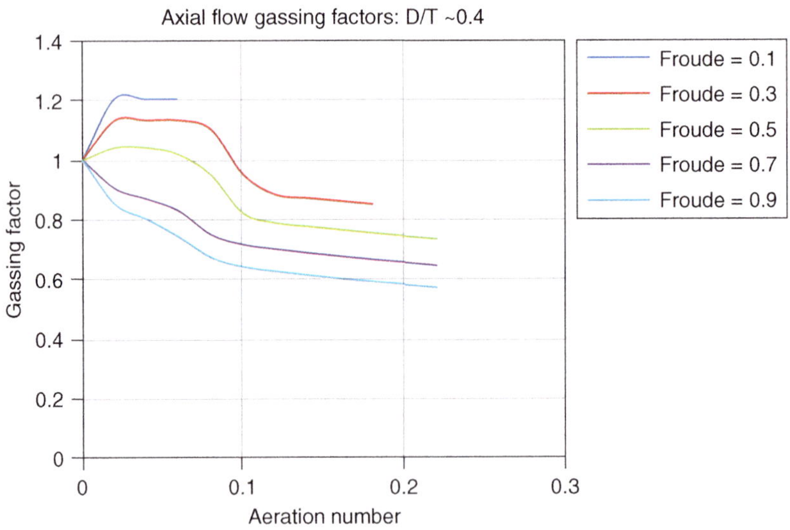

Figure 5.9 Axial flow gassing factors at D/T~0.4.

$$P = \rho * N_P * N^3 * D^5 = 1050\,kg/m^3 * 0.75 * (125/60\,seconds)^3 * (1.5\,m)^5 \\ = 54073\,W = 54.1\,kW \tag{5.1}$$

To calculate the gassing factor, we must first calculate the Aeration number and Froude number:

$$N_a = Q_g / ND^3 = (1\,m^3/s) / ((125/60\,seconds) * (1.5\,m^3)) = 0.142 \tag{5.2}$$

$$N_{Fr} = N^2 D / g = (125/60\,seconds)^2 * 1.5\,m / 9.8\,m/s^2 = 0.664 \tag{5.3}$$

Since our D/T value is 0.4, Figure 5.9 can be used to estimate the gassing factor. By inspection, it is about 0.7. Thus, the gassed power can be estimated as follows:

$$P_g = (P_g / P_u) * P_u = 0.7 * 54.1\,kW = 37.9\,kW \tag{5.4}$$

Up-pumping vs. Down Pumping

Since the 1990s, the use of up-pumping impellers has become more common in fermenters. There are a variety of reasons for this. If there were only a single impeller blending liquids, it would make little difference, except in very shallow tank geometries, where up-pumping would blend better because it would not draw a vortex into the impeller as readily. However, when gas is introduced, buoyancy forces will make the bubbles rise, and a great deal of liquid flow is induced by the rising gas bubbles. When the impeller is down-pumping, it will fight that flow, creating less flow overall, and a longer blend time. Some of the bubbles will bypass the impeller and not be dispersed. Where the upward velocities of the bubbles cancel the downward velocities of the impeller, gas coalescence can occur, creating poor dispersion, large bubbles and large mechanical loads. If there is a variable liquid level, when it passes through a down pumping impeller, it creates much larger loads than when it passes through an up-pumping impeller. It also splashes, creating more foam and increasing antifoam requirements. (One can demonstrate this by going to the beach and trying to splash with the hand pitched down versus pitched up. A down-pitch splashes much more.) Up-pumping impellers can sometimes pull foam back in at the walls, reducing the need for chemical antifoam. Some of my clients were able to reduce antifoam additions by as much as 50% when switching to up-pumping impellers.

When used as part of a multi-impeller system, with a radial impeller on the bottom, up-pumping produces a shorter blend time in the gassed condition. (The ungassed blend time would be longer, as the up-pumping flow pattern fights the lower radial flow pattern. But fighting the gas is a bigger issue).

Much of the credit for promoting up-pumping systems belongs to Professor Alvin Nienow.

There is one drawback: up-pumping impellers create thrust in the downward direction, adding that to the weight of the shaft and impellers. The net downward load is less with down pumping impellers, as their thrust counteracts the wetted parts weight. (In some cases, the thrust can exceed the wetted parts weight). These loads must be accounted for in the mechanical design of the equipment.

In summary, up-pumping compares to down pumping as follows:

- Shorter gassed blend time/more uniform dissolved gas concentration
- Slightly higher gassing factor
- Better gas dispersion: more uniform bubble size
- Less foaming/less antifoam usage
- Much lower shaft bending loads
- Higher net downward load on gear drive bearings

Examples of Radial Flow Impellers

There are many styles of radial flow impellers in the marketplace. Some are generic and some are proprietary. This section will explore the attributes of some of the better-known ones. For most proprietary impellers, several manufacturers have essentially similar products, so the characteristics may be close among them.

Straight Blade Impeller

This style has blades attached to an open-style hub (with no disc) at an angle of 90° to the plane of rotation. The most common style in use today has four blades, with a blade height one-sixth of the impeller diameter. It is illustrated in Figure 5.10.

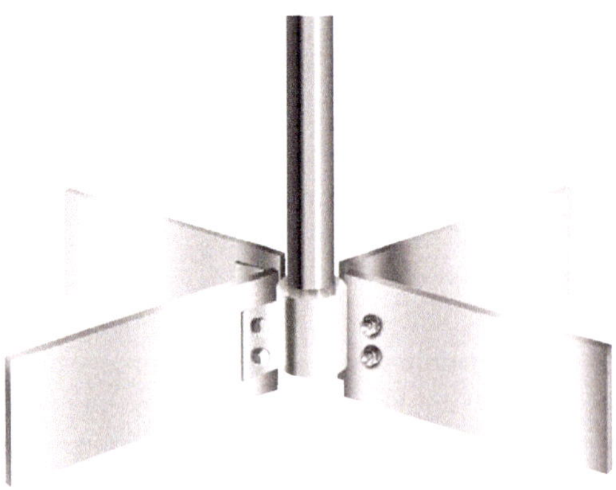

Figure 5.10 Straight blade turbine by Chemineer. *Source:* Photo courtesy Chemineer, a brand of NOV. Permission granted by NOV.

Mostly this style of impeller is used as a bottom turbine to avoid dead spots and keep solids in suspension, especially at low liquid levels. In that service it is often referred to as a "tickler" or "kicker" turbine. It is not normally used to disperse gas. Its power number is about 3, and pumping number about 1. Its gas dispersion characteristics, such as flooding and gassing factor, have not been studied much, so they will not be discussed here.

Though rare, there have been cases where a six-bladed version of the straight blade turbine has been used in fermenters. The primary reason is cleanability. The disc used on most radial turbines does not drain well. Though the gas handling capacity might be reduced by the absence of the disc, at high P/V levels there is little or no evidence that the gas bypasses the turbine and rises up straight through it. There has been little or no published information on gas dispersion characteristics of six or eight bladed versions of the straight blade turbine either, so there is little for us to say about it in such applications.

Disc, aka Rushton, Turbines

This impeller style is one of the oldest in industrial use. It has a series of flat blades mounted at 90° to the plane of rotation, attached to a flat disc. Most commonly, there are 6 blades, but impellers have been built with 4–12 blades. The most common six-blade version has a blade height of 20% of the diameter, and has a turbulent power number of about 5.5. The turbulent power number is proportional to the w/D ratio for the six-bladed version, and is largely independent of D/T in the range of 0.2–0.5.This impeller has been extensively studied in the literature. Indeed, many graduate students are still studying it, though not necessarily in optimum uses. It has become a generic impeller, made by virtually every agitator manufacturer in the world. A typical embodiment is shown in Figure 5.11.

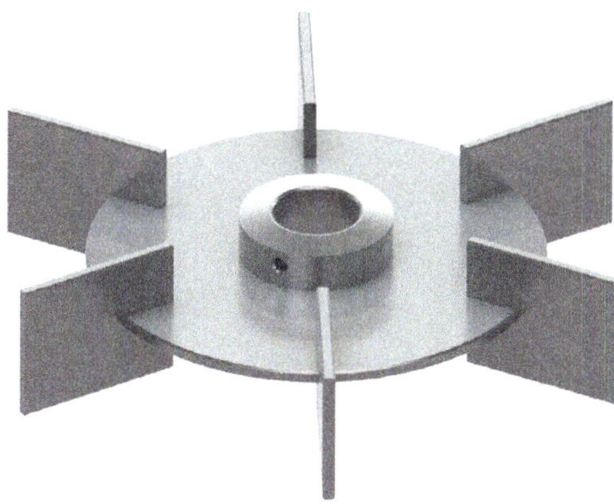

Figure 5.11 Rushton turbine. *Source:* Courtesy of Fusion Fluid Equipment; permission granted.

This was the impeller style first used in industrial fermentation, notably in the commercialization of penicillin in the 1940s. It remained popular well into the 1980s, but has been mostly replaced by more modern impellers in production-scale equipment, though, curiously, it is still the most common impeller supplied with bench-top and small pilot fermenters. The reasons why it has been largely replaced will become evident in this and in the sections describing various concave radial turbines.

Rushton impellers go through several changes in gas flow regimes around the impeller blades, as a function of gas loading. These regimes are illustrated in Figure 5.12.

There are roughly three different kinds of gas cavities or pockets that form behind a Rushton impeller blade, as a function of gas loading. The bottom of Figure 5.12 illustrates these. At low gas rates, the gas is collected in two counter-rotating vortex cavities, with rotation direction as shown, and with the vortices rotation shown by the curved arrows. At larger gas flow rates, clinging cavities form. These still have two distinct vortices, but a portion of the cavity clings to the backside of the blade, filling part of the space with gas. At high gas flow rates, large cavities are formed, which fill in the entire back of the blade and extend some distance behind it, after which the bubbles break off and are dispersed by vortex action.

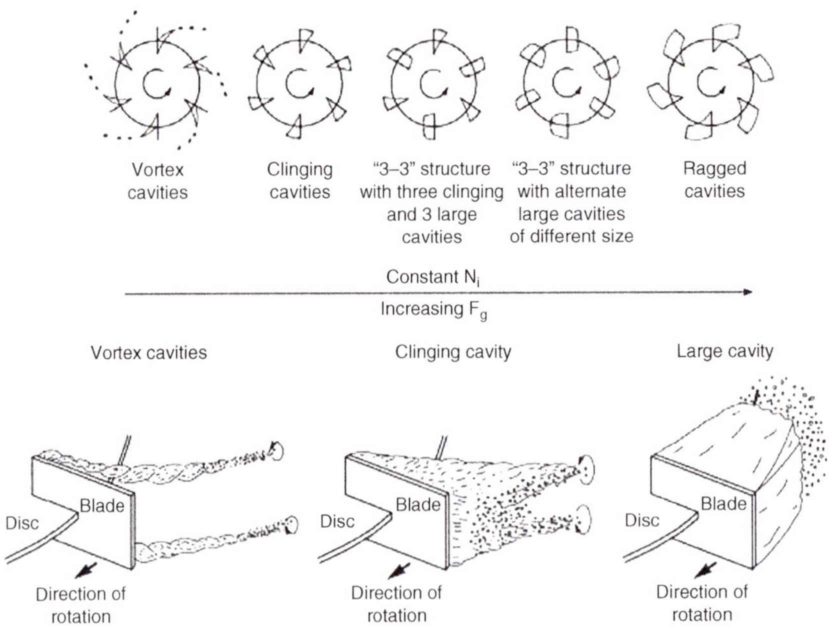

Figure 5.12 Rushton gas cavities. *Source:* Licensed by Elsevier.

These transitions do not happen to all blades simultaneously. As the upper part of Figure 5.12 shows, for a six-bladed turbine it is possible to have alternating large and clinging cavities, or alternating large cavities of different sizes. These conditions are called a "3-3" structure, and are a stable condition. Such a pattern cannot form on an impeller with an odd number of blades. For example, with a five-bladed impeller, one might develop a "3-2" pattern, which can move from blade to blade or change to a "2-3" pattern. This is hydraulically unstable, and can create large lateral hydraulic forces. The author recommends avoiding Rushton impellers with an odd number of blades for this reason.

At very high gas loading, "ragged" cavities can form. These indicate the impeller is close to flooding. In fact, one mechanism of flooding is when the cavities extend from the back of one blade to the front of the other. Essentially the impeller is running in air. Power draw under this condition varies periodically between a high and a low value, as the impeller "burps" the excess gas as it loses control of the flow pattern. At higher gas rates still, the gas basically rises through the impeller without being dispersed, as shown in Figure 4.3.

Figure 5.13 illustrates how these regimes were observed by Brujin et al. [1]. A camera rotating at the same speed as the shaft filmed the process from below the tank, allowing the appearance of a stationary impeller with gas being dispersed by it.

The consequence of the existence of these cavities is that they streamline the apparent shape of the impeller for drag creation purposes, reducing both pumping

Figure 5.13 Gas pocket photo. *Source:* Licensed from Elsevier.

and power draw considerably. It is not unusual for a Rushton impeller in an industrial fermenter to have a gassing factor of 0.4 or less. Since mass transfer depends on gassed power, this represents a considerable loss of mass transfer potential as the gas flow rate increases. Also, if the unit design is based on the gassed power draw, a loss of gas flow can create massive overloading, as mentioned in Chapter 4.

Quantitatively, the gassing factors are illustrated in Figure 5.14, based on a D/T ratio of 0.3. The figures end at each Froude number approximately at the flooding point, i.e., the maximum aeration number for that Froude number, or, in some cases where correlation data were not available. Note that the gassing factor can get as low as 0.3, although under typical conditions, it is usually around 0.4. These figures apply when the Rushton turbine is used as a lower impeller. When used as an upper impeller, the gassing factor can be estimated by the following:

$$(P_g / P_u)_{upper} = 0.9 * (P_g / P_u)_{lower} + 0.1 \tag{5.5}$$

Rushton turbines have a gas handling ability well-correlated by the following equation [2].

$$N_a(\text{maximum}) = 30 * (D / T)^{3.5} * N_{Fr} \tag{5.6}$$

As we will see later, that is a much lower gas handling capacity than is afforded by more recent impeller designs. In fact, we are listing the radial impellers in this section in order of increasing gas handling capacity.

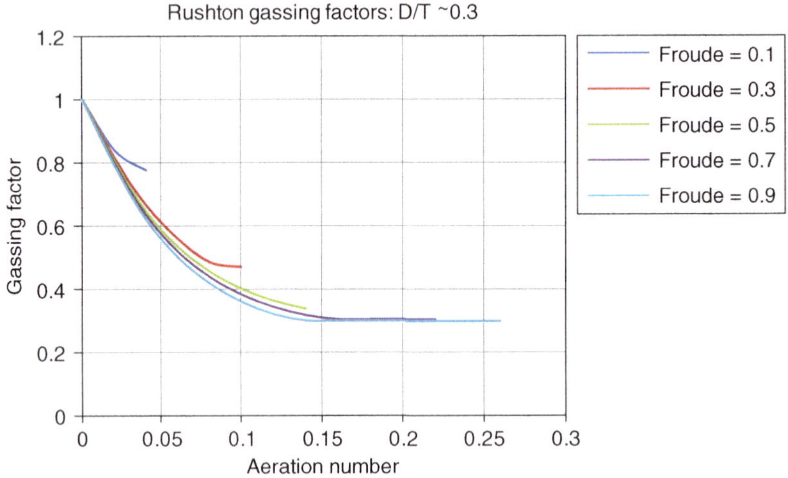

Figure 5.14 Rushton turbine gassing factors.

Note that the minimum Froude number for any degree of dispersion for a Rushton turbine is about 0.045.

In summary, the Rushton impeller has disadvantages at high gas flow rates due to potential flooding and substantial drop in power draw. But there are cases where it will perform as well as more modern impellers. Basically, these are cases where little gas flow is present, such as in micro-aeration (typical OTR < 10 mmol/l-h) or when a very high utilization of gasses is required due to either the expense of the gas or the cost of disposing of any gas that leaves the reactor. Examples include hydrogen, methane and oxygen-enriched gas streams. We will cover these cases in more detail in Chapter 11. Also, it has been noted that Rushton impellers are easier to clean with spray balls than concave impellers.

Example

How much gas flow can a 1.2 m diameter Rushton impeller handle without flooding when operated in water at 100 rpm in a 4 m diameter tank, and what will its gassed power draw be?

First, let us calculate the ungassed power:

$$\begin{aligned} P_u &= N_P * \rho * N^3 * D^5 = 5.5 * 1000\,kg/m^3 * \left(100/60\,seconds\right)^3 * \left(1.2\,m\right)^5 \\ &= 63360\,W = 63.4\,kW \end{aligned} \tag{5.7}$$

Next, let us calculate the Froude number:

$$N_{Fr} = N^2 D/g = \left(100/60\,seconds\right)^2 * 1.2\,m / \left(9.8\,m/s^2\right) = 0.34 \tag{5.8}$$

Now we can calculate the maximum aeration number:

$$N_a\left(max\right) = 30 * \left(D/T\right)^{3.5} * N_{Fr} = 30 * \left(1.2/4\right)^{3.5} * 0.34 = 0.15 \tag{5.9}$$

The maximum airflow is thus:

$$Q_g\left(max\right) = N_a\left(max\right) * ND^3 = 0.15 * 100/minutes * \left(1.2\,m\right)^3 = 25.92\,m^3/min. \tag{5.10}$$

Note that we did not convert N to revolutions per second, as we were looking for a dimensional result. By using revolutions per minute, we got the gas flow in cubic meters per minute. Had we uses revolutions per second, we would have just gotten gas flow in cubic meters per second.

Next, let us determine the gassing factor. Though Figure 5.14 does not have a recorded value for a Froude number of 0.34 and an aeration number of 0.15, the trend lines would lead one to estimate a value of 0.4 for the gassing factor. Thus, a reasonable estimate of the gassed power would be:

$$P_g = P_u * \left(P_g/P_u\right) = 63.4\,kW * 0.4 = 25.4\,kW \tag{5.11}$$

Smith Turbines

One of the pioneer researchers in the field of gas–liquid dispersion is Professor Emeritus John M. Smith. His works will be frequently referenced in this book. While at the University of Surry in the UK and Delft Technical University in the Netherlands in the 1980s, his team developed data on the mechanisms of dispersion, power draw and flooding for the Rushton turbine. In particular, upon observing the gas pockets that form behind the Rushton turbine blades, he reasoned that by using concave blades with the concave side being the leading side, more gas would be discharged from the front of the blade and any gas pockets formed would be smaller, increasing the gas handling capacity and the gassing factor. He tested his ideas and found that both were true. The particular impeller he came up with had blades with circular segments as cross sections, with about a 120° curvature. This design is properly called a Smith Turbine. We do not have an illustration of that design.

In that configuration, with 6 blades, its typical power number was about 3.8, its gassing factor was about 0.6 under highly aerated conditions, and its gas handling was about 50% more than that of the Rushton. So, it succeeded admirably.

This exact style of turbine has never been produced as a commercial product, to the author's knowledge. Instead, Chemineer, Inc. developed a modification based on this product, called the CD-6, and sold many of them starting about 1990. Other manufacturers soon started building essentially the same impeller but called it a Smith turbine, even though the geometry is based on the Chemineer CD-6 rather than the actual Smith turbine geometry. Since the original Smith turbine is not what manufacturers offer today, we will provide more detailed commentary on the CD-6 turbine, which will often be referred to as a Smith turbine, as that term is the generic version today. Below we describe how the CD-6 is different from the original Smith turbine.

CD-6 Turbine by Chemineer; aka Smith Turbine by Many Manufacturers

As stated above, Chemineer was likely the first to commercialize the impeller concepts developed by Smith. However, they did additional research before commercializing the impeller, which they named the CD-6 (presumably for Concave Disc-6 turbine). That research included detailed, proprietary power draw algorithms and flooding correlations, as well as some changes to the impeller geometry. The principal change is that the blade cross section is 180°, i.e., a semi-circle, rather than the 120° used by Smith. This change lowered the typical power number from 3.8 to about 3.2, further increased the gassing factor from about 0.6 to about 0.65 under highly gassed conditions, and increased gas handling to more than double that of the Rushton at higher D/T and Froude number conditions. The impeller is illustrated in Figure 5.15. This impeller has been widely used since the 1990s and is currently manufactured by multiple vendors under various names.

Unlike the Rushton impeller, the CD-6 does show a dependency of power number on D/T ratio. Myers et al. [3] provided the data shown in Table 5.2. Note that there is a 20% drop in ungassed power going from a D/T of 0.25 to 0.5.

As mentioned previously, the CD-6 unloads much less than the Rushton. This is because the curvature on the rear of the blade allows for less room for gas pockets to form, as shown in Figure 5.16, so that there is less streamlining of the flow. Figure 5.17 summarizes the gassing factor at a D/T of about 0.3. Note that the dependency on Froude number virtually vanishes above an aeration number of 0.15. Note also that the scale is different from the Rushton scale; it covers up to about triple the aeration number. That is because of the higher gas handling characteristics. The curves for each Froude number stop at the flood point or at the limits of the correlation.

Flooding behavior of the CD-6 is quite complex. The maximum aeration number depends on D/T and Froude number. At a given D/T ratio, the maximum aeration number is proportional to Froude number up to a certain Froude number. Above that critical Froude number, the maximum aeration number increases

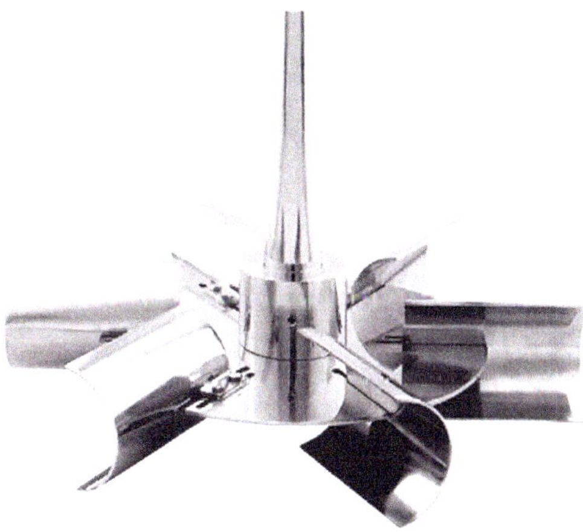

Figure 5.15 Chemineer CD-6 impeller. *Source:* Photo courtesy Chemineer, a brand of NOV. Permission granted by NOV.

Table 5.2 Ungassed power number of the CD-6 impeller.

D/T	0.25	0.32	0.39	0.45	0.50
N_P	3.21	2.89	2.77	2.56	2.55

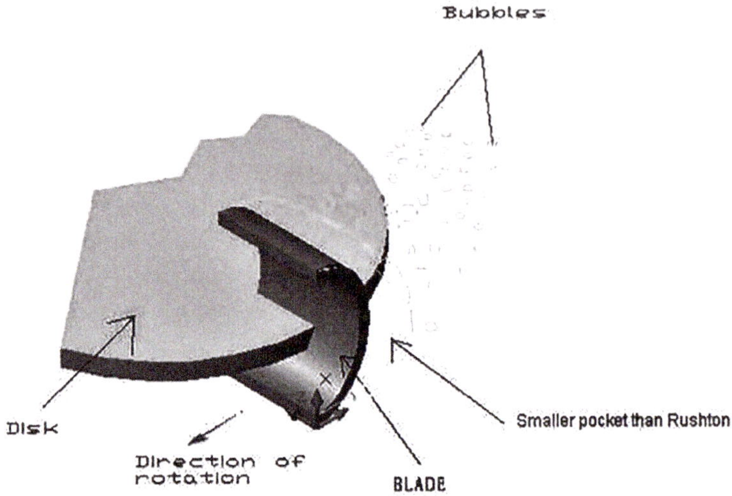

Figure 5.16 CD-6 gas pockets.

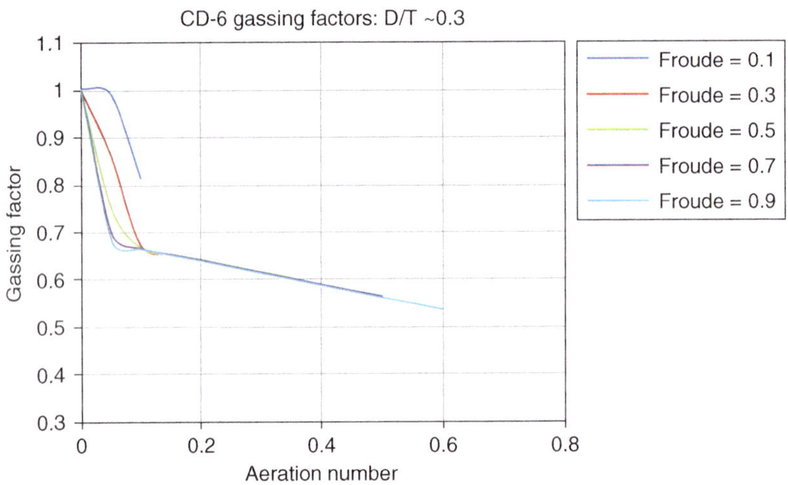

Figure 5.17 CD-6 gassing factors.

rapidly, sometimes by a factor of 3–5 within a 10–20% increase in Froude number. Reference [3] has considerable analysis of this phenomenon empirically.

If we denote the proportionality constant as K_{fl}, we can write this as:

$$N_a(\max) = K_{fl} * N_{Fr} \tag{5.12}$$

Table 5.3 Flooding constants of CD-6.

D/T	K_{fl}	Critical Froude number
0.25	0.3	0.98
0.32	0.5	0.5
0.39	1.0	0.3
0.45	2.5	0.15
0.50	5.4	0.06

Adding the caveat that there is a critical Froude number above which Eq. (5.12) no longer holds true, we can express this as a function of D/T in tabular form, as shown in Table 5.3.

Rushton turbines do not exhibit such behavior. They retain a basically linear relationship between maximum aeration number and Froude number. The end result is that at small D/T and Froude number values, the CD-6 has little or no gas handling advantage over a Rushton. However, at higher D/T ratios, the CD-6 has considerably more gas handling capacity, even below the critical Froude number. Above the critical Froude number, the gas handling capacity dwarfs that of the Rushton. And that critical Froude number decreases as the D/T increases. So, by judicious manipulating of D/T ratios, it should be possible to avoid flooding in most cases by using the CD-6 impeller.

Use the same problem statement as under Rushton turbines, but substitute a CD-6 impeller.

Example

First, let us calculate the ungassed power. Our D/T ratio is 0.3, so our power number based on Table 5.2 is about 2.87.

$$P_g = N_p * \rho * N^3 * D^5 = 2.87 * 1000\,kg/m^3 * \left(100/60\,seconds\right)^3 * \left(1.2\,m\right)^5$$
$$= 33062\,W = 33\,kW \tag{5.13}$$

Since we have the same diameter and speed as before, the Froude number remains at 0.34. From Table 5.3, we can see that we are below the critical Froude number, so we can use a linear assumption. The value of K_{fl}, by interpolation, is about 0.6–0.65. We will use 0.6. Thus:

$$N_a\left(max\right) = 0.6 * N_{Fr} = 0.6 * 0.34 = 0.204 \tag{5.14}$$

That is about 36% higher than for the Rushton. The resulting maximum air flow is:

$$Q_g(\max) = N_a(\max) * ND^3 = 0.204 * 100 / \text{minutes} * (1.2\,\text{m})^3 = 35.25\,\text{m}^3 / \text{min}. \tag{5.15}$$

Now let us calculate the gassed power. By inspection of Figure 5.16 at an aeration number of 0.204 and a Froude number of 0.34, the gassing factor is about 0.64. Thus:

$$P_g = P_u * (P_g / P_u) = 33\,\text{kW} * 0.64 = 21.1\,\text{kW}. \tag{5.16}$$

Though the CD-6 in this case has a ungassed power only 52% of the same-sized Rushton, and a gassed power 83% of that of the Rushton, it can handle 36% more gas flow before flooding. It would handle about double the gas flow at equal gassed power, which would require a larger impeller diameter.

In summary, the CD-6 is a substantial improvement over the Rushton in most cases. The only comparative drawbacks seem to be that the impeller weighs more and is more difficult to clean with spray balls. This is due to several reasons:

- Larger diameter required for equal power input
- More total metal in the blades (though they can be made of thinner material, as a circular cross section is stronger than a flat plate)
- An additional mechanical load is created: fluctuating loads normal to the disc. This requires a greater disc thickness for stability
- The curved shape makes spray ball access from above and below insufficient. Immersion in cleaning fluid may be required.

Deeply Concave Turbines

As noted above, Smith and similar turbines greatly reduce the size of the gas pockets, but they do not eliminate them. Smith himself has said that he was under the impression that strongly curved blades would be most effective. Therefore, several manufacturers embarked on development programs to create impellers that had a deeper degree of concavity than was possible using circular segments as cross sections. Examples of such impellers include the ICI radial impeller (Figure 5.18a), the SCABA 6SRGT (Figure 5.18b; Ref. [4]), the Lightnin R135 (Figure 5.18c), and the Ekato Phasejet (Figure 5.18d)

All of these impellers are mounted on a disc. All have deeper cross sections than a semicircle, and all are symmetric above and below the disc. As shown, all come to a sharp spine at the trailing edge of the impeller. However, it should be noted that the actual configuration of most Ekato Phasejet turbines sold for fermenters differs from the patent illustration in three ways:

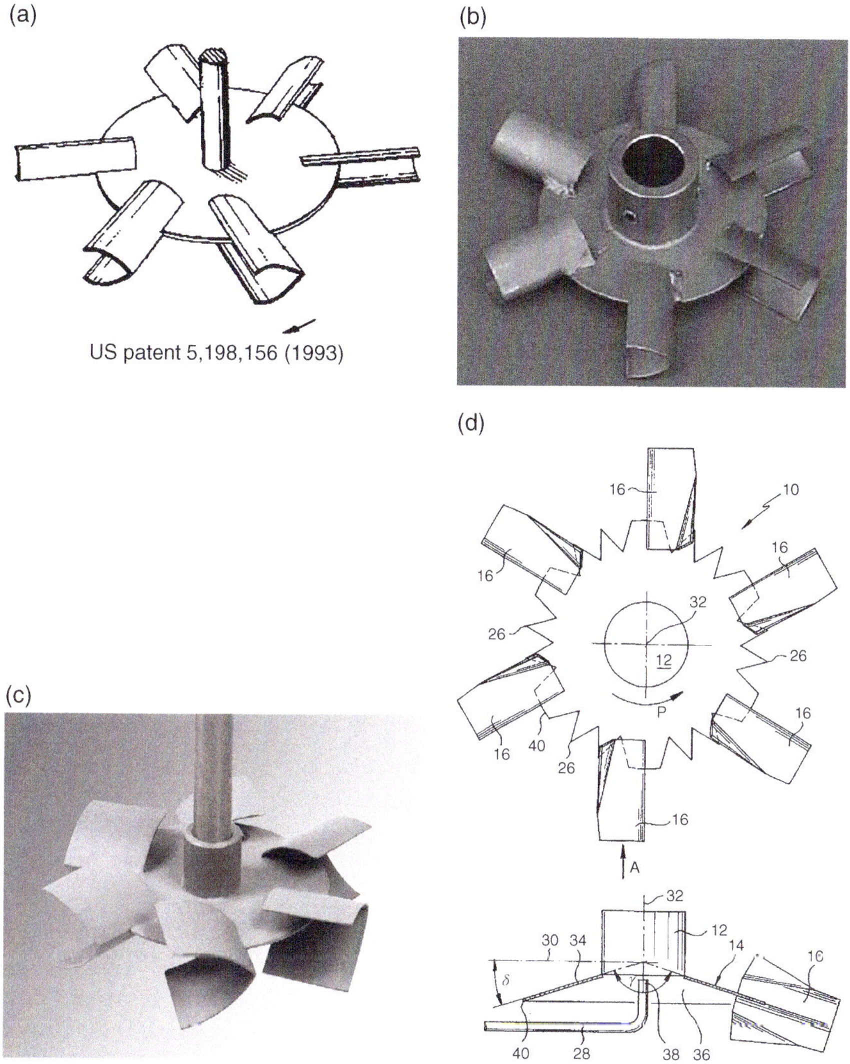

Figure 5.18 (a) ICI Patent drawing. *Source:* John C. Middleton and Colin Ramshaw, 30 March 1993, Agitators, US Patent 5198156. (b) SCABA 6SRGT. *Source:* Permission of Rob Van der Lans. (c) Lightnin R-135. *Source:* Courtesy of SPX FLOW, Inc. © 2019 SPX FLOW, Inc. (d) Ekato Phasejet Patent drawing.

1) The disc is flat, rather than conical or dished
2) The disc normally does not have serrations
3) The blades generally end in a curved section, rather than a sharp spine or ridge.

Because all of these are proprietary impellers, it is difficult to obtain reliable performance characteristics. Reference [4] reports the power number of the SCABA 6SRGT impeller as 1.5. Daub [5] reports the power number of the Phasejet as 1.6. Most likely, all are in the range of 1.3–1.8.

Likewise, little has been published about the gassing factor results. Some studies show a value of close to 1 for aeration numbers up to 0.05, but this is rather low for impellers operating in fermenters. Some fermenters have been operated with aeration numbers as high as 0.5, and many are in the range of 0.1–0.2. Ekato's literature on the Phasejet claims a gassing factor of 0.8 at an aeration number of 0.16 and a Froude number of 0.5. It is expected that the other impellers in this group will behave similarly. As expected, this gassing factor range is substantially higher than that obtained with Rushton and CD-6/Smith turbines. It is almost as high as just simply accounting for density reduction due to gas holdup.

Gas holdup is expected to be similar to the Rushton at equal gassed power and airflow.

There is also not much in the literature regarding flooding. However, it is expected that these impellers will generally have around twice the gas handling capacity of the CD-6 style impeller at equal invested power and shaft speed, in part because of studies done on the kind of impeller described in the section "Deep Asymmetric Concave Turbine with Overhang (BT-6)."

Summarizing, the impellers in this category have the following advantages over the CD-6 geometry:

- Much higher gas handling capacity before flooding
- Substantially smaller drop in power due to gassing.

They have the following disadvantages:

- Larger impeller diameter, due to lower power number
- Greater impeller weight, in part because of greater vertical oscillating loads, requiring a thicker disc
- Under overload conditions, they can open up (become taller in terms of blade height), which further increases loads and leads to failure. Sometimes a strut (illustrated in the section "Deep Asymmetric Concave Turbine with Overhang (BT-6)") is needed to prevent this.

Deep Asymmetric Concave Turbine with Overhang (BT-6)

More or less concurrently with the development of the impellers in section "Deeply Concave Turbines," or slightly later in some cases, engineers at Chemineer (notably Dr. André Bakker) reasoned that, since gas flow was inherently asymmetric with respect to the impeller (it always tries to flow upwards), why do the blades need to be symmetric about the disc plane? Their reasoning is that the upper section of the blade could be longer than the lower as a means to force discharge more of the gas on the front side of the impeller, leaving less to form pockets on the rear. They also studied the shape of the gas pockets, and designed the impeller curvature to match it, so that there would effectively be no pockets. The upper part of the blade has a different curvature from that of the lower part of the blade. The resultant impeller was dubbed the BT-6, and was patented, though the patent is now expired and several companies have copied the design. The impeller is depicted in Figure 5.19. A conceptual illustration of the lack of gas pockets is shown in Figure 5.20.

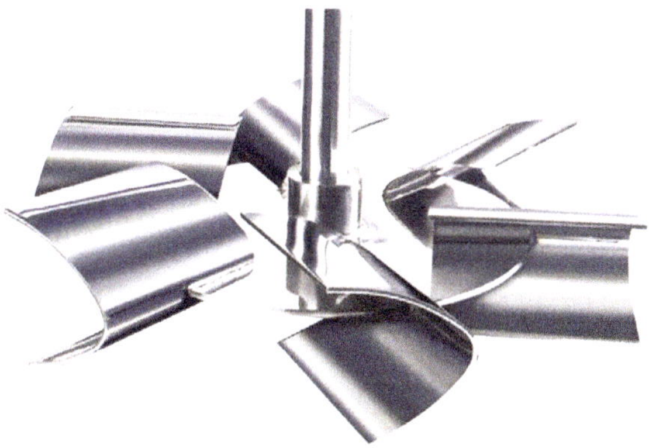

Figure 5.19 BT-6 impeller. *Source:* Photo courtesy Chemineer, a brand of NOV. Permission granted by NOV.

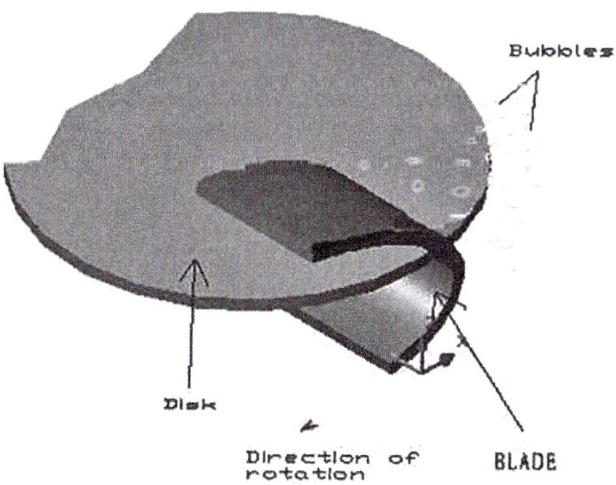

Figure 5.20 BT-6 no gas pockets.

The turbulent power number of the BT-6 is about 2.3 over the range of D/T from 0.25 to 0.5. It can be made lower by reducing the w/D ratio. The standard w/D ratio is 0.2. The power number is proportional to w/D with a six-bladed impeller.

As expected, according to the theory behind its design, its gassing factors are quite high. Figure 5.21 illustrates these for a D/T ratio of 0.3. If it seems that the results are basically independent of Froude number, it is because they are. The main difference Froude number makes is how much airflow can be handled before flooding. Note that the lowest gassing factor reported was about 0.82. One notable feature of these curves is that the gassing factor does not exceed 1 at low gas flows and high Froude numbers, as it can for some radial turbines such as the Rushton and CD-6 impellers. This reduces the risk of overloading when designed for ungassed conditions and operated at lightly gassed conditions.

How does this compare with the results in the previous category? Although we do not have gassing factors for all deeply concave turbines listed, Ekato was kind enough to publish values in their literature for a Froude number of 0.5 and for aeration numbers from 0 to 0.16. These are plotted along with gassing factors for the BT-6 over the same range in Figure 5.22 plus extending it to an aeration number of 0.25 for the BT-6. On average, except for zero gas flow, the BT-6 shows a 3% higher gassing factor up to an aeration number of 0.16, and a larger difference above that seems likely. In the absence of data for the other deep concave turbines, we would assume their gassing factors are similar to those of the Phasejet. These gassing factors are likely about the same for upper and lower impellers, as there are almost no gas pockets. The gassing factor is essentially the density reduction factor.

Flooding is more complicated than it is for a Rushton, similarly to the situation of the CD-6 impeller. This is shown in Figure 5.23, taken from Ref. [6]. Below an

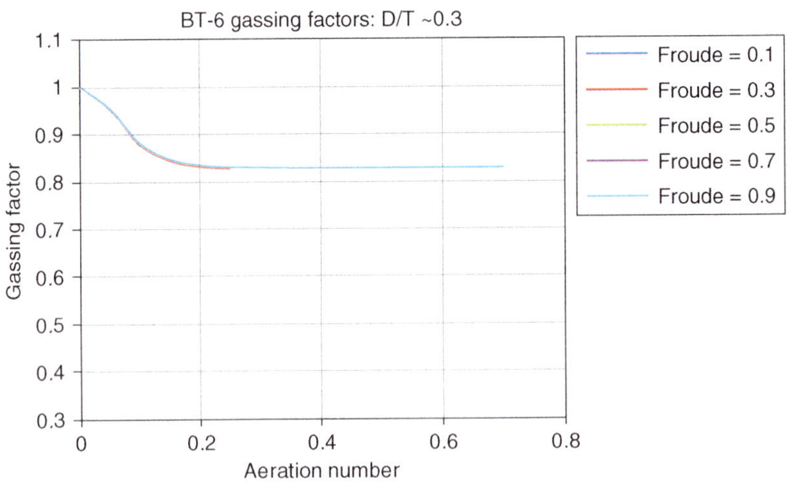

Figure 5.21 BT-6 gassing factors.

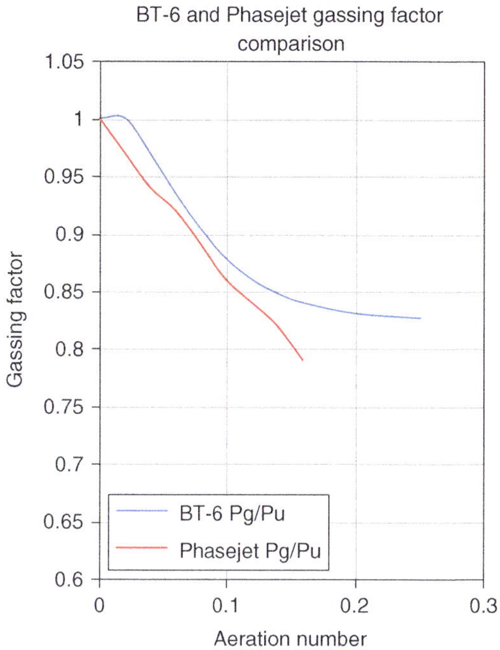

Figure 5.22 BT-6 vs. Phasejet gassing factors.

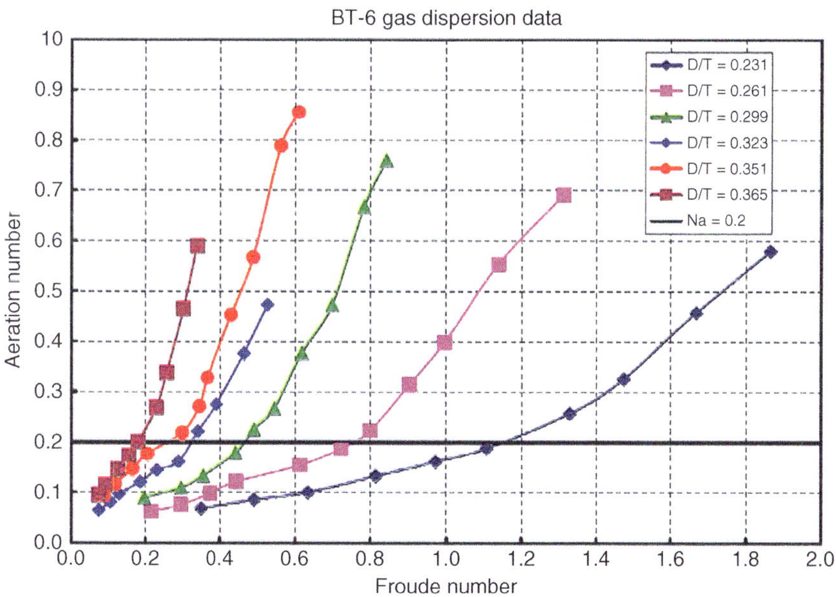

Figure 5.23 BT-6 gas dispersion data. *Source:* Myers et al. [6]. © 2006 American Institute of Chemical Engineers.

aeration number of 0.2, the maximum aeration number is roughly linear with respect to Froude number. The slope increases as D/T increases. Above an aeration number of 0.2 the slope increases even more, and it increases by a larger factor as D/T increases. Since the BT-6 impeller has on occasion been applied at larger D/T ratios than those shown, it would be ideal if more data were presented. But we expect that its gas handling ability would continue to improve at larger D/T ratios.

The behavior below an aeration number of 0.2 can be approximated by the following equation:

$$N_a(\max) = K_{fl} * N_{Fr}; N_a < 0.2 \tag{5.17}$$

At an aeration number of 0.2, the slope increases, and there is a critical Froude number associated with that change. The behavior above an aeration number of 0.2 and above the critical Froude number can be approximated by the following:

$$N_a(\max) = 0.2 + K_{fl2}\left(N_{Fr} - N_{Frc}\right); N_a \geq 0.2 \tag{5.18}$$

The constants are D/T dependent, as shown in Table 5.4. No correlation as a direct function of D/T has been developed by the author.

Regardless of these equations, it is not recommended to disperse gas below a Froude number of 0.035 for a BT-6 impeller.

Thus, the BT-6 impeller has the following advantages:

- Low sensitivity to viscosity; power number does not rise much until the Reynolds number drops below 600.
- Highest gassing factor of any impeller the author is aware of
- Highest gas handling capacity of any impeller the author is aware of (about 5–5.5 times that of the Rushton at equal gassed power)

It has a few disadvantages as well:

- Heavier than Rushton due to its low power number and increased vertical load fluctuation. Possibly lighter than other deeply concave turbines, however, as its power number is a bit higher.

Table 5.4 BT-6 flooding constants.

D/T	K_{fl}	K_{fl2}	N_{Frc}
0.231	0.17	0.53	1.17
0.261	0.26	0.91	0.77
0.299	0.43	1.43	0.47
0.323	0.62	1.45	0.325
0.351	0.8	1.75	0.25
0.365	1.1	2.27	0.18

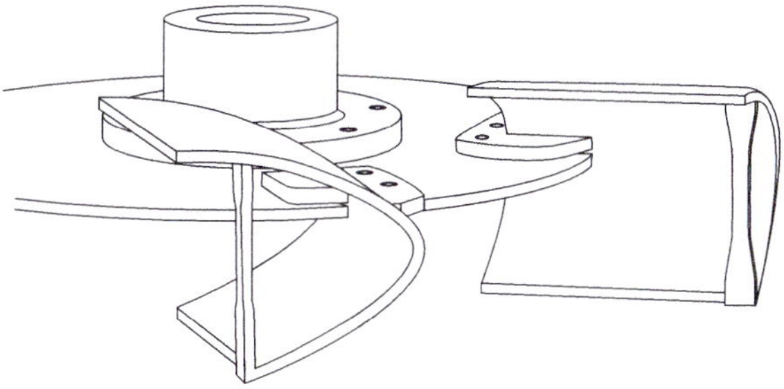

Figure 5.24 Blade struts.

- In common with other deeply concave turbines, if subjected to an overload, the blades can open up, increasing overload and leading to catastrophic failure. However, this can be prevented in the design phase by adding struts as shown in Figure 5.24. The strut design must avoid von Karman vortex shedding.

Examples of Mixed Flow Impellers

Though one could imagine a number of ways to get to a mixed flow pattern, the main way is to have a nominally axial flow impeller but with a blade angle greater than the boundary separation angle. In turbulent flow, that is usually somewhere around 40°. Up until hydrofoil impellers came to be popular, the standard 45° pitched blade turbine was called an axial flow turbine. Now we recognize it as a mixed-flow impeller. It is depicted in Figure 5.25.

The version shown in Figure 5.25 is pretty much the industry standard. Its blade w/D is 0.2, and has a turbulent power number of 1.37 (based on the flat-to-flat diameter) over a range of D/T ratios from 0.25 to 0.5.

Pitched blade turbines can be considered to be of mixed flow character with an angle ranging from 40 to 75°. Below 40, they should probably be considered axial flow, and above 75°, they are mostly radial.

Variants of the pitched blade turbine include numbers of blades from 2 to 8, and they can be built with a disc or open-style.

Though pitched blade turbines have been used in fermenters, their use is questionable. Like the Rushton, they coalesce incoming gas flow into pockets behind the blades, and then re-disperse it in the vortex trail. But the energy dissipation rate is less in the wake of a pitched blade turbine, so the bubble size discharged is larger. It may even be larger than the inlet bubble size if the gas is well-dispersed by a lower radial impeller.

Their best use is as a top impeller when the gas in the head space must be sucked in, as is commonly the case with hydrogenators. That concept would also apply

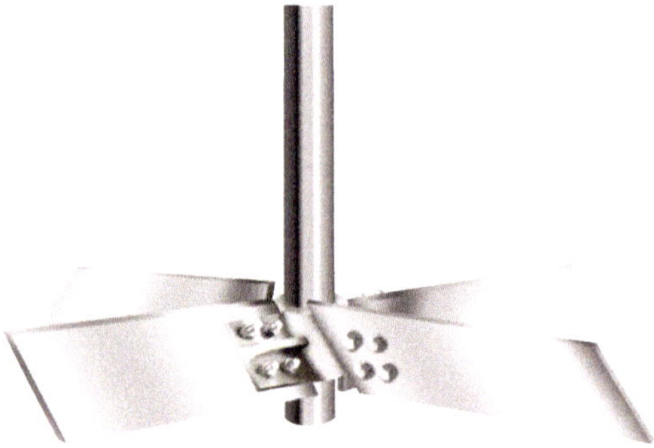

Figure 5.25 Pitched-blade turbine. *Source:* Photo courtesy Chemineer, a brand of NOV. Permission granted by NOV.

when the fed gas is more expensive than air and when there are costs associated with disposal of unreacted gas. See Chapter 11 for more detail.

Though not much has been studied about the gassing factors of pitched blade turbines, it is believed they are similar to hydrofoils.

Advantages include:

- Cheap
- Versatile

Disadvantages include:

- Not very effective at dispersion

Examples of Chaos Impellers

A number of manufacturers produce such impellers. They are mainly categorized by having an inner section that pumps one direction (usually down) and an outer section that pumps the opposite direction (usually up). Examples of such an impeller include the Ekato Intermig, Figure 5.26a, and the ProQuip Doubly Pitched impeller, Figure 5.26b. Another example, which puts a new "twist" on the design, is the Chemineer JT-2, Figure 5.27. This impeller is claimed to greatly reduce blend time compared to competitive products.

Reference [7] lists the power number of the Intermig (two-blades) as 0.23 Private communications report it as ranging anywhere from 0.21 to 0.32. Most likely, the power number is D/T dependent. The power number of the JT-2 is unknown to the author, but is probably higher. These impellers are almost always applied using more than one, and they are normally staggered 90° to each other.

Such impellers have a variable flow pattern that is D/T dependent. When the transition from the inner section to the outer section is set at 70.7% of the tank diameter, the area available for up flow and down flow is equal. This produces the

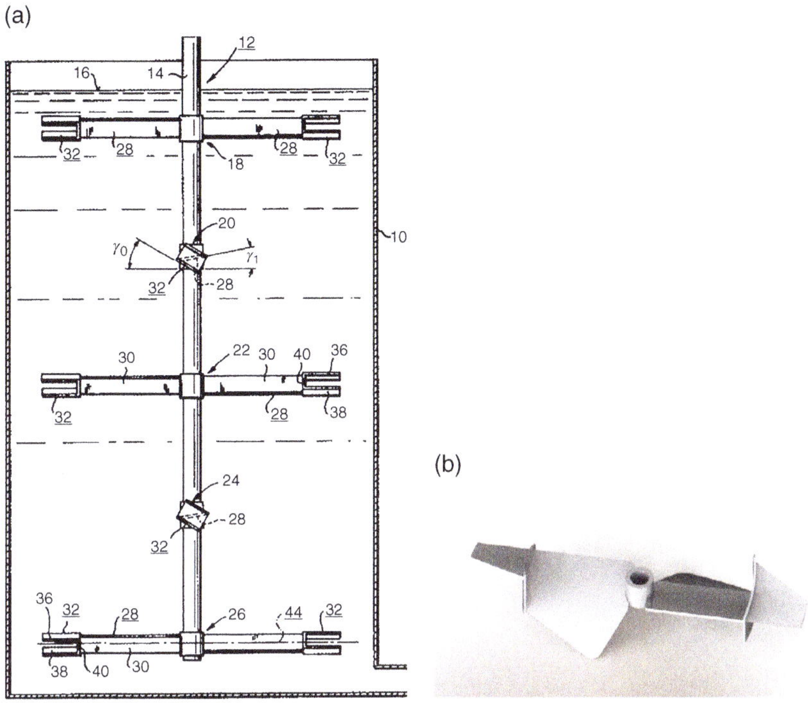

Figure 5.26 (a) Ekato Intermig patent drawing. (b) ProQuip Doubly Pitched impeller. *Source:* ProQuip Inc.; with permission.

most axial flow pattern, though there is still a lot of fluctuation. If the impeller is applied at a larger diameter, more swirl (tangential flow) occurs. Smaller diameters create a flow pattern that is both more chaotic and more radial.

Though these impellers are less effective than most others for blending in either laminar or turbulent flow, they can excel in transition flow. Normal axial, radial and pitched blade turbines can create torus-shaped dead zones which are very slow to blend, illustrated in Figure 5.28. These zones have a virtual surface which rotates around a circular core, and the entire torus slowly rotates around the shaft axis. There is usually one of these above and below each impeller when the Reynolds number is in the low hundreds. Under such conditions, the part of the tank volume outside the torus structure may blend in seconds or minutes, but the inside of the torus intermingles with the bulk fluid very little, so it may take hours to complete the blending. The cure to such phenomena is to introduce more irregularity to the flow pattern. Ways of doing that with conventional impellers include off center mounting, angle mounting, asymmetric baffling, unequal impeller diameters, different impeller types on the same shaft, ramping the shaft speed up

Figure 5.27 Chemineer JT-2. *Source:* Photo courtesy Chemineer, a brand of NOV. Permission granted by NOV.

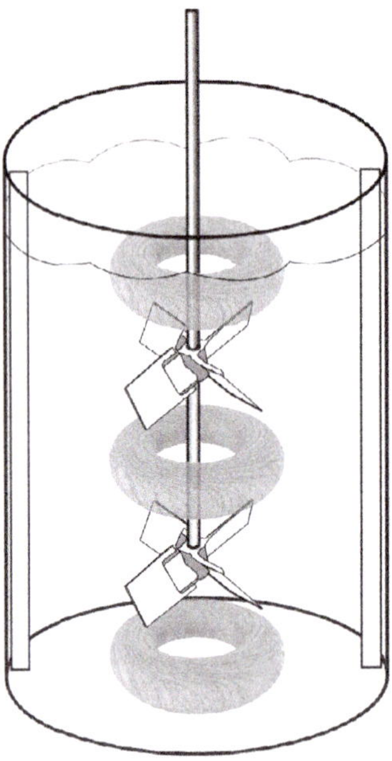

Figure 5.28 Torus zones.

and down to change the Reynolds number and staggering the impellers, among other things. Some of these methods increase the hydraulic force imposed on the shaft. However, chaos impellers introduce the needed irregularity to avoid formation of these torus structures, and can dramatically reduce blend time in the transition flow regime.

The above blending benefits are important in some applications. However, they would not seem to apply in a gas–liquid contactor, as the gas bubbles would break up any torus structures that tried to form. Moreover, most fermenters are operated in the turbulent regime, even for fairly high viscosities, due to the high power input.

Due to the low power number, it is difficult to invest the required power with these impellers if used as a primary disperser. The author has observed a multi-impeller lab test where all impellers were of this type, at normal power inputs. Generally, the bottom 2 or 3 impellers were flooded. When the total cumulative power in the vertical direction reached a sufficient level, the upper impellers dispersed the gas, but that means that gas dispersion in the bottom of the vessel was poor, and did not take advantage of the pressure driving force for mass transfer at the bottom of the vessel. When the bottom impeller was replaced with a radial gas-dispersion impeller, the dispersion was OK throughout the vessel. But is unclear to the author what advantage, if any, using such impellers as upper impellers would have over conventional axial flow impellers.

Shear Effects

Some microorganisms are sensitive to shear. Individual cells, such as mammalian cells and insect cells from complex organisms are often more so. How to minimize shear damage will be covered in more detail in the Chapter 6. Here we will describe different attributes of shear and shear damage.

There is a common misunderstanding that shear rate is somehow inextricably tied to the kind of impeller used. We will show that depending on conditions, a "low shear" impeller may in fact produce more shear and shear damage than a "high shear" impeller.

To explain this, we will cover average shear, maximum shear, and the real source of shear damage.

Conventionally, the average shear rate of an impeller in the volume swept by the impeller has been thought to be proportional to shaft speed, and mildly dependent on impeller type, using the Metzner–Otto concept [8]. This can be expressed as:

$$dv / dx = K_s * N \tag{5.19}$$

Table 5.5 gives commonly used values of K_S.

Reference [9] suggests that the value of K_S may be quite a bit lower for axial flow impellers, as this work indicates that K_S is proportional to N_Q, and may be as low as 3.5 in the turbulent regime, and even lower in laminar flow. Regardless of which set of values is used for K_S, the resulting calculated shear rates are quite low, especially in production-scale equipment, where the shaft speed is often less than 100 rpm. It is unlikely that many organisms would be damaged by the average shear rate in the impeller zone.

However, the average shear rate is not the maximum shear rate. References [10, 11] demonstrate that the maximum shear rate is proportional to impeller tip speed. The proportionality constant could possibly vary with impeller type. However, Ref. [10] studied a non-coalescing system and found that the ultimate mean droplet size was dependent on tip speed only and not impeller type. Specifically, a marine impeller and a saw tooth Cowles-type blade produced the same droplet size distribution at the same tip speed.

Table 5.5 Metzner–Otto constants for common impellers.

Impeller type	K_s
High efficiency hydrofoil	10
45° pitched blade	11
90° straight blade	11
Rushton turbine	11.5
Ekato Intermig, D/T = 0.8	40

How does this apply to a fermenter? With a given impeller type, the power can be invested by using a larger impeller at a slower speed or a smaller impeller at a higher speed. The latter case would have a higher tip speed and thus higher maximum shear, even though it is the same impeller type.

Most modern fermenter agitators use axial impellers in the upper positions. Since they have a lower power number than a Rushton impeller, they must be larger in diameter, and hence tip speed. So, for a given power input at a given shaft speed, the modern axial flow impellers actually produce a higher maximum shear rate than a Rushton turbine! So do the modern concave radial turbines. Thus, using marine propellers in cell culture bioreactors may make little difference based on impeller shear, and could even be worse.

However, impeller shear is far from the highest source of shear in a fermenter. Professor Jeff Chalmers of Ohio State University has studied shear effects in both biological and non-biological systems, and has written a number of papers on the topic, one of which is Ref. [12]. What he found is that the main source of shear damage in a gas–liquid bioreactor is not impeller shear. It is the rupturing of small bubbles at the liquid surface. Both the shear rate and the energy dissipation rate in these rupturing bubbles are orders of magnitude (sometimes more than 5 orders of magnitude) higher than the maximum found in the high shear zone of any impeller. Thus, shear damage caused by impeller shear is, for all practical purposes, negligible. This does not mean that a set of lab or pilot plant data from a fermenter that seems to indicate a dependence on impeller tip speed is entirely incorrect or bad data. In a given piece of apparatus, higher tip speed means higher shaft speed which means higher P/V which means more small bubbles to rupture at the surface. But this is a case of correlation truly not meaning causation. It is the small bubbles, not the tip speed per se, that are the mechanism causing the damage when they rupture. Some of Dr. Chalmers work included looking for shear damage in an ungassed system. He found almost none. But when the same system was gassed, shear damage was readily evident.

Now you might be asking, How do we avoid small bubbles yet keep good mass transfer? We will address that partially in the Chapter 6 on impeller system design, and also to some extent in Chapter 8.

Specialty Impellers

These would be impellers with combined characteristics. They would be difficult to classify under just one heading. So far, their use is rare in fermenters, so I will just give one example: the Ekato Combijet, as shown in Figure 5.29. This impeller is intended to produce a combination of axial and radial flow, and have gas dispersion characteristics.

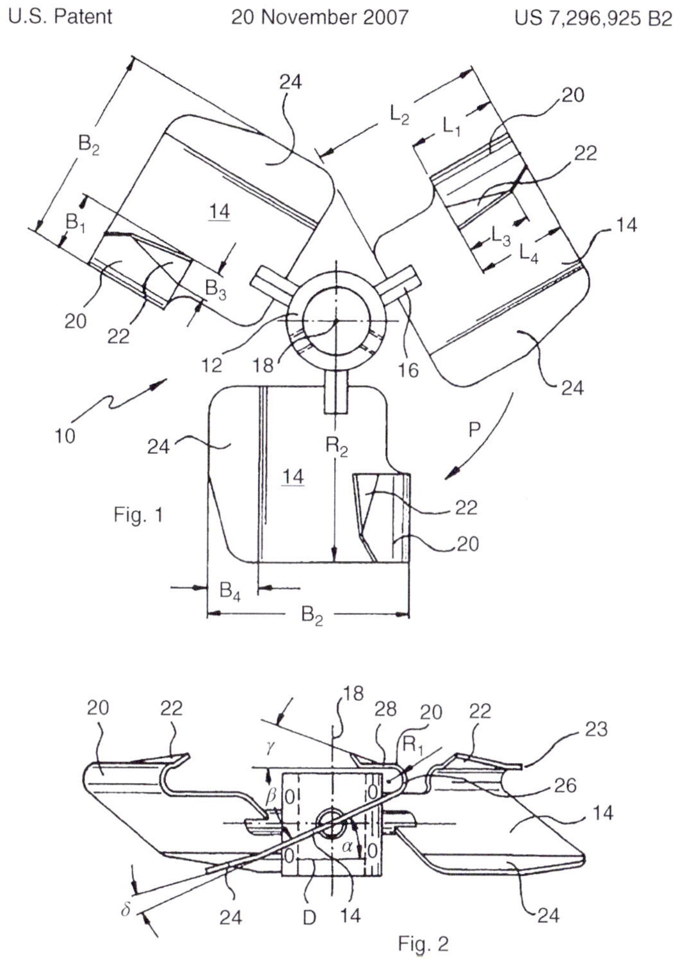

Figure 5.29 Ekato Combijet patent drawing. *Source:* Werner Himmelsbach and David Houlton, 20 November 2007, Agitator with improved blade configuration, US Patent 7296925B2.

Some performance testing of this impeller is presented in Ref. [13], written by Ekato employees. The tests were conducted for the mineral processing industry, and reached a maximum superficial gas velocity of 0.025 m/s, which is moderate by fermenter standards. Unfortunately, all of the results are plotted as comparative ratios, rather than as absolute numbers. Therefore, I was unable to discern such characteristics as power number, gassing factors or flooding characteristics as functions of the normal variables. It is hoped that more data will be presented for such impellers in the future.

Summary of Chapter

We have identified the major categories of impellers used in bioreactors and fermenters, and given examples of each. When possible, we have provided a limited amount of quantitative performance data for some example impellers. Such data may include power numbers, pumping numbers, gassing factors and flooding characteristics. Most such data is held as private information by various manufacturers, so we must depend on them to do final calculations. The data presented here should be sufficient for ballpark estimation of impeller sizes and gas handling ability.

Impeller development is a continuing process. More specialty impellers will be on the market as time progresses. My own personal desire would be to see the development of axial flow impellers with higher gassing factors, comparable to those of the best radial impellers. Regrettably, I have no idea how to make this happen, or if it is even possible.

List of Symbols

D	Impeller diameter
F_g	Alternate symbol for gas flow rate
g	Acceleration due to gravity
K_{fl}	Flooding proportionality constant
K_{fl2}	Flooding proportionality constant above $N_a = 0.2$ for BT-6
K_s	Shear rate constant
N	Shaft speed
N_a	Aeration number
N_{Fr}	Froude number
N_{Frc}	Froude number where slope changes for BT-6 flooding curve
N_i	Alternate symbol for shaft speed
N_P	Power number
P	Power draw
P_g	Gassed power draw
P_u	Ungassed power draw
Q_g	Gas flow rate
T	Tank diameter
V	Liquid volume
w	Blade width or height
ρ	Liquid density

References

1 Brujin, W., van't Riet, K., and Smith, J.M. (1974). Power consumption with aerated Rushton turbines. *Trans. IChemE* 52: 88–104.

2 Middleton, J.C. and Smith, J.M. (2004). Gas-liquid mixing in turbulent systems. In: *Handbook of Industrial Mixing: Science and Practice* (eds. E. Paul, V. Atiemo-Obeng and S. Kresta), 585–638. Hoboken, NJ: Wiley.

3 Myers, K.J., Janz, E.E., and Fasano, J.B. (2006). Gas dispersion capabilities of the CD-6 impeller. 12th European Conference on Mixing, Bologna (27–30 June 2006).

4 Vrabel, P., van der Lans, R.D., Luyben, K.C.A.M. et al. Mixing in large scale vessels stirred with multiple radial or radial and axial up-pumping impellers: modelling and measurements. *Chemical Engineering Science* 55: 5881–5896.

5 Daub, A. (2015). Hydromechanical stress in aerated stirred tanks. PhD thesis. Rhein-Westfalia Technical Institute, 28 May 2015.

6 Myers, K.J., Janz, E.E., and Fasano, J.B. (2006). Gas dispersion capabilities of the BT-6 impeller. Proceedings of the 2006 Annual AIChE Meeting, Multiphase Mixing I: Gas-Liquid Mixing, Paper 554b, San Francisco, CA (16 November 2006).

7 Hidalgo-Milan, A., Zenit, R., Palacios, C. et al. (2012). On the hydrodynamics characterization of the straight Maxblend® impeller with Newtonian fluids. *Chemical Engineering Research and Design* 90 (9): 1117–1128.

8 Metzner, A. and Otto, R. (1957). Agitation of non-Newtonian fluids. *AIChE Journal* 3: 3–10.

9 Wu, J., Graham, J., and Mehidi, N. (July 2006). Estimation of agitator flow shear rate. *AIChE Journal* 52 (7): 2323–2332.

10 Fondy, P. and Bates, R. (1963). Agitation of liquid systems requiring a high shear characteristic. *AIChE Journal*: 338–342.

11 Rodgers, T. and Cooke, M. (2012). Correlation of drop size with shear tip speed. Proceedings of the 14th European Conference on Mixing, Warzawa (10–13 September 2012).

12 Ma, N., Koelling, K., and Chalmers, J. (2002). Fabrication and use of a transient contractional flow device to quantify the sensitivity of mammalian and insect cells to hydrodynamic forces. *Biotechnology and Bioengineering* 80: 428–437.

13 Jung, J. and Keller, W. (2016). Process and cost optimized agitator solutions for hydrometallurgical base metals processing. Paper presented at the Lead-Zinc Conference, Pb-Zn 2015, June 14-17, Dusseldorf, Germany, and in World of Metallurgy, ERZMETALL 69, No. 2, pp. 108–118.

6

Impeller Systems

This chapter presents an overview of the main steps and logic required to achieve the best impeller system design. This mainly applies when the tank geometry is such that more than one impeller must be used. There is no one system design that is best for all applications, but the principles in this chapter should enable one to choose one that is close to the best for most fermenter/bioreactor applications when all process requirements are known.

This chapter assumes that the required power and gas flow have been calculated, using the principles of Chapter 8, at the design conditions required. We believe this chapter provides useful background information necessary to have prior to Chapter 8, though the reader will come back to several earlier chapters to complete the design.

Why Do We Need a System?

The rest of this question is, Why not just use the best gas dispersion impeller? If the tank is tall, why not just use as many as needed? To answer this question, we will delve into a bit of (qualitative) reaction engineering and then a bit of history.

Reaction Engineering

A bioreactor is a reactor. But unlike in a purely chemical reactor, the main "reactants," the biomass, can only thrive and produce the correct product over a narrow range of conditions, such as temperature, pH, and DO. If the conditions are outside this target range, several things can happen. The cells can make the wrong product. They can make product more slowly or with lower yield. They can go into a reduced metabolic state, which can take time to recover from. Or, in some cases, they can simply die.

Agitator Design for Gas–Liquid Fermenters and Bioreactors, First Edition. Gregory T. Benz.

Contrast this to a chemical reactor, especially one that does not have complex reactions such as competitive–consecutive reactions. In such a reactor, it is OK to have a high concentration of a reactant in one place and a low one in another, as long as the overall rate of reaction is as planned. This will be the case for a first-order irreversible reaction. One could even get a better overall rate with a second order reaction by having a higher concentration of dissolved gas in the bottom. So, in a gas–liquid system, placing most of the power in the bottom impeller for a purely chemical reaction may not harm the overall reaction rate or yield. In some cases, it might actually make it better. Such is not usually the case in a bioreactor. It is important to maintain the DO and other nutrients in the target range throughout the vessel.

So, the job of an agitator is somewhat different in a bioreactor versus a chemical reactor. Mass transfer is job number one, but distribution of nutrients is also very important. Insufficient performance in either area will result in an underperforming bioreactor or even a complete failure.

Fermenter History

When Penicillin was first made in commercial scale equipment in the 1940s, the agitators had multiple Rushton turbines. These did a good job of dispersing gas, but not such a good job of creating top-to-bottom movement, which is necessary to minimize the DO gradient. Nonetheless, use of multiple Rushton turbines remained the norm for almost all production fermenters into the early 1980s. Such facilities rarely had fermenters larger than 150 m^3 working volume. If more production capacity was needed, they just added more fermenters.

Even in that scale, there was evidence that large DO gradients affected production. So, some companies experimented with replacing the upper impellers with pitched blade turbines, and later, axial flow hydrofoils. Their reasoning was that once the gas was dispersed by the bottom impeller, the upper impellers just had to circulate the broth. Many of these installations increased the power in the bottom dispersing impeller and used upper impellers that, on paper, matched the pumping capacity of the lower impeller. That resulted in each upper impeller investing less than 10% of the power of the bottom impeller. Some installations even had the sum of the power in the upper impellers adding up to only about 10% of the total power. To the author's knowledge, all of the installations designed this way were failures. Productivity decreased.

The reason they failed was a flaw in the design premise: that the upper impellers need only circulate the liquid. That would be true if the gas bubbles, once formed, would never coalesce. However, they do coalesce, albeit to varying degrees. Though fermenter broth is often called "non-coalescing" and tap or distilled water is called "coalescing," the reality is that fermenter broths do allow coalescence, but at a slower rate than tap water. As the gas bubbles rise, some or many of them will collide with each other and coalesce into larger bubbles, diminishing mass transfer. Therefore, the job of the upper impellers is not just to circulate. It must also maintain the dispersion and even redisperse the gas that has coalesced. To make this happen, the upper impellers need more power invested than just that required to match the pumping capacity of the lower turbine. In fact, there is an optimum power split, which will be discussed in the detailed impeller procedure logic.

When the optimum power split is applied, mass transfer may improve due to more metabolic activity, as the organisms are not cycled through unfavorable conditions. In one case I personally worked on, a customer had a 150 m^3 fermenter with 500 Hp installed, and four Rushton turbines. They had a DO probe at the top and the bottom. Under peak OTR conditions, they measured a DO value of 3 mg/l at the bottom and 0 at the top. I recommended they increase the diameter of the bottom impeller and replace the upper impellers with hydrofoils. The power distribution was based on what I had personally determined to be optimum, which also agreed with the work of others in my company at the time. After the retrofit, the DO probes read 3 and 2, respectively, the OTR was the same, and undesirable byproduct formation was reduced. I call that a success.

Later in the 80s, a trend started that has continued to this day. Rather than add capacity by just using more fermenters, many large production facilities today build fewer and larger fermenters. These save both capital and operating costs. Some are larger than 450 m^3 working volume. Such large fermenters are made possible in part by more vendor experience building large agitators (some over 1200 kW motor size) and in part by the better DO gradient control afforded by properly designed, modern impeller systems.

Steps to Impeller System Design

We list the major steps here and then go into detail on each one afterwards. The details can be extensive.

- Choose number of impellers
- Choose placement of impellers
- Choose type(s) of impellers
- Choose power split or distribution
- Choose D/T and shaft speed (they go hand in hand, as we will see)

Choose Number of Impellers

To assure that bubble distribution is relatively even throughout the fermenter, the number of impellers should follow the guidelines presented in Table 6.1. These are based mainly on the author's experience and observations in transparent vessels. The number of impellers depends on viscosity and whether the impellers are axial or radial flow. A single impeller can handle up to the Z_g/T ratio in the second column. Each additional impeller can handle an incremental section as listed in the third column. Round up the number for increments larger than these. For example, for a 50 cP liquid and axial/radial combinations, with a Z_g/T ratio of 2.5, we need one impeller for the first 0.8T increment, leaving a 1.7T increment for the upper impellers. The axial upper impellers can handle up to 1.0T each. 1.7 needs to be rounded up to 2, so we need a total of three impellers: one radial at the bottom and two axial on top.

Mathematically, this can be expressed as:

$$\text{No. of impellers} = 1 + \text{Roundup}\Big(\big(Z_g / T - Z_g / T(\text{first impeller})\big) / Z_g / T(\text{increment})\Big),\ 0\,\text{digits} \tag{6.1}$$

Choose Placement of Impellers

Impeller placement is based on the gassed liquid level, Z_g. In systems with high gas holdup, this sometimes means the upper impeller is located above the maximum ungassed liquid level, Z. The traditional rule of thumb places the bottom radial impeller at T/6 off bottom, measured from the lower edge of the impeller blades. In practice, in many fermenters, by the time room is made for the sparger and the steady bearing, it is not possible to place the lower impeller that close to the bottom. It is OK to place it higher if needed to clear these items. A small loss

Table 6.1 Quantity of impellers.

Viscosity, cP ($n' > 0.5$)	Z_g/T for first impeller (radial)	Z_g/T increment: radial/axial
<500	0.8	0.6/1.0
500–5000	0.6	0.5/0.8
Xanthan, Gellan, or similar gums	0.5	NR/0.6

in driving force and mass transfer potential is the only consequence, which can be compensated for by a very slight increase in airflow. If there is enough agitation to drive bubbles to the tank bottom, the loss in performance will be negligible.

Upper impellers should normally be evenly spaced between the lower impeller and the gassed liquid surface. This is not an iron clad rule and can be adjusted for specific circumstances.

One example is when a fermenter has a relatively constant gassed liquid level and a strong foaming tendency. If up-pumping axial impellers are used, placing one about D/2 below the surface, or slightly lower, will help reduce foam generation and even pull in some of the foam. If the Z_g/T value is near the maximum for the number of impellers chosen, it is OK to add 1 more impeller.

Another example is when a fermenter is operated in fed-batch mode (variable liquid level), and the peak OUR occurs at a partial level such that it is just below one of the upper impellers or barely covering it. In such a case, it can be useful to lower one of the impellers to make sure it is fully covered, so the power draw is increased, aiding mass transfer when it is needed the most. It is good practice to do design calculations for every significant liquid level in a fed-batch fermentation.

Choose Type(s) of Impellers

This topic is likely to always be a work in progress, as impeller research is a never-ending story, and there is a tendency to create new impeller styles for special purposes, sometimes without doing the necessary experimental testing for the whole range of conditions over which they may be applied. In this section, we will only discuss proper choice of impellers from the list of kinds we have discussed in Chapter 5.

There may be some disagreements within the fluid agitation community about which impeller system is best for various fermenter conditions. But there is an almost universal consensus, at least, that multiple Rushton or other radial turbines is not the way to go in large scale fermenters, due to the staged flow pattern and resultant poor top-to-bottom mixing and large DO gradients. Today, almost all agree that some axial flow is needed, which can be gotten by a combination of a lower radial impeller plus upper axial impellers (either up or down pumping), or sometimes, all-axial (generally all up-pumping). Figures 6.1 and 6.2 qualitatively compare the flow patterns of an all-radial system with one comprised of a lower radial plus upper up-pumping axial impellers. Both systems have four impellers.

Oddly, virtually no pilot fermenters are sold with modern axial and radial impeller combinations unless the customer asks for it. We comment more on that in Chapter 7.

Figure 6.1 Multiple radial flow pattern.

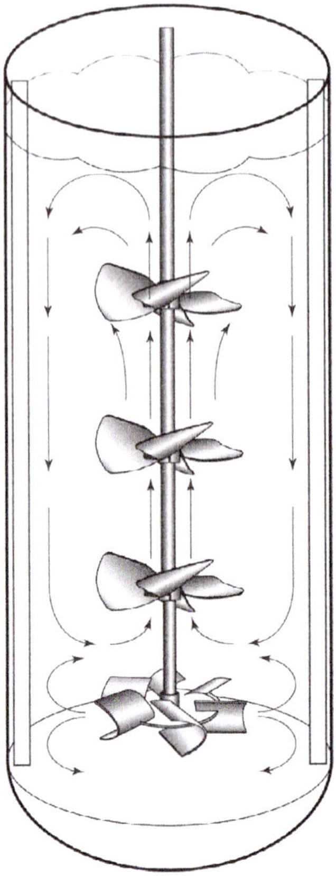

Figure 6.2 Axial–radial flow pattern.

At this point, we can probably conclude that we need an axial radial combination for best results. But what specific kinds of impellers should be used? The answer to that depends on whether we are designing to load the machine fully in the gassed condition or the ungassed condition and, if in the gassed condition, whether there is a variable gas flow requirement. That makes three main cases for design, broadly speaking.

Let us deal first with the case of designing to be able to operate ungassed at full speed and motor load (as an aside, the author does not normally recommend designing this way. The reasons will become evident shortly). When gas is applied, the power draw will go down. And since the mass transfer coefficient depends on the gassed power input, impellers with the highest gassing factor will give the highest value of $k_l a$. Based on that criterion alone, the best impeller system would be multiple deeply concave turbines, especially with the top overhang, as exemplified by the Chemineer BT-6. However, the staging resulting from such a system would be counterproductive in a biological system. So, the best system would have a lower deeply concave turbine with upper axial impellers having the highest gassing factor possible. In general, high-solidity axial impellers have a slightly higher gassing factor than low-solidity, and up-pumping has a slightly higher gassing factor than down pumping, plus other advantages. So, for this design requirement, the best impeller system would be a lower deeply concave radial with an upper blade overhang plus upper up-pumping high-solidity axial impellers. Still, the net system gassing factor will probably be about 0.75–0.8 for a large system, under loading the motor. That is OK if the net gassed power is still sufficient for mass transfer, but may be more costly and the motor will be operating at a less efficient condition.

The second case designs for loading the motor in the gassed condition, with a fixed or only slightly variable gas flowrate. Such an approach is uncommon, as most fermenters do, in fact, have a variable gas flowrate. As we will see in the third case, that can make a huge difference. But this design case may be used in a continuous-flow fermentation, where the conditions are steady after start-up.

This case is not very straightforward, as it is comparing impellers on the basis of equal gassed power. Many researchers believe that there is little difference between impellers at equal gassed power input, in terms of the mass transfer coefficient. However, in reference [1], (which is misnamed; the authors studied more than two impeller systems), the authors report some $k_l a$ differences even at equal gassed power input. In total, they actually studied four systems: a Chemineer CD-6/HE-3 down-pumping combination (equivalent to a Smith turbine and low-solidity hydrofoils), a Chemineer Maxflo-W4 (abbreviated as MW-4) up-pumping system (high-solidity up-pumping), a Lightnin A-340 up-pumping system (high-solidity up-pumping), and a Chemineer BT-6/Maxflo W4 up pumping (deeply concave asymmetric radial plus high-solidity up-pumping.

The differences in these systems were a bit confounding, in that none of these systems outperformed the rest across the board. Not surprisingly, there was no statistically significant difference between the two all-axial high-solidity up-pumping systems tested. Therefore, we can lump the comparisons together and just discuss the three unique systems.

At low gas velocities (U_S ~0.03 – 0.04 m/s) and low P/V (about 0.25 W/l), in a 1 cP fluid, the mass transfer coefficient was about the same between the BT-6/MW-4 system and the up-pumping high-solidity system, and the CD-6/HE-3 system had about a16% advantage. At the same gas flow, in a P/V range from 0.5 to 1.0 W/l, the CD-6/HE-3 system and the all-axial system were essentially equal, but the BT-6/MW-4 system gained a 10–12% advantage. As the P/V approached 2 W/l, the BT-6/MW-4 retained that advantage over the all-axial system, which in turn gained about a 13–14% advantage over the CD-6/HE-3 system.

So, at low P/V and low gas flow, it seems to make little difference, but the HE-3/CD-6 system came out ahead. At middle to high P/V, the BT-6/ Maxflo system was the clear winner, with the all-Maxflo system surprisingly coming out ahead of the CD-6/HE-3 system.

When the gas flow was increased by a factor of 10 ($U_S \sim 0.3 - 0.4\,m/s$), at low P/V (0.75 W/l), the CD-6/HE-3 system had about a 10% advantage over the other two, which were statistically equal. The advantage diminished at P/V = 1.5 W/l. At P/V levels of 3 and 6 W/l, all three systems had the same mass transfer coefficient within statistical accuracy.

The above two data sets were in an air–water system. A third data set was taken in a 31 cP fluid, with low airflow and the same P/V range as the first data set. The results were a bit odd. The best system across the board was the CD-6/HE-3 system, though it had its greatest advantage at the lowest and highest P/V values, with up to 40% higher k_la than the MW-4 system and 11% higher than the BT-6/MW-4 system at low P/V (0.25 W/l) and a 10%/5% advantage, respectively, at a P/V of 2W/l. The all-MW-4 system had the lowest k_la across the board. It is not clear whether absolute viscosity or Reynolds number contributes to the performance differences between 1 and 31 cP. Many fermenter broths have a higher viscosity than 31 cP. More study in larger scale equipment is needed. But, from a mass transfer standpoint, the axial–radial combinations look pretty good over a range of conditions. The ranking may change with broth properties. It should also be noted that, in actual broth, it is hard to get repeatable k_la values with the same precision as reported in this paper.

To add to the potential confusion, the same paper studied holdup. In general, under the same gas flow and P/V conditions, the CD-6/HE-3 system had the lowest holdup, yet sometimes the highest k_la. While not a prime design variable, lower holdup means that potentially a higher broth volume could be used, increasing annual production by several percent. This could be a significant additional source of revenue, especially given the well-known tendency of plant production personnel to increase broth volume until broth starts coming out the gas discharge opening. On the other hand, the blend time is much shorter with an up-pumping system, which leads to a more favorable DO gradient. It is unfortunate that the paper did not study a CD-6/Up-pumping low-solidity hydrofoil system.

The third case would be the most common: design for variable airflow and sometimes variable batch volume. For such a case, the normal protocol would be to design the impeller system to load the motor at the minimum target airflow that must be handled at full shaft speed. This approach favors impeller systems that have intrinsically high gassing factors. When gas flow increases above the minimum, such impeller systems do not lose power draw as quickly as others, and therefore, they have a higher mass transfer potential at the higher gas flowrates. They are also less likely to flood. Also, if gas flow is reduced below the design minimum, they are less likely to overload the motor, and, if variable speed drive is used, they do not have to be slowed down as much to avoid overload, so they are capable of investing more power at reduced airflow due to the constant torque characteristic of most variable speed drives. This was covered in some detail in Chapter 4, especially in Figure 4.4.

To illustrate the point, we will create a hypothetical batch fermentation cycle, with an airflow profile as shown in Figure 6.3. In this profile, the airflow starts out very low (0.1 VVM) during the initial inoculation, when the cell mass concentration is low and the OUR is low. As the fermenter enters the growth phase, the airflow is increased as needed. It remains constant during the stationary phase and then falls off at the end as nutrients are depleted or cells die off, or any other reason productivity slows down.

To keep things simple, rather than have a variable speed agitator, we simulated one with a two-speed motor, where the low speed is ½ of the high speed. At the beginning of the batch cycle, the agitator is operated at low speed. Five hours into the batch, the airflow is increased to 0.2 VVM and the agitator switches to high speed. That is the design condition: load the motor at high speed at the minimum airflow required at that speed.

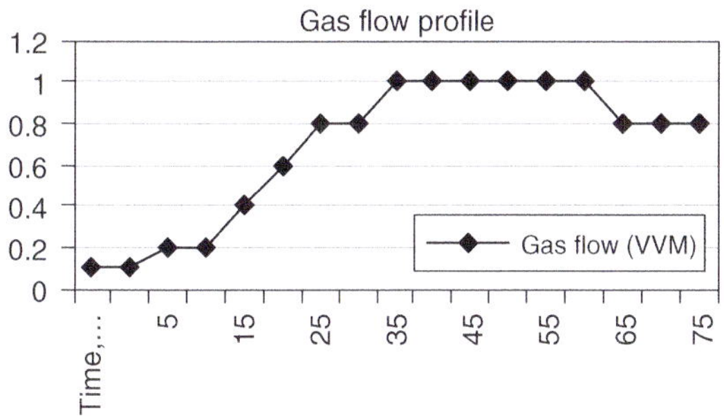

Figure 6.3 Gas flow profile.

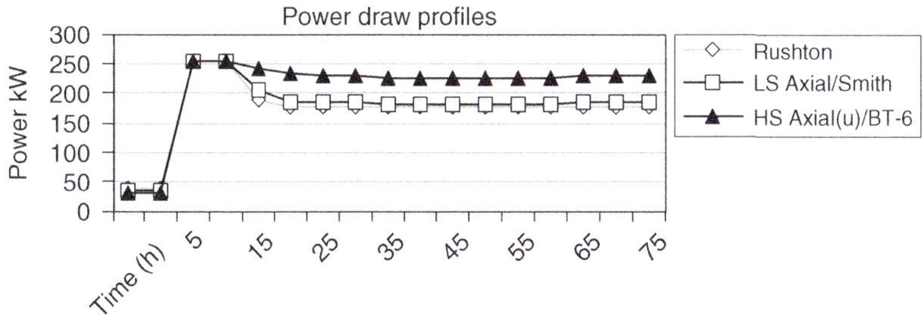

Figure 6.4 Power draw profile.

We model three impeller systems: an all-Rushton system, a down-pumping low-solidity axial plus a Smith turbine (essentially, the CD-6/HE-3 system described above), and an up-pumping high-solidity axial (such as the Maxflo W-4) with a lower BT-6 deeply concave asymmetric turbine. Since the design basis is the same for all systems, they all have the same gassed power draw at 0.2 VVM, as shown in Figure 6.4. However, as the airflow increases, the power draw decreases for all three systems. But it decreases the most for the Rushtons, followed by the Smith/LS hydrofoil system and then the BT-6/HS up-pumping system. At peak airflow, the Rushton system power draw drops 32%. The Smith/LS axial system drops 30%, and the BT-6/HS system drops about 11.5% In other words, the BT-6-based system draws 25–28% more power at peak airflow than the other two systems.

What effect does this have on mass transfer potential? Different broths have different $k_l a$ sensitivities to P/V, but most are in the range of an exponent of 0.2–0.8. Klaas van't Riet [2] reports an exponent of 0.7 for P/V in ionic, non-coalescing liquids (which fermenter broths normally are), so we used that exponent to prepare Figure 6.5. Under peak conditions, the BT-6/HS axial system has a 20% potential mass transfer advantage over the Rushton system and a 13% advantage over the Smith/LS axial, even allowing for a factor of 1.1 on $k_l a$ for that system based on the work in reference [1].

Clearly, for the normal case of variable airflow, a system with a high gassing factor is best overall. A lower deeply concave radial impeller combined with upper high-solidity up-pumping impellers seems to be the best among systems available today. But there is no good substitute for doing your own carefully planned pilot work in order to do a quantitative assessment of impellers and, indeed, to be able to predict and optimize performance. This is especially true when new impellers come out with new marketing claims.

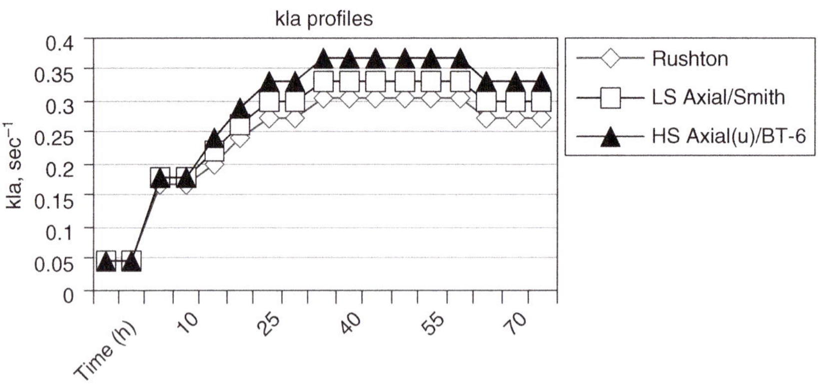

Figure 6.5 $k_l a$ profile.

Choose Power Split or Distribution Among Impellers

Though not explicitly stated, the above comparisons were based on what is believed to be an optimum power distribution among multiple impellers, except for the Rushton systems, which used equal diameter impellers. So just what is this optimum distribution? Based on visual observation and holdup measurements, the following gassed power draw ratios are recommended by the author:

For multiple Rushtons, the lower impeller power draw should be about 15–20% higher than the rest. For axial–radial combinations, the lower radial impeller should have about 2–2.5 times the power draw of each of the axial impellers, with the optimum being about a 7/3 ratio. Note that this split results in the upper impellers each pumping several times as much as the lower radial. This is necessary to avoid coalescence. For an all-axial system, the optimum split has not been studied, but most people use equal diameter impellers, which is pretty much equal power in each. Probably, the lower impeller should be a little larger if it is directly gassed.

The above recommendations are based on tests done in an air/water system, which is considered a coalescing system. It is possible that even more power could be placed in the bottom impeller if the broth is non-coalescing. Ionic broths are often classified as non-coalescing, as are viscous broths. However, this author does not believe any fluid is truly non-coalescing; some just do it more slowly than others. And, remember, early experiments with ~90% of the power in the bottom impeller were usually failures. I would recommend sticking to the guidelines above unless a careful test program shows that a different split will work.

The above guidelines can be tweaked a bit, just as impeller placement can, when a fed-batch fermentation has a peak OUR at an intermediate liquid level. It is OK to alter the power distribution by 10–15% on a given impeller to help get through the peak condition.

Choose D/T and/or Shaft Speed

This is another complex choice, and there is no one fixed answer. We will discuss the results of different D/T ratios here. But first, let us note that gear drives from a given manufacturer have fixed ratios; it is not normally possible to get a custom ratio to get an exact speed needed to get the desired power draw at a given D/T. So, choosing a D/T ratio is just a first approximation. It is really a four-step procedure, as shown below:

1) Choose a preliminary D/T ratio
2) Calculate the shaft speed needed to draw the desired power at that D/T ratio

3) Choose a gear drive with ample service factor at roughly the shaft speed chosen, and look up the exact ratio that gives the closest actual shaft speed, the use that shaft speed as the design basis
4) Recalculate impeller sizes (D/T ratios) at the actual shaft speed available.

Most, though not all, gear drive manufacturers try to have standard speeds within ±4% of the standard AGMA (American Gear Manufacturers' Association) speeds when using standard motor speeds as inputs. Those speeds, in the range likely to be found in production fermenters, are as follows, in units of RPM: 25, 30, 37, 45, 56, 68, 84, 100, 125, 155, 190, 230, 280, and 350. Benchtop and small pilot units usually have much higher speeds than those; many run direct drive at motor speeds.

The basic idea is that it is possible to invest the same power with an array of combinations of impeller size (D/T) and shaft speed. Much of the following discussion came from reference [3], which was also written by this author.

The effects of D/T on fermenter performance are often calculated, for comparative purposes, under the conditions of constant vs. variable gas flowrates. The outcomes can differ significantly. For the case of constant airflow, several specific observations may be made:

1) If impellers of different sizes but equal type and power draw are used, the impellers' shaft speed is obviously lower when bigger impellers are used. In addition as an impeller system's D/T increases at constant P, the impeller tip speed, S, decreases, resulting in lower maximum shear rate imposed by the agitator. Ignoring changes in the power number ($N_P = P/\rho N^3 D^5$) and gassing factor (P_g/P_u), S is proportional to D/T to the minus 2/3 exponent ($S \propto D/T^{-2/3}$). If the organism is shear sensitive, the effect of lower shear may lead to a lower incidence of cell death, though bubble size distribution is likely more important. If the organism tends to form clumpy colonies, added shear associated with the higher tip speed/smaller D/T designs may actually help mass transfer by breaking the colonies into smaller units.
2) As stated above, higher D/T ratios result in a lower agitator-shaft rotational speed, N. Ignoring changes in N_P and P_g/P_u, the agitator–shaft rotational speed, N, will be proportional to D/T to the minus 0.6 exponent ($N \propto D/T^{-0.6}$). This lower shaft speed results in lower average shear (different from the maximum shear rate, which depends on tip speed) imparted to the fluid, but that is likely irrelevant. What is relevant is that a lower shaft speed at constant P means higher torque. This means a more expensive gear drive, a larger shaft and impellers, and a larger support requirement. Agitator cost is roughly proportional to torque raised to the 0.8 exponent in large fermenter sizes. "Large" as used here means roughly above 200 m^3 working volume or above a 200-kW motor.

3) Larger impellers pump more. Higher turnover may aid in discharge of CO_2 at the surface. The pumping number, N_q, of impellers reduces with increasing D/T, so the net effect is roughly that Q is proportional to D/T to the 1.9 exponent ($Q \propto D/T^{1.9}$).
4) Blend time will be slightly shorter at larger D/T ratios. However, this is a minor effect. Blend time correlations are specific to impeller type and so is the D/T effect. However, over the range of D/T usually found in fermenters (0.2–0.6), blending at constant P will generally be faster by 10% or less using large D/T ratios than blend times found at the smaller D/T ratios.
5) Heat transfer coefficient also increases with D/T. How much will depend on the kind of heat transfer surface (such as helical coils, jacket, vertical tubes, or panel coils), but it will generally be in the range of a 0.95–1.05 exponent. The reader is cautioned against designing the agitator specifically for heat transfer. It is better to add more area or a bigger temperature differential if needed. Moreover, large D/T ratios may prevent use of extensive coils or vertical tube bundles in the vessel, due to mechanical interference. Heat transfer is discussed in more detail in Chapter 10.
6) Gas-handling capacity, measured as the gas volumetric flowrate (Q_g), increases with D/T. Sensel's work [4] suggests that gas-handling capacity before flooding, at constant P. is proportional to D/T to the 0.4 exponent ($Q_g \propto D/T^{-0.4}$). If large gas flowrates are anticipated, the flowrates may dictate the required D/T.
7) Mechanical parts such as gears, roller bearings, steady bearings, and shaft seals will all last longer at the lower shaft speeds associated with large D/T ratios, assuming the same service factors and stresses are used.
8) On the other hand, small D/T and high shaft speed can result in design challenges due to shaft natural frequency. In some cases, a larger shaft is required at a higher shaft speed, even though the torque is less. Hollow shafts and multiple steady bearings (ugh!) are possible solutions.
9) As we will see later in this chapter, large D/T and other techniques can be used to create a more narrow bubble size distribution, thereby reducing the number of small bubbles rupturing at the liquid surface.
10) Large D/T may generate less foam.

From the above, one may observe that most performance comparisons favor large D/T ratios at constant gas flow. Higher cost is the main drawback. However, most real fermenters have variable gas flowrates, to avoid wasting compressor power during the early stages of fermentation, while the cell population is low. The next section deals with what happens when one must allow for variable gas flowrates.

D/T Effects with Variable Gas Flowrates

In most fermenters, the air flowrate is not constant. That is because most are batch or fed-batch, rather than continuous. Typically, the airflow begins at a low value, such as when the cell population is low and not much mass transfer is needed. As the cell population increases, the airflow is ramped up and then remains constant for some time. Late in the batch, it may be reduced for various reasons, including control of product distribution. A change in the air flowrate directly affects impeller power draw and mass transfer performance.

The ratio of gassed to ungassed power draw, P_g/P_u, (a ratio also known as the gassing factor) is a function of impeller Reynolds number ($N_{Re} = D^2N\rho/\mu$), aeration number ($N_a = Q_g/ND^3$), and Froude number ($N_{Fr} = N^2D/g$). As D/T increases at constant power, P, values of N_a and N_{Fr} decrease. At such conditions, P_g/P_u changes more rapidly as a function of Q_g than it does at high N_a and N_{Fr} conditions. This means that the power draw of the impeller is more constant at low D/T ratios, which can affect mass transfer performance and required motor power. This concept is best illustrated by example. We will provide two cases. In the first case, motor power is specified. In the second case, shaft power is specified under peak gassed conditions.

Example 6.1

The model fermenter has a tank diameter, T, of 5 m and a working liquid volume, V, of 200 m^3. Using ASME heads, this yields an ungassed liquid level of 10.5 m. Fluid-specific gravity, *s.g.*, is 1.02, and viscosity, μ, is 5 mPa-s. The fluid has sufficient ionic strength so as to be essentially non-coalescing for mass transfer calculation purposes. Process temperature is 40 °C, and back pressure is 1.5×10^5 N/m^2 abs. The impeller system will consist of a lower concave blade disc turbine (180° concavity; CD-6 equivalent), plus two up-pumping, high-solidity hydrofoil impellers. For simplicity, the impeller diameters are set as equal, though in practice they will not always be the same diameter. Although the D/T value found in real fermenters ranges from 0.2 to 0.6, this example will compare more typical values of 0.3 and 0.5.

In this first case, motor power is specified. For this problem, the motor size is set at a typical 420 kW. Q_g is specified as a range of 0.1–0.7 VVM. For this problem, then, 0.1 VVM translates to 20 m^3/min airflow at standard conditions (1 atm. and 0 °C), and 0.7 VVM translates to 140. Power is set at 90% loading at 0.1 VVM for D/T values of 0.3 and 0.5 by using the appropriate shaft speed, then, with this speed fixed, calculated power and k_la will be figured at 0.1, 0.3, 0.5, and 0.7 VVM. The results are tabulated and plotted in Table 6.2 and Figures 6.6 and 6.7.

Performance is the same at the minimum airflow, as expected by setting the actual gassed power essentially the same. However, at 0.7 VVM, the power drops more for the larger impellers, so that the power draw is only 82% of that drawn by the smaller ones, and k_la is about 91% of the value of the smaller impellers. Here

Table 6.2 Example 6.1 data.

D/T effects on fermenter k_la
Case 1: motor power specified (420 kW, 90% loading at 0.1 VVM gas flow)

	D/T = 0.3		D/T = 0.5	
VVM	**P (kW)**	**k_la (1/s)**	**P (kW)**	**k_la (1/s)**
0.1	379	0.172	378	0.173
0.3	355	0.29	308	0.271
0.5	339	0.365	278	0.33
0.7	328	0.424	270	0.385

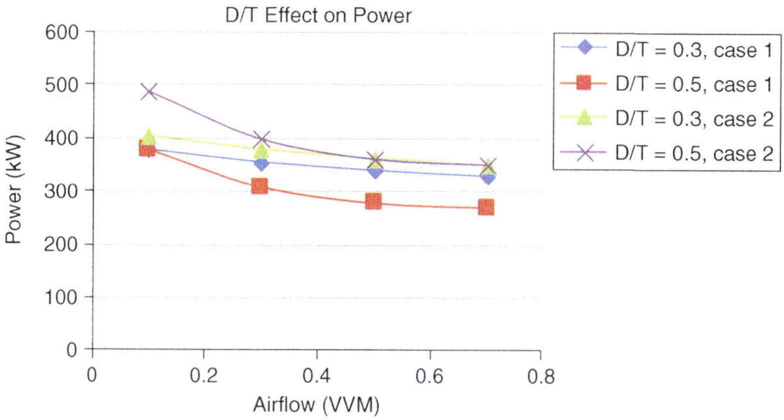

Figure 6.6 Example graphic comparison of power draw.

is clearly a case where smaller may be better, at least in terms of mass transfer performance. Moreover, the larger D/T design agitator costs about 2–2.5 times as much as the smaller design. Better performance for less money! Imagine that!

Example 6.2

The second case is one in which the shaft power is specified at the maximum Q_g, and the motor power must be chosen based on the minimum Q_g (in the author's opinion, this is the "right" way to do it in most cases). Shaft power is specified as 350 kW at 0.7 m^3/min, and P_g has been calculated at 0.1, 0.3, 0.5, and 0.7 VVM, at D/T values of 0.3 and 0.5. Table 6.3 and Figures 6.6 and 6.7 summarize and display the results.

As expected, the values of P and k_la are the same at the design condition of 0.7 VVM. However, the motor size required at 0.1 VVM is 450 kW for the case of D/T = 0.3 and 550 kW for the larger impeller system. Alternatively, the larger system

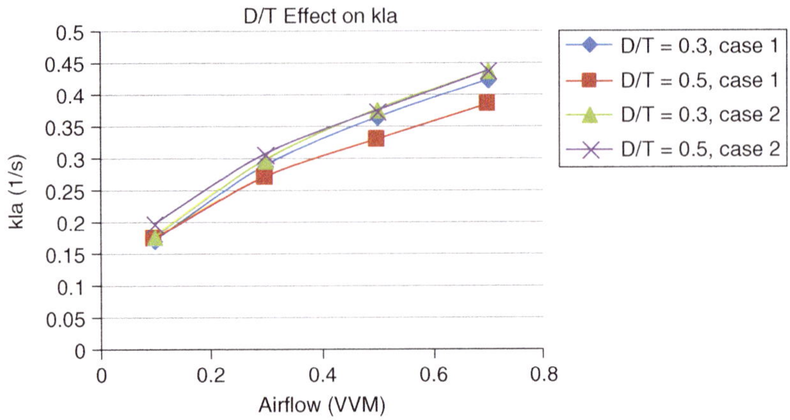

Figure 6.7 Example graphic comparison of k_la.

Table 6.3 Example 6.2 data.

D/T effects on fermenter k_la
Case 2: shaft power specified (350 kW at 0.7 VVM gas flow)

	D/T = 0.3		D/T = 0.5	
VVM	P (kW)	k_la (1/s)	P (kW)	k_la (1/s)
0.1	403	0.177	487	0.197
0.3	378	0.298	398	0.307
0.5	362	0.377	360	0.375
0.7	350	0.438	350	0.438

could use a 450 kW motor if the minimum Q_g were reset to 0.3 VVM. This, of course, would use more compressor power during the early growth stage of the batch. The larger D/T agitator would also cost two to three times as much as the smaller one if a 550-kW motor were used or 1.5–2.5 times as much if a 450-kW motor were used.

Conclusions on D/T Ratio

Agitator power may be invested using high shaft speed and small impellers (low D/T) or low shaft speed and large impellers (high D/T). Although the high D/T case has some advantages at fixed gas flow conditions, in a typical fermenter with variable gas flowrates, a more constant power draw, which is characteristic of a smaller impeller system, may lead to higher mass transfer performance or energy savings. Therefore, one should seriously consider a small D/T design if the following criteria are met:

- Maximum S is within acceptable limits for the organism.
- Organism is not sensitive to small bubble rupturing (high speed and small D/T create a broader bubble size distribution, therefore more fine bubbles)
- Q is sufficient
- Heat transfer is sufficient (hint: add more area if it is not, rather than change the agitator design)
- Shaft design is feasible.
- Gas handling before flooding has adequate safety margin (20% or more).

If the above criteria are met, higher performance can be realized at a lower equipment cost.

Design to Minimize Shear Damage

As stated in Chapter 5, direct shear created by the impeller system is not the main source of shear damage. Instead, rupture of small bubbles at the liquid surface does most of the damage, as the shear rates in a rupturing film may be 100 000 or more times those produced by an agitator impeller. This does not mean agitator design has no effect on shear damage. In fact, it has a major indirect effect, by how it affects bubble size distribution.

Normally, we desire small bubbles for mass transfer, as these can easily give us large interfacial areas. However, we can get just as much area by having more bubbles of larger size. Also, if we could get all of our bubbles to be large enough to avoid the threshold of damage, that would be ideal.

Unfortunately, we cannot create a perfectly uniform bubble size distribution. But we can make the distribution narrower and favor a larger quantity of bubbles by using the following tips:

- As we will see in Chapter 8, we can get the same mass transfer rate using high agitation power and low airflow or high airflow and low agitation power, with the total system power going through a minimum at some airflow rate. However, the high airflow, low agitator power approach produces about the same interfacial area by having more volume of gas in contact with the liquid at a larger mean bubble size. So, to minimize shear damage, use a high airflow and low agitation design, but be sure the impeller is not flooded.
- At a given power input, a large D/T design invests its power over a larger volume in the tank, thus creating a more uniform bubble size distribution, with fewer small bubbles. So, a large D/T is preferred for minimizing bubble damage.

- Similar to the above principle, using more impellers than normal distributes power more evenly, creating a more uniform bubble size distribution. An impeller spacing of T/2 should be considered for shear sensitive organisms.
- Using a large ring sparger (near the wall) allows fewer bubbles to pass through the impeller at low agitation intensity, thereby resulting in larger bubbles (it might not matter as much at high agitation intensity). But mass transfer will be lower. Mostly, this technique is used for sensitive mammalian cell fermentations where the OUR is very low anyway.
- Placing the top impeller further than normal from the surface will allow more coalescence, reducing fine bubble rupture. However, this may lead to a low DO zone.

These tips can lead to much less shear damage. However, those pertaining to larger D/T and more impellers will tend to increase the cost of the agitator. And, operating at high airflow, low agitator power may not be optimum from an energy cost standpoint. If the bubble size corresponding to the threshold of damage is known, it is possible to use CFD techniques to predict a bubble size distribution, which could be very helpful.

Sparger Design

In general, the purpose of a gas sparger is to uniformly distribute gas below the impeller, so as to minimize mechanical asymmetry of loads and allow efficient dispersion. There are many variations of sparger design. Even with the common ring sparger, there are many variants. Such may include location (above or below the impeller; in the discharge plane of a radial turbine if $D_S > D$), relative diameter to the impeller, number and size of holes, etc. Some of these variants are discussed in reference [5]. Reference [2] notes no difference in mass transfer performance for any sparger placed below the impeller and smaller than the impeller. That includes a simple open pipe. This makes some sense, as he was studying Rushton turbines, which coalesce gas before redispersing it. We will discuss three major sparge systems here, bearing in mind that there are a lot of discussions still going on about this topic, and the results may depend on impeller type. The three types are ring sparger, pre-dispersion with a motionless mixer combined with a recirculating line, and a fine bubble diffuser.

Ring Sparger

This is by far the most common way to distribute gas. To keep the ring from filling up with broth, it is recommended that the holes be on the bottom of the sparger, as shown in Figure 6.8. The shape of the ring need not always be circular. Indeed,

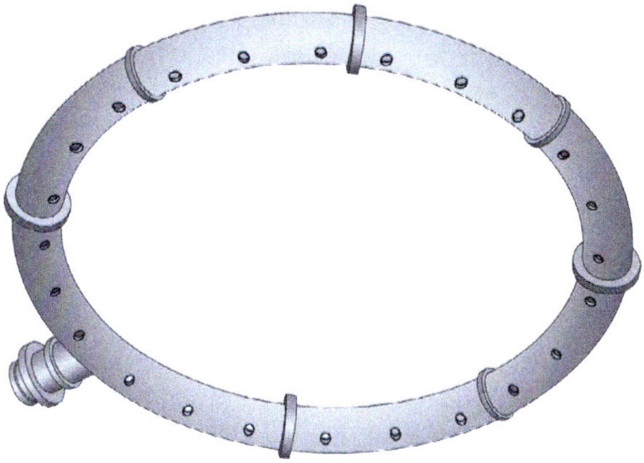

Figure 6.8 Ring sparger.

at high gas flowrates, the pipe size required to carry the flow may be difficult to bend into a circle. It is acceptable to make an octagon out of straight sections and 45° elbows and even a square with 90° elbows. It is not necessary to use small holes or lots of them, particularly with Rushton impellers, which will coalesce and then redisperse the gas. For deeply concave turbines, or axial turbines, it could be that there is a minor difference, but the author is unaware of this being studied. In general, there should be at least eight holes, except in laboratory scale equipment where four holes are OK for impellers less than 150 mm diameter. For large spargers (say, >1.5 m diameter), up to 36 holes may make sense just to avoid having to drill holes larger than 25 mm diameter.

To avoid gas coming out from only some of the holes, the sparger should be level and the pressure drop at peak airflow should be about 0.3 bar. Higher pressure drop just wastes power.

Allowable flow per hole may be calculated by solving the appropriate equation below, according to the ratio of pressure drop to incoming pressure, compared to the product of the specific heat factor times the pressure differential ratio factor.

Case 1, $(p1 - p2)/p1 < F_y{*}X_t$; $X_t = 0.72$. For air, use Eq. (6.2)

$$Q_a = 0.0695 * C * (d_o / 4.654)^2 * p1 * \left(1 - ((p1 - p2)/p1)/(3 * F_y * X_t)\right) * \left(((p1 - p2)/p1)/T_a\right)^{0.5} \quad (6.2)$$

Case 2, $(p1 - p2)/p1 \geq F_y{*}X_t$; $X_t = 0.72$. For air, use Eq. (6.3)

$$Q_a = 0.0464 * C * (d_o / 4.651)^2 * p1 * (F_y * X_t / T_a)^{0.5} \quad (6.3)$$

To assure the gas is evenly distributed, the pressure drop through the holes should be about 10 times as high as the pressure drop through the ring. So, if the total design pressure drop is 0.3 bar, the pressure drop through the holes should be 90% of that, or 0.27 bar. That will be achieved if the cross-sectional area of the ring pipe is greater than or equal to the total hole area. So, for example, if we have (24) 10 mm holes, the total hole area is 1885 mm^2, and therefore, the minimum ring pipe size should be about 49 mm ID. Since the feeder pipe splits the air into two paths in the ring, it should have twice the area of the ring pipe, or, for this example, about 69.3 mm ID.

But the crux of the design problem is really deciding how big the sparge ring should be, and where we should put it. The most common location is below the lower (dispersing) impeller, at a diameter less than or equal to the impeller diameter. In reference [2], the sparger type seems to be irrelevant as long as it is below the impeller, quoting ". . .no indication is given for the influence of the sparger. This should indeed be absent because the bubbles are sucked up and dispersed by the cavity immediately after sparging. Also, the stirrer type does not appear in the relations. The literature shows that only the energy dissipated in the vessel is determining k_la, independent of stirrer type." The latter statement contradicts some of the conclusions in references [1, 4]. Reference [5] says that a sparger placed above the impeller increases its gas-handling ability and power draw. It also says that a large ring sparger placed in the plane of the impeller increases gas handling, power draw, and mass transfer coefficient. On the other hand, reference [6], among others, shows that, while sparging outside of the impeller diameter (or anywhere but below the impeller) may increase gas handling and power draw, it reduces the mass transfer rate. In my own opinion, it is necessary to directly gas a dispersing impeller to get the best dispersion and mass transfer. The main reason to consider a large ring sparger is to avoid formation of fine bubbles so as to reduce shear damage. This should only be necessary in cell culture fermentations. The OUR is so low in those fermentations that an increase in airflow to increase OTR is a viable approach.

There are some differences between dispersing impellers in terms of the best ratio of sparge diameter to impeller diameter. The recommended ranges are given in Table 6.4.

Table 6.4 Sparge ring diameter ratios.

Impeller type	Rushton	Concave radial	Axial up-pumping
Minimum d_s/D	0.75	0.5	0.5
Maximum d_s/D	1.0	0.7	0.75

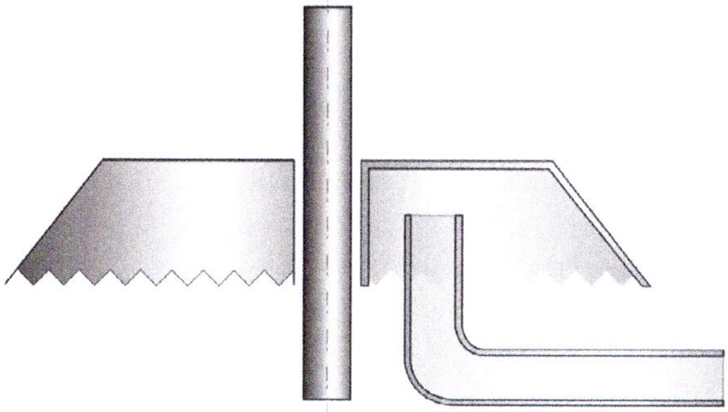

Figure 6.9 Pie-plate sparger.

An alternative to a sparge ring, mainly used in the mining industry, but equally effective, is the serrated cone over an open pipe configured to act as an airlock. This is sometimes referred to as a "pie-plate sparger." It is illustrated in Figure 6.9. As long as a spray ball can be configured to spray the underside, it should be cleanable. There is a possibility that its relatively large footprint may affect the flow produced by the impeller above it and thereby affect flooding. However, the author is not aware of any studies on this.

Pre-dispersion

This system employs a motionless mixer in a pipe, located inside the tank such that it discharges directly below the lower dispersing impeller, normally aimed upwards. The dispersion is accomplished by means of a recirculating liquid line pumped from a location near the liquid surface. This approach requires a lot of liquid, as the motionless mixer cannot exceed 70% volume fraction of gas, or the phases will invert and the gas will become the continuous phase, essentially delivering a liquid spray to the tank. The best design point appears to be about 50% volume fraction or equal gas and liquid volumetric flow. For example, if a fermenter is operating at 1 VVM airflow (standard conditions) and the actual airflow adjusted for temperature and pressure at the bottom of the tank is 0.5 VVM, then we need half the liquid volume of the tank delivered to the recirculating line per minute. That is a two-minute turnover time.

But how well does it work? From a gas-handling standpoint, it increases the gas handling of the bottom impeller enough so that impellers just barely flooded are no longer flooded. But a large ring sparger does that, too.

From a mass transfer viewpoint, it increases it slightly; about the same as increasing the agitator power input as much as the incremental power added by the recirculating pump system. And it adds a high speed submerged seal pump to the system, which must be able to operated pumping a gassed liquid, as the recirculating inlet line will pick up aerated liquid. The maintenance cost on a pump is generally much higher than on an agitator. Therefore, this author does not think such a system makes sense on a new installation; it would be better to just buy a bigger agitator. There may be some cases where it is viable as a retrofit where flooding is a problem. However, deeply concave turbines have so much gas-handling capacity that an impeller retrofit might be the better alternative.

The above observations are based on the author's personal experience piloting such a system and analyzing the data. The work is unpublished.

Fine Bubble Diffuser

This method uses sintered metal or ceramic diffusers to disperse the gas. They can be in a pipeline, in a ring (acting as segments of the ring), or as an array of individual diffusers mounted in the tank. In many ways, they resemble an aquarium fritted stone bubbler, though the pore sizes are usually smaller.

Manufacturers of these devices talk a lot about pore size, but in fact, pore size has little or nothing to do with the bubble size generated. The bubbles are typically much larger than the pore size. This is because interfacial tension keeps the bubble intact until a combination of buoyancy and fluid drag forces exceed the interfacial tension, causing the bubble to leave the diffuser. One can readily see this in an aquarium; the typical bubble size is 1–3 mm, whereas the pore size is on the order of magnitude of 0.01 mm. If too much airflow is forced into a unit, the whole surface will be blanketed in gas, and the bubbles will be quite coarse. Moreover, in a coalescing system such as water, eventually the bubbles will coalesce into a stable bubble size distribution on the way to the surface and thus act like a bubble column in a tall vessel. The equilibrium bubble size in water is generally about 6 mm [2]. It is likely to be smaller in a non- or less-coalescing system.

When these are used in an agitated tank, the flow created by the agitator helps to dislodge the bubbles from the surface of the diffuser, producing finer bubbles and enhancing mass transfer. Though there would be no point in discharging such a diffuser directly below a Rushton or similar turbine, as it would just coalesce the bubbles before redispersing them, the author has successfully used an array of such diffusers set near the tank wall, with a down-pumping axial turbine. The axial turbine is not designed to disperse gas; it is there only to circulate the tank contents. As such, it is usually designed to provide a characteristic fluid velocity of 0.1–0.2 m/s or a scale of agitation of 3–6. Such systems are generally employed at the micro-aeration level, i.e., an OTR of less than 10 mmol/l-h., though some have been used up to 20 mmol/l-h. The best location for the array is at 85% of the tank diameter; this creates the strongest reinforcement of the primary flow created by the agitator. However, it may be more convenient to mount

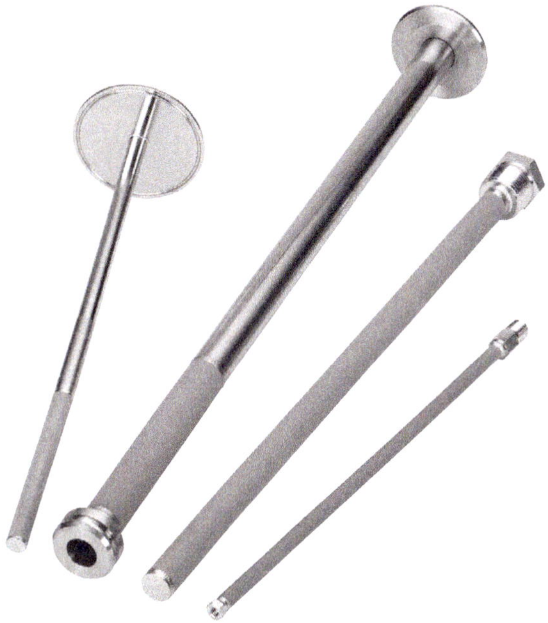

Figure 6.10 Fine bubble diffusers. *Source:* Courtesy of Mott Corporation.

them on nozzles through the tank wall, thus eliminating in-tank piping and allowing units to be removed without entering the tank. Several configurations of such spargers are illustrated in Figure 6.10.

Though such systems have been successful, there is little quantitative information about their design or sizing. The manufacturers do not seem to have any mass transfer correlations. So, it comes down to in-plant experimentation in the full scale equipment, adding more units and more gas if needed to get the required mass transfer. Fortunately, such modifications are not prohibitively expensive.

Summary of Chapter

We have discussed impellers as a system, including such attributes as gas handling, mass transfer, and blending. We have treated the sparger as part of the system. While research and modeling are always ongoing, for many fermentation processes, a system of upper high-solidity up-pumping axial impellers above a lower deeply concave impeller with asymmetric blades, combined with an appropriate sparger and an appropriate power split, seems to be near optimum.

List of Symbols

C	discharge coefficient of sparge (~0.65)
d_o	diameter of sparge hole (mm)
d_s	sparge ring diameter
D	impeller diameter
D_s	sparge ring diameter
DO	dissolved oxygen concentration
F_y	specific heat ratio factor (specific heat ratio/1.4; 1.4 is the value for air)
k	thermal conductivity
k_la	overall mass transfer coefficient
n'	power law exponent
N	shaft speed
N_{Fr}	Froude number
N_{Re}	Reynolds number
OTR	oxygen transfer rate
OUR	oxygen uptake rate
p1	primary pressure feeding sparger (kPa abs)
p2	sparger exit pressure (kPa abs)
P	agitator power draw
P_{motor}	motor nameplate power
P_g	gassed power draw
P_u	ungassed power draw
Q	impeller pumping rate
Q_a	gas flowrate as Nm^3/min
Q_g	gas flowrate
S	impeller tip speed
T	tank diameter
T_a	air temperature (K)
U_s	superficial gas velocity
X_t	pressure differential ratio factor (=0.72)
Z	liquid height
Z_g	gassed liquid height
ρ	liquid density
μ	liquid viscosity

References

1 Fasano, J.B., Reeder, M.F., and Myers, K.J. A comparison of two gas-liquid, up-pumping impeller systems. In: *AIChE Annual Meeting*, Los Angeles, CA, November 12–17, 2000, Paper 178d.

2 van't Riet, K. and Tramper, J. (1991). *Basic Bioreactor Design*. Marcel Dekker Publishing.

3 Benz, G.T. (February 2004). Sizing impellers for agitated aerobic fermenters. *Chemical Engineering Progress* 100: 18S–24S.

4 Sensel, M. Gas dispersion at high aeration rates in low to moderately viscous newtonian liquids. M. Sensel, Master's Thesis, University of Dayton, April 1992.

5 Sardeing, R., Aubin, J., Poux, M., and Xuereb, C. (2004). Gas-liquid mass transfer: influence of sparger location. *Trans IChemE* 82 (A9): 1161–1168.

6 Ineguez, A., Kehn, R. The effect of sparge type on oxygen mass transfer and gas dispersion for minerals processing gassed reactors. Poser Session, Mixing XXVI, San Juan PR June 24–29, 2018.

7

Piloting for Mass Transfer

This chapter deals with the why and how of piloting for mass transfer. This chapter expands on some of the principles in Ref. [1]. The "how" is not so much about detailed methodology, though an overview is given of the common methods of determining $k_l a$. Rather, it is about experimental strategy and protocol to come up with a $k_l a$ correlation that covers the range of conditions likely to exist full scale, and therefore, will be usable for design and optimization of the full-scale system.

This chapter will be divided into four main sections, with subtopics as needed. The four topics are as follows:

1) Why pilot for mass transfer
2) Overview of methods for determining $k_l a$
3) Equipment needed for proper scalable data
4) Experimental protocol; to match full-scale operating conditions

Why Pilot for Mass Transfer

It almost goes without saying that it is impossible to accurately design a system without proper mathematical equations describing that system. Additionally, the ability to minimize power consumption, as described in reference [2], is severely constrained or even impossible. Although we will cover the detailed procedure for minimizing power consumption in the next chapter, it is worth summarizing the concept here as a way to illustrate why piloting for mass transfer can save operating costs.

Figure 7.1 conceptually plots agitator power, compressor power, and the sum (total) power resulting from adding these two together, as a function of airflow, for a given mass transfer rate. In other words, it is possible to achieve the same mass transfer rate with different combinations of airflow and agitator power. One can

Agitator Design for Gas–Liquid Fermenters and Bioreactors, First Edition. Gregory T. Benz.

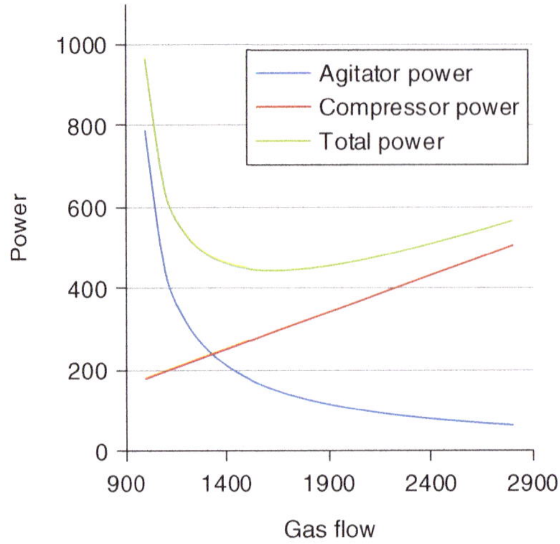

Figure 7.1 Power optimization curve.

readily see that the sum goes through a minimum. Designing for a condition substantially removed from the minimum will result in increased energy costs. For a large installation with multiple fermenters, those increased costs could amount to hundreds of thousands of dollars per year. It is probably worthwhile to pilot specifically for mass transfer for this reason alone, though other reasons are also compelling. However, there may be additional constraints, such as a maximum allowable partial pressure of a gas in the exhaust stream that may govern the choices.

One may ask, Why not use published correlations? There are many things which affect $k_l a$. A partial list of fluid properties that may affect it would include liquid density, liquid viscosity, gas diffusivity in the liquid, ionic strength, interfacial tension, presence of solids or cells at the gas–liquid interface, presence of an oil film at the gas–liquid interface, and perhaps other things not listed here or even known about. There are no uniform correlations that take all such properties into account. The result is that even published correlations can differ markedly from each other, even in tap water or distilled water. Actual broth correlations may also differ considerably from published correlations. Using the relationship $k_l a$ (1/s) = A*(P/V, W/kg)^B (U_S, m/s)^C, Table 7.1 lists the constants A, B, and C for three published correlations and two actual broth correlations developed by Benz Technology International ("Benztech") using data from an experimental protocol designed by Benztech.

One can readily see that there are significant differences in the constants for these various correlations. Figure 7.2 compares total power required as a function of airflow for each of these correlations, for a hypothetical problem. One can note that the airflow corresponding to the minimum total power is different for each

Table 7.1 Mass transfer correlation constants.

Source	A	B	C
Bakker [3]	0.946	0.6	0.6
Coalescing [4]	0.41	0.4	0.5
Noncoalescing [4]	0.25	0.7	0.2
Broth 1	0.67	0.55	0.6
Broth 2	3.73	0.542	0.741

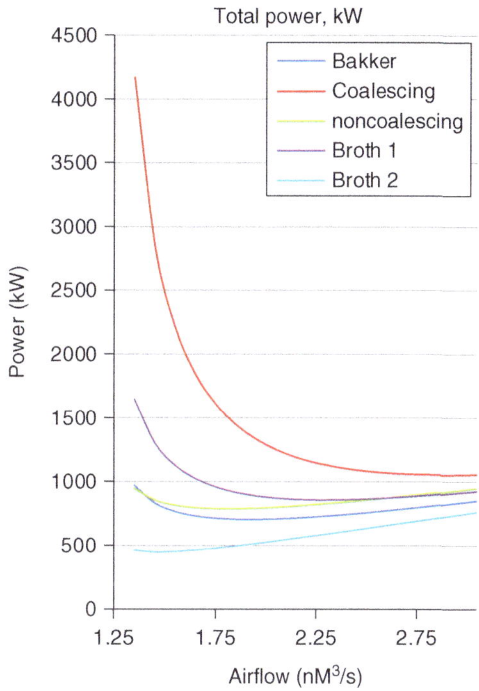

Figure 7.2 Comparing correlations.

one. Also, the difference in total power at the minimum point differs by more than a factor of 2:1. The differences are less pronounced at higher airflow rates, but still significant. The V'ant Riet noncoalescing (ionic) correlation is the closest to the actual broths at higher gas rates.

So, it is definitely important to use the best correlation possible. For feasibility studies, reasonable results can be obtained by using the noncoalescing correlation. But a broth-specific correlation is recommended for actual design.

Methods for Determining k_la

There are a lot of methods available. This section is not intended to delve into details, as there are many references on this subject. Rather, it is to give an overview of the more common methods and the pros and cons of each. Reference [5] describes the more common methods covered below.

Sulfite Method

In the presence of a copper ion catalyst, sulfite ions will rapidly oxidize to sulfate when exposed to oxygen in solution. Essentially, the dissolved oxygen is zero, so the driving force is C_{sat}, calculated at the midpoint of the liquid. In vessels taller than about 1–2 m, it is probably better to use the log mean. Basically, the sulfate concentration is titrated or otherwise monitored versus time, and the mass transfer coefficient can be calculated thusly:

$$k_l a = 0.5 * \left(dC_{sulfate} / dt\right) / C_{sat} \tag{7.1}$$

The concentrations must be on a molar basis. The coefficient of 0.5 reflects the fact that 0.5 mol of oxygen oxidizes 1 mol of sulfite. The denominator can be changed to a log mean form if the data are taken in a taller vessel. The tests are repeated for different gas flow rates and different power inputs to come up with a correlation.

Although this method is fairly simple and quite reproducible, the results are not realistic for fermenter broths. The fact that oxygen is consumed almost at the surface of the bubble results in a very steep local concentration gradient, compared to fermentation where the consumption takes place within the cells. This results in a much higher k_la value than would actually be achieved in most broths. Its use is not recommended for designing fermenters.

Dynamic Method; aka Dynamic Gassing/Degassing Method

This method strips dissolved oxygen by using nitrogen or other inert gas and then introduces air or oxygen. The dissolved oxygen concentration (DO) is monitored versus time, possibly at more than one location for tall vessels so that a log mean concentration driving force may be obtained. The log of the driving force decreases with time linearly. A plot of log of driving force vs time will have a slope of $-k_la$. This is illustrated in Figure 7.3. However, the DO probe has a finite response time. Using τ as the probe first-order time constant, a corrected form of response is as follows:

$$C(t) = C_{sat} + \left(\left(C_{sat} - C_0\right) / \left(1 - k_l a * \tau\right)\right) * \left(k_l a * \tau e^{-t/\tau} - e^{-k_l a * t}\right) \tag{7.2}$$

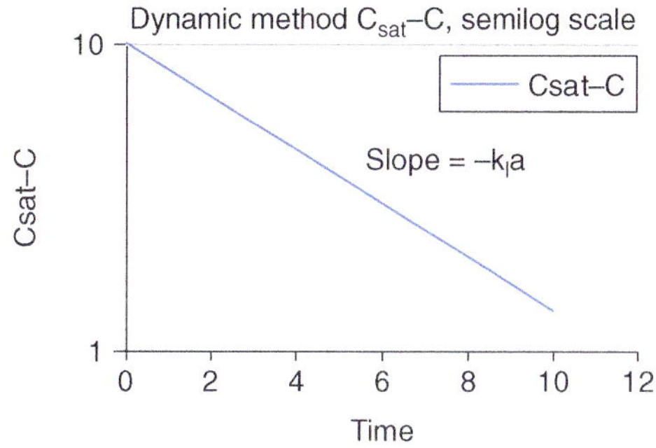

Figure 7.3 Dynamic k_la method.

This method can be used for water, cell-free broth, or broth with dead cells, as long as there are no consumption agents present. For broth, the results are more accurate than either the sulfite method, or the dynamic method applied to distilled or tap water. It can also be formulated to match conditions where the OUR is at its peak.

Disadvantages include probe response time issues and the possibility that the results of a dead broth may not match those of a live broth. Also, it has been shown that the presence of live cells consuming oxygen affects the k_la value [6]. Usually, the result is a higher k_la, but sometimes a lower value results.

Steady-State Method; aka Mass Balance Method

This method typically involves monitoring an actual running fermenter under quasi steady-state conditions, which are varied over the course of the run to cover a range of P/V and U_S values.

The mass balance part is done by monitoring flow and concentration of the gas at the inlet and outlet. OTR can be determined thusly:

$$\text{OTR} = \left(\text{moles gas in / time} - \text{moles gas out / time}\right) / \text{V} \tag{7.3}$$

Then, the mass transfer coefficient can be determined by dividing OTR by the driving force:

$$k_la = \text{OTR} / \left(C_{sat} - C\right) \tag{7.4}$$

The driving force in the denominator should be replaced by the log mean form for taller vessels. Using mid-liquid level results will be sufficient for liquid depths of less than 1 m.

This method has two strong advantages: It takes data under the exact conditions desired. It also has no DO probe response time issues.

Its main disadvantage is that conditions may be changing as the fermentation progresses, making it hard to get a good correlation for a given state of the process.

Also, to get a good correlation, it is necessary to vary P/V and U_S over a fairly wide range. This could lead to very low DO values when both P/V and U_S are low, possibly resulting in cell death or dormancy. A workaround is to try to get such data when the cell population is still low.

Combined Dynamic and Steady-State Method

This technique uses the dynamic method described above in a live fermentation, where oxygen is being consumed. The way it works is that the air is shut off for a short period of time, long enough for all bubbles to exit. Then, the DO value will fall. It must not be allowed to fall below the minimum required for the organisms to thrive. The value of the slope of its concentration vs time equals the OUR, times −1. Then, gas is turned back on, and the rate of DO increase vs time equals the OTR minus the OUR. These may be expressed mathematically as follows:

$$\text{Gas off}: \text{OUR} = -dC/dt \tag{7.5}$$

$$\text{Gas resumed}: dC/dt = \text{OTR} - \text{OUR} \tag{7.6}$$

Then, by calculating the driving force and substituting, we can calculate $k_l a$:

$$dC/dt = k_l a\left(C_{sat} - C\right) - \text{OUR} \tag{7.7}$$

The advantage this method has is that it does not require an off-gas recorder. However, there is a risk of killing the organisms without careful operator attention, and it is hard to know when the vessel is fully de-aerated, especially for viscous broths. Moreover, in tall vessels where the log mean driving force would be best, one can run into different slopes in different parts of the vessel. Absorption of gas from the head space can also be a problem. Over all, this seems like a rather difficult way to measure $k_l a$ in practice.

Equipment Needed for Scalable Data

Here, we are primarily talking about agitator and vessel needs, although we will briefly go into data requirements, which will demand their own set of equipment.

When it comes to piloting gas–liquid contacting equipment, size matters! For mass transfer studies, the gas dispersion mechanism must be the same in the pilot scale as it is in the production scale. Size matters in two ways: The bubbles must be smaller than the vortex trail created by the dispersing impeller, and mass transfer directly to the free liquid surface at the liquid level should not be more than about 10% of the total mass transfer. These two factors mean that the minimum size for piloting mass transfer is about 50 l working volume, with a dispersing impeller diameter of at least 125 mm diameter. One hundred liters or more would be even better. Small benchtop fermenters are OK for studying microorganism response to various inputs but cannot be relied upon for scalable mass transfer data.

Although most production fermenters have aspect ratios of 2–3, it is OK for the pilot scale to be a bit shorter, as long as it has at least two impellers so it can match full-scale impeller types.

Most pilot fermenters operate over a very different range of P/V and U_S compared to production equipment. Normally, mass transfer correlations are presented in the form $k_la = A(P/V)^B(U_S)^C$. Though this form works reasonably well, it is not perfect and should not be used for extrapolation. Rather, it should be used for interpolation. What this means in practice is that the values of P/V and U_S should cover the same range in the pilot scale as anticipated in the production scale. Table 7.2 illustrates how the range of these variables can vary with scale. It is based on a common mass transfer correlation and an OTR of 200 mmol/l-h, optimized using the principles of Chapter 8.

The general trend is for higher P/V and lower U_S in the pilot scale than in the production scale. However, for proper piloting of mass transfer, we need to develop a test program that brackets the anticipated full-scale values in the pilot scale [2]. To do that, we must get some idea of what the production scale values might be.

Table 7.2 Range of variables from pilot to full scale.

Approximate OTR = 200 mmol/l-h				
Scale	**Volume (L)**	**U_S (m/s)**	**P/V (W/L)**	**Gas flow (VVM)**
Pilot	1	0.0065	9.8	3
	5	0.009	7.1	2.5
	20	0.011	6.1	2
	100	0.015	4.5	1.6
	300	0.019	3.9	1.4
Production	5000	0.032	2.4	0.96
	20 000	0.044	1.8	0.85
	150 000	0.064	1.3	0.68
	350 000	0.074	1.1	0.62
	1 000 000	0.091	0.88	0.57

Table 7.3 illustrates some hypothetical values for vessel dimensions, volumes, and other parameters for an array of production volumes and oxygen transfer rates. What does this mean for modeling in the pilot scale? It would take up too much space to illustrate the scaled-down parameters for all the production vessel sizes shown, so we will illustrate the concept by scaling down the $200\,m^3$ production fermenter.

Table 7.4 shows the required power and airflow in several pilot fermenter volumes as a function of full-scale OTR based on a $200\,m^3$ production scale. The requirement for equal U_S results in a higher VVM gas flow in the small scale than in the production scale. Most off-the-shelf pilot fermenters are not set up to deliver that much air, so they must be modified to do so if proper pilot data are to be obtained. The author has actually run as high as 30 VVM in a 100 l test vessel to study hydrodynamics.

Most pilot fermenters come as packages from well-known manufacturers. They do not normally have sufficient air supplies for proper piloting of mass transfer, as shown above. They also normally come standard with Rushton turbines, even though most production fermenters built since 1980 have used other impellers. Presently, the preferred impeller system for most production fermenters is a combination of 1 lower deeply concave radial turbine +1 or more up-pumping high-solidity axial impellers. Though perhaps not absolutely essential, confidence in the derived correlations may be higher if the pilot scale uses the same general types of impellers as those that will be used in the production scale. But, since the modern impellers mostly seem to outperform Rushtons, use of Rushton impellers will often lead to a slightly conservative k_la correlation, and thus a slightly conservative full-scale design. However, aside from not matching production equipment, there is a very practical reason why Rushton impellers are not the best choice for pilot equipment. When setting up an array or grid of P/V and U_S values to test, some combinations of low P/V and high U_S may lead to a flooded impeller, and such test conditions must be avoided. Concave impellers such as the Smith turbine or Chemineer CD-6 have at least double the gas handling capacity of a Rushton. The best deeply concave radial impellers have more than five times the gas handling capacity of a Rushton. This means that more possible combinations of gas flow and agitator power can be tested, improving the range of validity of the resulting correlation.

How do we translate the parameters into test equipment? Table 7.5 lists impellers and speeds that would match production scale P/V and U_S values, as well as the nearest standard motor sizes and shaft speeds. This table is based on the best-practice design concept as of 2019, per the author's experience.

For general purpose pilot equipment, at least double the maximum anticipated U_S and P/V should be possible. Table 7.6 lists guidelines for purchase of a few sizes of pilot equipment. Specific recommendations can be made on a case-by-case

Table 7.3 Full-scale parameters (typ) using a common k_La correlation.

Temp = 37 °C, head pressure 1.7 bar abs. aspect ratio nominally 2.5

Volume (m^3)	OTR (mmol/l-h)	Tank diam (m)	Ungassed liquid level (m)	Power (kW)	Airflow (nm^3/min)	Superficial gas velocity (m/s)	Airflow expressed as VVM
40	50	2.73	7.17	17.7	5.89	0.0096	0.15
40	100	2.73	7.17	39.1	11.8	0.019	0.30
40	150	2.73	7.17	59.7	18.2	0.03	0.46
40	200	2.73	7.17	80.4	24.9	0.041	0.62
200	50	4.67	12.26	77.3	26.5	0.0138	0.13
200	100	4.67	12.26	161	54.9	0.028	0.27
200	150	4.67	12.26	251	83.6	0.043	0.42
200	200	4.67	12.26	344	113	0.058	0.57
400	50	5.88	15.45	137	52.2	0.016	0.13
400	100	5.88	15.45	300	105	0.033	0.26
400	150	5.88	15.45	473	158	0.05	0.40
400	200	5.88	15.45	640	214	0.068	0.54

Table 7.4 Pilot-scale parameters to match 200 m^3 production parameters.

Aspect ratio nominally 1.5								
Pilot volume (l)	Full-scale OTR (mmol/l-h)	Tank diam (mm)	Ungassed liquid level (mm)	Power (W)	P/V (W/l)	Airflow (slpm)	Superficial gas velocity (m/s)	Airflow expressed as VVM
50	50	349.00	523.50	19.3	0.386	79	0.0138	1.58
50	100	349.00	523.50	40.25	0.805	161	0.028	3.21
50	150	349.00	523.50	62.75	1.255	247	0.043	4.94
50	200	349.00	523.50	86	1.72	333	0.058	6.66
100	50	439.00	658.50	38.6	0.386	125	0.0138	1.25
100	100	439.00	658.50	80.5	0.805	254	0.028	2.54
100	150	439.00	658.50	125.5	1.255	391	0.043	3.91
100	200	439.00	658.50	172	1.72	527	0.058	5.27
500	50	751.00	1126.50	193	0.386	367	0.0138	0.73
500	100	751.00	1126.50	402.5	0.805	744	0.028	1.49
500	150	751.00	1126.50	627.5	1.255	1143	0.043	2.29
500	200	751.00	1126.50	860	1.72	1542	0.058	3.08

For k_la studies, pilot volumes less than 50l are not recommended

Table 7.5 Pilot-scale impellers and speed to match 200 m^3 production parameters.

Lower impeller deep asymmetric concave turbine. Upper impeller up-pumping high solidity.							
Pilot volume (l)	**Full-scale OTR (mmol/l-h)**	**Power (W)**	**Lower impeller size (mm)**	**Upper impeller size (mm)**	**Shaft speed (rpm)**	**Standard motor size (Hp(kW))**	**Standard shaft speed (rpm)**
50	50	19.3	140	140	301	0.25(0.19)	350
50	100	40.25	150	150	350	0.25(0.19)	350
50	150	62.75	140	140	430	0.25(0.19)	430
50	200	86	140	140	495	0.25(0.19)	600
100	50	38.6	160	160	301	0.25(0.19)	350
100	100	80.5	170	170	350	0.25(0.19)	350
100	150	125.5	170	170	414	0.25(0.19)	430
100	200	172	175	175	430	0.33(0.25)	430
500	50	193	255	255	230	0.33(0.25)	230
500	100	402.5	263	263	280	0.75(0.55)	280
500	150	627.5	255	255	350	1.0(0.75)	350
500	200	860	270	270	350	1.5(1.11)	350

Table 7.6 Pilot-scale buying recommendations.

Pilot volume (l)	Motor size (Hp(kW))	Shaft speed (rpm)	Lower impeller size (mm)	Upper impeller size (mm)	Air supply (slpm)
50	0.5(0.37)	600	150	150	600
100	0.75(0.55)	600	175	175	1000
500	3(2.2)	430	270	270	3000

basis as well. High OUR fermentations (say, >300 mmol/l-h) may need even more power and airflow.

Data Gathering Needs

In order to develop a correlation for $k_l a$, the following data are required:

- Inlet gas flow rate under standard conditions. (The reference standard conditions must be stated, as there are several "standard" conditions.)
- Liquid temperature
- Barometric pressure
- Back pressure in head space
- Outlet gas flow rate. If a neutral gas such as nitrogen is present in the inlet, it can be used as a tie element to calculate outlet flow. Outlet flow = inlet flow*(X_{in}/X_{out}) where X is the mole fraction of the inert gas
- Ungassed liquid volume and liquid level
- Density of the liquid
- Viscosity, especially if it will vary during the test. (May have to add it as a parameter in the correlation.)
- C_{sat} value at a reference gas composition and pressure at operating temperature. This must be expressed in actual concentration units, such as mg/l or mmol/l, not simply %
- Off-gas analyzer measuring oxygen, nitrogen, and CO_2 at a minimum for aerobic fermentations
- Dissolved oxygen (DO) at top and bottom of fermenter, or in middle if only one is available. The readings must be converted from % to concentration units, so the actual concentration at 100% must be known. Merely stating % saturation does not allow for mass transfer calculations
- Agitator shaft speed
- Agitator power draw, if possible. If vessel has standard baffles and the impellers are of a well-studied type, the manufacturer's power draw calculations may be substituted in lieu of measured power. It is difficult to measure power draw of

an agitator; a whole chapter could easily be devoted to that. We will not cover that topic in this book.

The above data will allow calculation of OTR, driving force, superficial gas velocity, and P/V, all of which are necessary to develop a correlation.

Experimental Protocol

Because of the need to avoid extrapolation, the pilot studies for mass transfer should employ runs with the same range of P/V and U_S as full scale. This means lower agitator shaft speed and higher airflow than typical pilot studies. Combinations of high airflow and low rpm may pose a danger of flooding the impeller, so this should be checked. The pilot air supply and sparge system may need to be modified for the higher-than-normal airflow. The concepts below are also outlined in reference [7].

The pilot protocol can be organized into four steps:

Step 1. Start with an estimated $k_l a$ correlation.
If any pilot data are available and sufficient to fit a mass transfer correlation to them, use these for the preliminary full-scale design, even if the variable values are different from full scale. It at least is a place to start. If no such data are available, data or correlations for similar systems, either from published sources or available internally, may be used as a basis, as we will bracket the conditions in the protocol.
One such correlation for an ionic air/water system, taken from reference 4, is as follows:

$$k_l a = 0.25(P/V)^{0.7}(U_S)^{0.2} \tag{7.8}$$

This is a dimensional equation. In it, $k_l a$ has units of 1/s, P/V is in W/kg and U_S is in m/s.

Step 2. Estimate a reasonable full-scale design by calculating airflow and power at the full scale to give the required OTR.

a) Start with the peak oxygen uptake rate (OUR) requirement and assume that the OTR must match it.
b) Set the airflow based on 50% oxygen transfer. This is based on the author's experience that the lowest total power consumption (agitator plus compressor) often occurs roughly at this condition. 1 ft^3 of air at 60 °F and 1 atm. pressure contains 1.2 g-mol of gas or 0.25 g-mol of oxygen. Similarly, 1 standard m^3 at 1 atm and 0 °C contains 44.06 g-mol of gas.

c) Calculate a mass balance and determine the mass transfer driving force.
d) Solve the k_la correlation for agitator power.
e) Now an approximate full-scale P/V and U_S are known.

Step 3. Design a series of pilot runs using shaft speeds and airflow rates to bracket the values of P/V and U_S calculated for the full-scale design of Step 2.

a) Calculate the pilot airflow based on using the same U_S as full scale.
b) Calculate the agitator speed to give the same P/V as full scale.
c) Prepare an experimental plan or grid of values that go above and below the values calculated in Step 2 by at least factors of 0.5–2 for airflow and 0.75–1.3 for agitator speed. This will give a 4: 1 range on both U_S and P/V.
d) Check for impeller flooding and eliminate flooded runs. To avoid flooding, limit the maximum aeration number, Q_g/ND^3, to $K(N^2D/g)(D/T)^{3.5}$, where the correlation constant, K, is 30 for a Rushton turbine, 70 for a 180° concave turbine (such as the Chemineer CD-6), and 170 for the Chemineer BT-6. Several studies, such as some mentioned in Chapter 4, have suggested that the maximum aeration number for concave turbines increases faster than linearly as a function of the Froude number, N_2D/g, so the above flooding relationship may be conservative for such turbines in some cases. This chapter does not address the effect of viscosity on flooding. If the viscosity exceeds 100 cp, consult other sources or visually check for flooding.

Some runs with combinations of low rpm and low gas flow may not provide sufficient OTR for the organisms to thrive at full biomass concentration. Yet, such data are needed for full-scale design. There are several tactics to cope with this problem. One is to reduce the biomass. Sometimes, this can be done by taking the data for such runs in the early stages of fermentation, while the organism growth is just beginning. Another method is to increase the back pressure at equal actual airflow. This provides higher mass transfer capability for a given superficial gas velocity. However, it also increases the concentration of CO_2 in the liquid and the off-gas, which may be a problem for some organisms. A third method is to use an oxygen-enriched feed gas. While expensive, it may be useful if the other methods do not work.

Step 4. Fit a correlation to the new data.

With good measurements of driving force, OTR, and other parameters, it is now possible to correlate the data. The most common correlation takes the form:

$$k_la = A(P/V)^B(Us)^C \tag{7.9}$$

Other forms may also be used if they fit the data better or make mechanistic sense. Typically, the data fit is ±30%. This new model may now be used for full-scale design and optimization, as will be described in Chapter 8. However, a reasonable safety factor should be applied both to P/V and U_S to allow for experimental error and possible process changes in the future. Alternatively, the design OTR can be factored upward.

Example Problem: Experimental Protocol for Steady-State Analysis

Suppose we have a pilot fermenter, with a vessel of 700 mm diameter, flat bottomed, with a working volume of 300 l, corresponding to an ungassed liquid height of 800 mm. It is equipped with a lower concave radial turbine with a power number of 3.2, and an upper up-pumping, high-solidity axial impeller with a power number of 1. We wish to develop a $k_l a$ correlation that will be used to design a production fermenter of 200 000 l capacity, tentatively choosing a production vessel diameter of 4.67 m, with an ungassed liquid level of 12.2 m. The production process will operate at a back pressure of 0.3 bar-g, at a location where the barometric pressure is 1 bar, or 0.986 atm. The operating temperature is 37 °C, and the target OTR at peak conditions is 150 mmol/l-h. We stipulate that 0.5 mol of CO_2 is respired for each mole of oxygen consumed. We require a minimum DO of 2 mg/l. To choose the test parameters for the pilot scale, we will first calculate an approximate design for the production scale.

For this example, we do not have a broth-specific correlation, so we will use the v'ant Riet noncoalescing (ionic) correlation, repeated from above:

$$k_l a = 0.25\left(P / V\right)^{0.7}\left(U_S\right)^{0.2} \tag{7.8}$$

Since this equation was normalized for ionic water solutions at a temperature of 20 °C, it is probably conservative for 37 °C. The value of $k_l a$ normally increases with temperature. We will not number the equations used for the calculations below, as they are specific solutions to this problem, rather than general equations.

Our first step will be to see how much air we will use, both for calculating superficial velocity and for driving force calculations. The first step in doing this is to convert the OTR into total oxygen demand (g-mol/time) required. With a volume of 200 000 l and an OTR of 150 mmol/l-h, we get a total oxygen demand of:

$$\begin{aligned}\text{Oxygen Demand} &= 200\,000\,\text{l} * 150\,\text{mmol} / \text{l} - \text{h} * 1\,\text{g} - \text{mol} / 1000\,\text{mmol}\\ &= 30\,000\,\text{g} - \text{mol} / \text{h}\end{aligned}$$

Noting that, at standard conditions of 1 atm. and 0 °C, 1 nm^3 of air contains 44.06 g-mol of gasses, and at 21% oxygen, it contains 9.25 g-mol of oxygen, we can calculate the minimum airflow based on stoichiometry, i.e., 100% utilization of the oxygen:

$$\text{Minimum airflow} = \left(30\,000\,\text{g} - \text{mol} / \text{h}\right) / \left(9.25\,\text{g} - \text{mol} / \text{nm}^3\right) = 3243\,\text{nm}^3 / \text{h}$$

As noted in an earlier section of thus chapter, the author's experience is that the minimum power usage often occurs approximately at an airflow corresponding to 50% oxygen utilization. Since our purpose is to develop an experimental protocol to bracket full-scale conditions, this assumption will be sufficient to set an approximate midpoint for the pilot protocol. Therefore, we will double the airflow to 6486 nm^3/h.

Now we will do a mass balance. We will divide the gasses into three components: oxygen, nitrogen, and carbon dioxide. We will lump all nonreactive gasses in with nitrogen and call them "other."

Our inlet airflow corresponds to a total molar rate of 6486*44.06, or 285 700 g-mol/h. At 21% oxygen, this consists of 60 000 g-mol/h of oxygen and 225 700 g-mol/h of "other." During the fermentation, based on the target OTR of 150 mmol/l-h, the process consumes 30 000 g-mol/h of oxygen, leaving 30 000 in the outlet stream. Since 0.5 g-mol of CO_2 is produced per mole of oxygen consumed, we have 0.5*30 000 or 15 000 g-mol/h of CO_2 added to the outlet gas. The "other" contribution remains constant. Therefore, the outlet molar flow rate is as follows:

$$30000\,\text{g}-\text{mol}/\text{h}\,\text{O}_2+15000\,\text{g}-\text{mol}/\text{h}\,\text{CO}_2+225700\,\text{g}-\text{mol}/\text{h}\\ \text{of other, totaling}\,270700\,\text{g}-\text{mol}/\text{h}.$$

This means less gas is leaving than is coming in. In terms of gas flow, this corresponds to 6144 nm^3/h. The inlet oxygen mole fraction is 0.21. The outlet mole fraction is 30 000/270 700, or 0.111. These mass transfer results are summarized in Table 7.7.

Now we can calculate the average superficial gas velocity. First, we must calculate the average flow rate, since it changes due to consumption and respiration:

$$Q_g(\text{avg})=(6486+6144)/2=6315\,\text{nm}^3/\text{h}.$$

Now we must convert this to the actual flow rate at the middle liquid height in the vessel. First, we apply a temperature correction factor based on absolute temperature: °K = °C + 273:

$$\text{Temperature factor}=(273+37)/273=1.136$$

Next, we apply a correction for the absolute pressure at the midpoint of the liquid level. This consists of barometric pressure plus half of the liquid head plus the head space pressure:

$$\text{Pressure}(\text{abs})=1\,\text{bar}*0.986\,\text{atm}/\text{bar}+0.3\,\text{bar}*0.986\,\text{atm}/\text{bar}+1.02*\\ (12.2/2)\text{m}(\text{liquid head})*0.0967\,\text{atm}/\text{m}=1.88\,\text{atm absolute}.\\ \text{The correction factor is}\,1\,\text{atm std}/1.88\,\text{atm actual}=0.5319$$

Table 7.7 Mass transfer summary.

	In	Out
Gas flow (nm^3/min)	6486	6144
Mole fraction oxygen	0.21	0.111
Mole fraction carbon dioxide	0	0.055
Mole fraction other	0.79	0.834
Moles/h oxygen	60 000	30 000
Moles/h carbon dioxide	0	15 000
Moles/h other	225 700	225 700

Since flow was stated at standard conditions of 1 atm and 0 °C, we apply both factors to get actual volumetric flow:

$$Q_g(\text{avg,actual}) = 6315 * 1.136 * 0.5319 = 3815\,\text{m}^3\,/\,\text{h}.$$

To find superficial gas velocity, we divide this figure by the tank cross-sectional area, based on the 4.67 m tank diameter. The area is π*4.67^2/4, or 17.13 m^2. Thus, the superficial gas velocity is as follows:

$$U_S = 3815\,/\,17.13 = 222.7\,\text{m}\,/\,\text{h}, \text{which would more normally be stated as } 0.0619\,\text{m}\,/\,\text{s}.$$

So, we have a midpoint value of superficial gas velocity for our pilot protocol. Next, we must establish a midpoint P/V value. To do that, we solve for $k_l a$ using OTR = $k_l a$*(driving force) and then solving Eq. (7.8) for P/V.

Since this is a tall vessel, we must use the log mean driving force. To do this, we must calculate C_{sat} at the top and bottom of the vessel, and estimate C at the top and bottom. To estimate C_{sat}, we start with a value for 37 °C in water at 1 atm. We found a value of 6.85 mg/l using reference [8]. For this problem, we are adjusting this downward to 6.7 to account for dissolved solids lowering the saturation value. (Good pilot protocol would include comparing C_{sat} in water and the broth to account for the effect of dissolved solids, etc.) The absolute pressure at the bottom is equal to the barometric pressure plus the entire liquid head pressure plus the vessel head pressure:

$$\text{Bottom pressure} = 0.986 + 1.02 * 12.2 * 0.0967 + 0.3 * 0.986 = 2.485\,\text{atm}.$$

Thus, C_{sat}(in) = 6.7*2.485 = 16.65 mg/l.

On the outlet, we must adjust for pressure and change in mole fraction. The outlet pressure is just the barometric pressure plus the head pressure:

$$\text{Outlet pressure} = 0.986 + 0.3 * 0.986 = 1.2818$$

But the outlet mole fraction has changed from 0.21 to 0.111, so the outlet C_{sat} is as follows:

$$C_{sat}(\text{out}) = 6.7 * 1.2818 * (0.111/0.21) = 4.539\,\text{mg/l}.$$

Our problem statement required that we have a minimum DO of 2 mg/l. That would need to occur at the top of the vessel. It will be higher at the bottom. Though we do not know exactly how much higher it will be (we would need a CFD model for that, and we would need a broth-specific $k_l a$ correlation to input to the CFD model for it to be accurate), we will simply assume a value of 2.5 mg/l at the bottom. Thus, our log mean driving force becomes:

$$\text{log mean concentration driving force} = ((16.65 - 2.5) - (4.539 - 2)) / \ln((16.65 - 2.5)/(4.539 - 2)) = 6.759\,\text{mg/l}$$

Solving for required $k_l a$:

$$k_l a = \text{OTR / driving force} = 150\,\text{mmol/l-h} * 32\,\text{mg/mmol} / 6.759\,\text{mg/l} = 710/\text{h}; \text{more commonly stated as } 0.197/\text{s}.$$

Now we can solve Eq. (7.8), re-arranging it to solve for P/V:

$$(P/V)^{0.7} = 0.197 / \left(0.25 * (0.0618)^{0.2}\right) = 1.375$$

$$P/V = 1.58\,\text{W/kg}.$$

For the production fermenter, this corresponds to 322 kW power required. (200 000*1.02*1.58/1000)

We would seek to achieve this same P/V in the pilot fermenter. Since the pilot unit has a 300 l or 306 kg batch volume, we seek a power input of 306*1.58, or 483 W. Since we will bracket the result, we will use a simplified calculation to target shaft speeds. We will assume the gassing factor at the conditions in the pilot fermenter is 0.7 for both impellers. Thus, we may use the following:

$$P = (P_g/P_u)(N_{P1} + N_{P2}) * \rho * N^3 D^5; \text{in the MKS unit system, for our pilot unit, this becomes:}$$

$$483\,\text{W} = 0.7 * (3.2 + 1) * 1020\,\text{kg/m}^3 * N^3 * (0.25\,\text{m})^5$$

Solving for N gives 5.48/s, more commonly stated as 329 rpm. To adequately bracket this figure, I suggest using 250, 300, 350, 400, 450, and 500 rpm. Bearing in

Table 7.8 Example pilot protocol.

	Gas flow rates (actual liters/min)				
Shaft speed (rpm)	**750**	**1125**	**1500**	**2250**	**3000**
250	k_la value	k_la value	k_la value	k_la value	k_la value
300	k_la value	k_la value	k_la value	k_la value	k_la value
350	k_la value	k_la value	k_la value	k_la value	k_la value
400	k_la value	k_la value	k_la value	k_la value	k_la value
450	k_la value	k_la value	k_la value	k_la value	k_la value
500	k_la value	k_la value	k_la value	k_la value	k_la value

Basis: 700 mm tank diameter, 800 mm liquid level, 300 l working liquid volume.
Dual 250 mm impellers; lower is concave, upper is up-pumping high solidity.

mind that power is related to speed cubed, this array gives a range of about 8:1 on P/V.

Similarly, scaling on U_s gives a midpoint flow rate of:

Q_g(actual, pilot) = $0.0618\,m/s*\pi 8(0.7\,m)^2/4 = 0.024\,m^3/s$, more commonly expressed as 1427 actual liters/minute, not to be confused with SLPM. We suggest an array of 750, 1125, 1500, 2250, and 3000 l/min flows, giving a 4:1 range. The mass transfer protocol array is summarized in Table 7.8. The value of k_la should be determined at each of the intersecting grid points so that a correlation may be fitted. Also, the actual power input at each grid point should be either calculated or measured. Each grid point should be checked for flooding. We did that for this problem using the correlation:

$$N_a(\max) = 70 * (D/T)^{3.5} * N_{Fr} \tag{7.10}$$

None of the conditions were flooded. Sometimes, Rushton turbines will be flooded at low RPM and high gas flow. Avoid using data from flooded conditions!

It is important to cover the range with actual gas flow rates, even if the pressure changes. In fact, the pilot fermenter may be operated at different pressures to keep the cells in an aerobic state.

It is not necessary to operate at the exact values in the array when running the protocol. In fact, that may be almost impossible. The important thing is to cover the range and record sufficient data. There will be some noise in the data and natural process variability. Results from a number of runs can be taken and averaged, though statistical techniques should be used to compare from run to run and look for outliers.

At the end of this process, one should have a correlation that will facilitate the techniques used in Chapter 8.

Summary of Chapter

In this chapter, both the why and the how of piloting for mass transfer were discussed. Using these techniques is an essential starting point for accurate fermenter design and allows energy cost minimization, as we will see in the next chapter.

List of Symbols

a	gas–liquid interfacial area per volume
C	gas concentration in liquid; also correlation constant in k_la correlation
C_0	gas concentration in liquid at time 0
C_{sat}	saturation concentration of gas in liquid
$C_{sulfate}$	concentration of sulfate ion in liquid
$C(t)$	dissolved concentration at time t
D	impeller diameter
DO	dissolved oxygen concentration
e	natural log base
g	gravitational acceleration
k_l	liquid film mass transfer coefficient
k_la	overall mass transfer coefficient
OTR	oxygen transfer rate
OUR	oxygen uptake rate
P	agitator power input
P_g	gassed agitator power
P_u	ungassed agitator power
T	tank diameter
Q_g	gas flow rate
t	time
U_s	superficial gas velocity
V	liquid volume
X	mole fraction
τ	DO probe time constant

References

1 Benz, G. (January/February 2013). Why conduct pilot studies for agitated gas-liquid mass transfer? *Pharmaceutical Engineering* 33 (1): 1–3.

2 Benz, G. (May 2003). Optimize power consumption in aerobic fermenters. *Chemical Engineering Progress* 99: 100–103.

3 Bakker, A., Smith, J., and Myers, K. (1994). How to disperse gases in liquids. *Chemical Engineering Magazine* 12 (1): 98–114.

4 v'ant Riet, K. and Tramper, J. (1991). *Basic Bioreactor Design*. Marcel Dekker, Inc.

5 Shuler, M. and Kargi, F. (2002). *Bioprocess Engineering Basic Concepts*, 2e. Prentice Hall.

6 Garcia-Ochoa, F. and Gomez, E. (2009). Bioreactor scale-up and oxygen transfer rate in microbial processes: an overview. *Elsevier, Biotechnology Advances* 27: 153–176.

7 Benz, G. (February 2008). Piloting bioreactors for agitation scale-up. *Chemical Engineering Progress* 104: 32–34.

8 https://antoine.frostburg.edu/chem/senese/101/solutions/faq/predicting-DO.shtml (accessed 9 October 2020).

8

Power and Gas Flow Design and Optimization

What This Chapter Is about

This chapter will show how to choose power and gas flow for a given mass transfer rate, and if full data are available, to optimize power consumption.

Where We Are in Terms of Design

Up to this point, we have introduced the concepts and details of a fermenter design procedure. We have provided background on impeller styles and systems, and how to pilot in such a way as to develop sufficient information for design and optimization. Now we are ready to undertake the first major step of actual design: choosing the combination of airflow and agitator power to be used. Once this is done, choosing the impeller system and shaft speed will complete the process design.

But what if we do not have the necessary pilot data and must design a full-scale system anyway? For talking purposes, we can divide design cases into three categories: no data, limited data, and full data. Subsequent sections will analyze these cases.

Design with no Data

Obviously, this is not a recommended practice. But sometimes, it is forced upon the design engineer by management decisions. The most common reason for this is that there is a great desire to get a product to market, and management is willing to risk the possibility that production rate and yield may be less than anticipated,

Agitator Design for Gas–Liquid Fermenters and Bioreactors, First Edition. Gregory T. Benz.

and that the design may not be optimum. But the design engineer should at least endeavor to classify the general nature of the fermentation into "easy," "average," or "difficult."

"Easy" corresponds, roughly, to a required OTR of 100 mmol/l-h. Such fermentations normally have less than the maximum possible cell population density or relatively low metabolic rates. Examples include some amino acid fermentations such as lysine or tryptophan, acids such as citric, etc. Fermentation time is often four to six days.

"Average" corresponds to an OTR of about 170 mmol/l-h. Examples include most antibiotic fermentations such as penicillin, yeast production, and other single cell proteins. Fermentation time is usually two to four days.

"Difficult" corresponds to an OTR of roughly 300 mmol/l-h. Some difficult fermentations are even over 400. Generally, such fermentations use "hot" organisms such as *Escherichia coli* or *Pichia pastoris*, at a relatively high cell population density. Fermentation times are often less than one day.

Based on a common correlation for noncoalescing liquids (Ref. [1]) and a back pressure of about 0.3 bar, Table 8.1 is suggested as a basis for design.

Based on the author's experience, there is considerable risk designing on this basis. Though some fermentations may actually result in higher OTR values than suggested by this table when the suggested P/V and U_S are used, some will be lower. Only the piloting procedures developed in Chapter 7 will ameliorate the risk.

If a plant is built based on these guidelines and fails to meet expectations, there is usually a way out. Though increasing agitator power is rarely an option (usu-

Table 8.1 Designing with no data.

	Easy	Average	Difficult
	OTR ≈ 100 mmol/l-h	OTR ≈ 170 mmol/l-h	OTR ≈ 300 mmol/l-h
Volume (m^3)	P (kW)/U_s (m/s)	P (kW)/U_s (m/s)	P (kW)/U_s (m/s)
10	30/0.05	55/0.08	110/0.14
20	60/0.055	100/0.1	180/0.18
40	100/0.067	200/0.11	330/0.21
80	180/0.08	350/0.13	600/0.25
160	350/0.09	600/0.16	1200/0.29
320	650/0.11	1200/0.19	2000/0.33
640	1200/0.12	2000/0.21	4000/0.38

ally a whole new agitator assembly would be required), increasing airflow is often an option, and sometimes increased pressure is an option. To make these possibilities easier, several things can be done in the initial design phase, such as:

- Put in air supply piping and filtration to allow at least double the design airflow without excessive pressure drop
- Design the sparger with feed pipe and header sizes appropriate for double the airflow, so that adding holes will be the only sparger modification required to increase airflow
- Provide a place for connecting another compressor if needed, or just buy a larger compressor in the first place
- Design the vessel for higher pressure than anticipated. If 0.3 bar is the planned back pressure, consider a code stamp for 1.5–2 bar.
- To allow for a larger agitator as a last resort, choose a larger mounting flange size than required for the initial agitator design and design for about double the design nozzle loads stated by the original agitator manufacturer.

Example of Procedure

Suppose it is desired to make lysine, an amino acid often used as a feed supplement, in a 100 m^3 working volume fermenter. This is considered an "easy" fermentation. Interpolating from Table 8.1, we would need approximately a 250 kW motor and a superficial gas velocity of 0.14 m/s. The resultant standard airflow would depend on the aspect ratio and back pressure. For an aspect ratio of 2.5 and 0.3 bar, that would be about 100 standard m^3/min or 1 VVM. Expected OTR would be about 100 mmol/l-h.

Design with Limited Pilot Data

Often, there is a limited amount of data available from pilot operations. Ideally, these should be from a pilot scale vessel of at least 100 l working volume or 10% of the production volume, whichever is greater. If the only data available are that the product was able to be consistently made at the rate and yield desired, and the airflow and power draw are known, it is still possible to scale-up to a successful operation from a mass transfer point of view, though the design may not be optimum.

A fairly popular way to do so is to maintain equal P/V and equal VVM in the production scale, while keeping the back pressure and other operating parameters the same. This is actually a rather conservative way to scale-up, even when geometric similarity is not maintained. We can see why when we re-examine the typical mass transfer equations:

$$\mathrm{OTR} = k_l a * (\text{driving force}) \tag{8.1}$$

$$k_l a = A(P/V)^B (U_S)^C \tag{8.2}$$

Scaling up on equal P/V and U_S will actually increase the OTR in two ways. When the absolute height of the liquid broth is greater, the absolute pressure at the bottom of the fermenter is greater. Therefore, the C_{sat} value at the bottom of the fermenter is greater. Even if the C_{sat} at the top is the same or a little less, the overall driving force still usually increases, thus increasing OTR, as per Eq. (8.1), unless the increase in OTR due to increased superficial gas velocity actually results in a decrease in driving force by significantly reducing the C_{sat} value at the outlet. Likewise, unless the scale-up is of equal liquid height (increasing diameter only), the cross-sectional area per volume will decrease upon scale-up. This means, at equal VVM, the superficial gas velocity will increase, and thus, $k_l a$ will increase even though the power/volume remains the same.

To illustrate this concept, we have prepared Table 8.2. For this table, we assumed a back pressure of 0.35 bar, a temperature of 30 °C, a DO value of 2 mg/l and constants of A, B, and C in the $k_l a$ correlation of 0.9, 0.6, and 0.5, respectively. We used volumes of 20 and 120 m^3. The larger volume has a higher aspect ratio, which is often the case due to heat transfer and other considerations. (See Chapter 19 for more detail on aspect ratio). For Table 8.2, we scaled up on equal P/V and equal VVM. The value for U_S almost doubled. At equal OTR, the exit C_{sat} would have been the same in the large and small scales, resulting in a higher driving force. However, the increased superficial gas velocity resulted in an increase in OTR, reducing the exit C_{sat} and slightly reducing the overall driving force. But the OTR increased 35%!

Table 8.2 Sample P/V, VVM scale-up.

	Small scale	Large scale
Aspect ratio	2	2.5
Tank diameter (m)	2.33	3.94
Ungassed liquid level (m)	4.68	9.85
Liquid volume (m^3)	20	120
P/V (kW/m^3)	2	2
Power (kW)	40	240
VVM at standard conditions	1	1
Mean superficial gas velocity (m/s)	0.057	0.108
Log mean driving force (mg/l)	6.89	6.74
OTR (mmol/l-h)	249	335
Relative OTR	1	1.35

The degree of improvement will vary case by case. But the OTR capability will always increase when using this method of scale-up.

Sometimes, the known data can include saturation values, DO values, and the actual OTR, as well as temperature, pressure, and airflow. In such a case, it is possible to calculate the $k_l a$ value in the small scale. It may then be possible to check to see whether any published correlations reasonably fit the data and, if so, use the best correlation to design the production equipment using the techniques in section "Design with Full Data." However, this can be risky. Sometimes, one data point can seem to fit a correlation, but the actual exponents for a particular broth may be different. In such a case, it may be wise to refrain from decreasing P/V and VVM by more than 10% or so, even if the correlation says this is possible.

Design with Full Data

Obviously, more knowledge can lead to better design. With "full" data, not only can design be done more accurately, but energy cost can be minimized, as per the next subsection. But what is "full data"? For agitator design/mass transfer purposes, the following constitute full data:

- Saturation concentration of gas in liquid at a reference pressure and concentration (mole fraction or mole %) of the active gas species in the gas and at process operating temperature. This must be in actual concentration units such as mg/l or mmol/l, not %. If Henry's law constants are known for the actual gas/broth combination, this can be used. If such data are not available, it is reasonable to use about 95–98% of the saturation values in water. One source of such data is The Engineering Toolbox website, Ref. [2]. For example, at 30 °C, it reports the saturation concentration of oxygen from air in fresh water to be 7.6 mg/l at a pressure of 760 mmHg. Likely, an actual broth would have a C_{sat} of 7.2–7.5 mg/l due to the presence of dissolved solids.
- Required oxygen transfer rate or other gas transfer rate.
- Required minimum DO for the organisms to thrive. Again, this must be stated in concentration terms, such as mg/l or mmol/l, not % saturation. A probe calibrated to 100% in water will have a different saturation at 100% from one calibrated in actual broth. Also, the result will vary according to the barometric pressure where it was calibrated. A biolab in Tibet will get a noticeably different true C_{sat} from one taken in Death Valley! If reported as 100% in both cases, serious errors can occur.
- Gas concentration at the inlet (mole % is typical for this measurement)
- Most importantly, a broth-specific $k_l a$ correlation, taken under conditions bracketing production conditions, per the procedures in Chapter 7. Without

this, accurate design and optimization are impossible. If viscosity varies, that may need to be included as another correlating variable

- Liquid density
- Liquid viscosity
- Barometric pressure at the site
- If off-gases are produced, ratio of moles of off-gas created per mole of target gas consumed. Most commonly, this is moles of CO_2 per mole of oxygen consumed. This affects outlet gas flow and concentration, affecting both driving force and superficial gas velocity.
- Maximum allowable partial pressure of any organism-inhibiting gases in the off-gas. Most commonly, this is CO_2.
- For design of a particular system, a tank diameter, back pressure, volume, and liquid level will have to be chosen. There can be additional optimization steps by changing these parameters. For now, we will focus on what can be done with a fixed geometry and back pressure. Chapter 19 addresses the effect of aspect ratio on design and optimization.

Choose Minimum Combined Power

For any given mass transfer rate, it is possible to achieve it with a spectrum of airflow rates and agitator power inputs as described in Ref. [3]. Low airflow rates require more agitator power than high airflow rates, but high airflow rates require more power from the compressor. The sum of agitator power and compressor power will go through a minimum at some airflow rate, as illustrated in Figure 8.1. This section will detail a procedure for how to arrive at the minimum total power.

There are lower and upper bounds to the airflow. The lower bound is stoichiometric. At standard conditions of 1 atm pressure and a temperature of 0 °C, 1 m^3 contains 44.64 g-mol. Since it is not possible to transfer more gas than is present in the gas stream, the minimum gas flow at 100% utilization, in units of standard cubic meters/time, is as follows:

$$\text{Minimum gas flow} = V * OTR / \left(44.64\ \text{g}-\text{mol} * x_a\right),$$
$$\text{where } x_a \text{ is the mole fraction of gas in the air stream.} \tag{8.3}$$

In the above equation, consistent units must be used. For example, if liquid volume is stated in m^3, OTR would need to be in units of g-mol/m^3-time. Although we stated mass transfer as OTR, Eq. (8.3) is equally applicable to other gases on a mole basis.

This is a "hard" lower limit. It is impossible to get the required OTR at a lower gas flow rate. The upper limit is a "soft" limit. Although there is probably a gas flow rate that would literally dry out the tank by transporting its contents out the

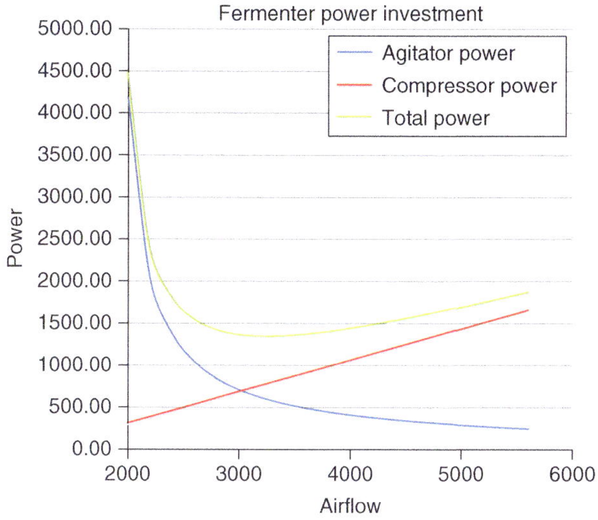

Figure 8.1 Optimum power.

gas vent, practical operation is limited to a much lower flow rate than that. The author's experience suggests that velocities above 0.6 m/s will result in excessive entrainment of liquid in the vent stream, though he is aware of some chemical reactors operating at around 1 m/s.

Required agitator power is calculated by re-arranging the k_la correlation and OTR formulas in Eqs. (8.1) and (8.2):

$$k_la = \text{OTR/driving force} \tag{8.4}$$

$$P = V * \left(k_la/\left(A * U_S^{\ C}\right)\right)^{1/B} \tag{8.5}$$

For design purposes, the maximum system pressure drop is held to be a constant independent of airflow chosen. It is assumed that the piping, sparger, and air filtration systems will be sized to provide that maximum pressure drop. With this assumption, the compressor power is directly proportional to airflow and may be calculated by Eq. (8.6), taken from Perry's Handbook [4]:

$$P_c(\text{kW}) = 5.97(\text{inlet pressure, atm})(\text{inlet flow, m}^3/\text{min}) \left(\left(\text{pressure ratio}\right)^{0.283} - 1\right)/\text{efficiency} \tag{8.6}$$

It is necessary to set the DO at the top of the vessel to the minimum required. The DO at the bottom will be higher. It may be possible at this point to model a fermenter and calculate the DO at the bottom using CFD techniques, though this is both costly and unverified by production-scale measurements. An estimate of the ratio of bottom DO to top DO, based on a very limited amount of data, may be made by using Table 8.3.

With the above information, it is possible to follow a rational, step-by-step procedure to minimize power consumption at peak mass transfer conditions:

1) Based on peak OTR, calculate the minimum airflow required at 100% utilization. (That is, all of the oxygen in the feed gas must be transferred in order to meet the required oxygen transfer rate.)
2) Choose a beginning airflow (greater than the theoretical minimum calculated in step 1) for the first design calculation. I recommend beginning about 20% above the minimum.
3) Perform a mass balance and determine the exit gas flow rate and composition, including CO_2 or other gases respired.
4) Calculate the driving force for mass transfer. Because in small-scale equipment, there will not be significant variation in the DO value, it is OK to use a simple form for driving force for the mass transfer relationship. However, for full-scale equipment there may be significant DO gradients, so use of a log mean driving force is recommended. The log mean driving force is as follows:

   $$((C_{sat} - C_L)_2 - (C_{sat} - C_L)_1)/\ln((C_{sat} - C_L)_2/(C_{sat} - C_L)_1).$$

   The subscripts 1 and 2 refer to inlet and outlet conditions. Accurate saturation data are needed.
5) Calculate the required k_la.
6) Calculate the actual volumetric airflow at the midpoint of the unaerated broth height, being sure to account for temperature, composition, back pressure, and liquid head. This is equivalent to averaging the actual volumetric flow rates at the inlet and outlet.
7) Calculate the superficial gas velocity at the midpoint of the unaerated broth height using the above gas flow rate.
8) Using the mass transfer relationship developed from pilot scale data, solve Eq. (8.5) for agitator power. Add about 5%, or use 95% mechanical efficiency, to account for gear drive and seal losses, to get motor power draw required.
9) Calculate compressor power. Include back pressure, liquid head pressure, and losses from the sparge ring, piping, and filtration systems. Also include compressor efficiency (typically 50–80%) to calculate motor power draw required. For an existing plant, these should be calculated based on the actual piping system. For a new plant optimization study, a simplifying assumption can be made: The piping system will be sized based on a certain maximum pressure drop at the final airflow chosen. Use Eq. (8.6) or a similar equation from an air handling handbook. Usually, the line losses are less than 2 bar-g

10) Calculate total power (agitator + compressor).
11) Repeat steps 1–10 for an increased airflow, typically incremented by 20% of the theoretical minimum.
12) Repeat step 11 until a minimum total power is found. Add safety factors to the results (or to the design OTR) to account for data error and to allow some flexibility for future process changes. The factors depend on how much data scatter is found in the correlation and how much confidence is needed. It is common for the error band in the k_la correlation to be ±30%, and it can easily be as high as 60%. It may easily require doubling agitator power to be certain. However, flexibility in airflow may also be provided. Future process changes are the most uncertain of all. There is no substitute for past experience with process changes as a guide for judgment. There are no fixed guidelines.

Example Problem

Suppose we have a production fermenter with a liquid volume of 150 m^3, with a vessel diameter of 4.0 m, with an ungassed liquid height of 12.2 m (the tank has dished heads, making the height a little more than it would be with a flat-bottomed cylinder). The broth has an unaerated specific gravity of 1.05. Temperature is 30 °C. It has been found that at 1 atm pressure the saturation value of dissolved oxygen is 7.5 mg/l, based on an air composition of 21% oxygen. When oxygen is consumed, 90% of it is returned in the form of respired carbon dioxide. The OTR target is 3200 mg/l-h or 100 mmol/l-h. Minimum desired oxygen concentration in the broth is 2.1 mg/l. The fermenter will be operated at a back pressure of 0.68 atm gage pressure, or 0.69 bar-g. The site location has a mean barometric pressure of 0.95 atm. Line losses from the compressor to the vessel will by design be set at 2.0 atm. We will assume 95% mechanical efficiency for the agitator and 70% for the compressor. We also have a broth-specific k_la correlation of the form we have used throughout this book, with constants A = 0.35, B = 0.6, and C = 0.45. Now we will follow the procedure step-by-step.

1) Total OTR required is 3200 mg/l-h × 150 m^3 × 1000 l/m^3 = 4.8E8 mg oxygen per hour. Based on a molecular weight of 32, or an OTR of 100 mmol/l-h, this equates to 15 000 g-mol/h. One cubic meter of gas at standard conditions contains 44.64 g-mol. Air at 21% oxygen contains 9.374 g-mol of oxygen, so we need a minimum airflow rate of 15000/9.374 = 1600 std. m^3/h, or 26.67 std. m^3/min.
2) While normally we would begin our iteration at roughly 20% higher airflow than the above value, for this sample problem, just to illustrate the calculation steps, we will do only one iteration, and we will use about twice the above value for illustration, or 53.4 std. m^3/min.
3) Mass balance, on a per-minute basis: Inlet airflow of 53.4 std. m^3/min times 44.64 gmol/m^3 contains a total of 2384 g-mol, of which 501 g-mol (21%) are oxygen, and the balance of 1883 is other gases, which will be carried through

the fermenter unchanged. Oxygen consumed is 100 * 150000 * 0.001 = 15000 g-mol/h, or 250 g-mol/min, leaving (501–250) or 251 g-mol/min of oxygen in the outlet gas. In addition, carbon dioxide is generated in the amount of (250 * 0.9) or 225 g-mol/min (90% of the value of oxygen consumed.) So, the outlet gas totals 251 + 225 + 1883 = 2359 g-mol/min. The oxygen content in the outlet gas is thus a mole fraction of 251/2359 = 0.106 or 10.6%. This corresponds to a gas flow at the outlet of 2359/44.64 or 52.8 std. m^3/min, slightly less than we started with.

4) To calculate driving force, we need to know absolute pressure and concentration of the air stream at the top and bottom, as well as the DO concentration in the broth at the top and bottom. Bottom pressure equals back pressure plus liquid head pressure plus barometric pressure. The liquid head pressure, based on 12.2 M liquid height and a density of 1050 kg/m^3 is 12.2 * 1050 * 9.81 = 1.257E5 kg/m-s^2 = 1.24 atm. Barometric pressure is 0.95 atm and back pressure is 0.68 atm. Thus, the absolute pressure at the bottom is 0.95 + 0.68 + 1.24 = 2.87 atm. Since the saturation value at 1 atm and 21% oxygen is 7.5 mg/l, the saturation concentration at the bottom of the fermenter is 7.5 × 2.87 = 21.53 mg/l. The pressure at the top is 0.68 + 0.95 = 1.63 atm. Absolute, and the concentration of oxygen is 10.6%, so saturation at the outlet is 7.5 mg/l × 1.63 × (10.6/21) = 6.17 mg/l. Now that we have established saturation values, we must calculate driving force. To do this, we must also know liquid DO concentrations. We have established that we need a minimum of 2.1 mg/l for the process to work effectively. However, real fermenters are not perfectly mixed, so there will be a concentration gradient. The author has seen cases using multiple radial impellers wherein the DO varied from bottom to top by more than an order of magnitude. It is the author's experience that if a combination of axial and radial impellers is used in fermenters of this size, the DO at the bottom will be about 50% higher than at the top, as per Table 8.3. Using this basis, we would assign a value of 2.1 mg/l at the top and 3.15 mg/l at the bottom. Our log mean driving force can be calculated thusly: Driving force = ((6.17 − 2.1) − (21.53 − 3.15))/ ln((6.17 − 2.1)/(21.53 − 3.15)) = 9.49 mg/l.
5) Required k_la = OTR/driving force = (3200 mg/l-h)/9.49 mg/l = 332.7 per hour or 0.0937 per second.
6) Average molar flow of gas at the middle is (2384 + 2359)/2 = 2371.5 g-mol/min, or 53.03 std. m^3/min. Absolute pressure at the midpoint equals back

Table 8.3 DO ratio guidelines.

Volume (m^3)	**20**	**40**	**80**	**160**	**320**	**640**
DO ratio, bottom/top	1.1	1.2	1.3	1.5	2	3

pressure plus ½ of the liquid head pressure, or 0.68 + 0.95 + 1.24/2 = 2.25 at m. Temperature is 30 °C or 303 K. So based on ideal gas law adjustments, actual gas flow is 53.03 * (1/2.25) * (303/273) = 26.16 m^3/min.

7) Based on a tank diameter of 4.0 m, which has a cross-sectional area of 12.57 m^2, the above airflow yields a superficial gas velocity of 26.16/12.57 = 2.082 m/min or 0.0347 m/s.
8) For this example, the constants A, B, and C in Eq. (8.5) are based on units of 1/s for $k_l a$, kW/m^3 for P/V and m/s for U_S. Ideally, such constants should be derived from experimental data taken in the actual broth and impeller system that will be used. Based on a required $k_l a$ of 0.0937 per second and a known superficial gas velocity of 0.0347 m/s, we calculate a required agitator power using Eq. (8.5):

$$P = V * \left(k_l a / \left(A * U_S^{\,C}\right)\right)^{1/B} = 150 * \left(0.0937 / \left(0.35 * 0.0347^{0.45}\right)\right)^{1/0.6}$$

= 207.5 kW. Allowing for 95% gear drive and seal efficiency, the actual motor power draw would be 207.5 / 0.95 = 218.4 kW.

9) To calculate compressor power, we need to know inlet airflow and pressure, and outlet pressure, assuming a roughly adiabatic operation. More detailed calculations can be made after the final system is chosen. Adjusting for units of kW for power, atmospheres for pressure and airflow in actual m^3/min., and using a specific heat ratio for air of 1.394, the required compressor power is 5.97(inlet pressure, atm) (inlet flow, m^3/min)((pressure ratio)$^{0.283}$ – 1). Our inlet pressure is the barometric pressure of 0.95 atm. The airflow is adjusted to 20 °C or 293 K and to the actual barometric pressure for design purposes as 54.36 m^3/min. Outlet pressure is equal to the back pressure plus liquid head plus lines losses, or 0.95 + 0.68 + 1.24 + 2.0 = 4.87 atm absolute. This means the pressure ratio is 4.87/0.95 = 5.13. So, compressor power is 5.97 × 0.95 × 54.36 × ([5.13]$^{0.283}$ – 1) = 181.3 kW. Allowing 70% efficiency, the compressor motor power draw is 259 kW.
10) Total power of the agitator plus compressor is 218.4 + 259 = 477.4 kW.
11) We will not repeat the optimization steps here, but by varying airflow, a point of minimum power requirement can be found, as shown graphically in Figure 8.1 Note, that figure was not generated using this problem as an example, so the numbers do not match.

State of Design Completion

After going through this chapter, we should be able to select the combination of airflow and agitator power required. After that is completed, impeller selection and shaft speed follow, per Chapters 5 and 6. Then, mechanical design would need to be completed, per Chapter 17.

Additional Considerations

The previous analysis was designed to minimize total power needed. However, some organisms are sensitive to partial pressure of CO_2 in the exhaust gas. In our sample problem above, the CO_2 mole fraction in the exhaust gas was 225/2359, or 0.095. The partial pressure would thus be 1.63 * 0.095, or 0.155 atm. If this were a problem, a higher gas flow would need to be used. Such a problem is more likely at low gas flows/high oxygen utilization.

Summary of Chapter

We have learned several ways to select a working combination of agitator power and airflow for a fermenter design basis. Only with "full" data can we do the best job possible and also minimize power consumption.

List of Symbols

A	Constant in k_la correlation
B	Exponent on P/V
C	Exponent on U_S
C_L	Concentration of dissolved species in liquid
C_{sat}	Saturation concentration of gas in liquid
DO	Dissolved oxygen concentration in liquid
k_la	Overall mass transfer coefficient
OTR	Oxygen transfer rate
P	Agitator power
P_c	Compressor power (usually in kW)
U_S	Superficial gas velocity
V	Liquid volume
VVM	Volume of gas per volume of liquid per minute
x_a	Mole fraction of gas "a" in gas stream

References

1 V'ant Riet, K. and Tramper, J. (1991). *Basic Bioreactor Design*. Marcel Dekker, chapter 11, equation 11.41.

2 The Engineering Toolbox website. *Oxygen – Solubility in Fresh Water and Seawater*. https://www.engineeringtoolbox.com/oxygen-solubility-water-d_841.html (accessed 10 October 2020).
3 Benz, G. (2003). Optimize power consumption in aerobic fermenters. *Chemical Engineering Progress* 99: 100–103.
4 Perry, R., Green, D., and Maloney, J. (eds.) (1997). *Perry's Chemical Engineering Handbook*, 7e. McGraw-Hill.

9

Optimizing Operation for Minimum Energy Consumption per Batch

Purpose of This Chapter

The purpose of this chapter is to apply the principles we have developed so far to the goal of minimizing total energy consumption per batch, in terms of power applied to the agitator and compressor. Though not expressly calculated here, this procedure also minimizes total cooling duty, as any heat transferred to the broth from agitation and compressor power must be removed by the cooling system, whether that takes the form of cooling the vessel contents as per Chapter 10, or cooling the hot compressed air before it enters the vessel. Since most fermenters require chilled water for heat transfer, that is a significant additional energy cost, though it will not be explicitly covered in this book.

The concepts and some of the calculations are based on reference [1] and are similar to those contained in reference [2].

Prerequisite

As we shall see when going through the detailed procedure, it is impossible to achieve this goal unless we have a broth-specific k_la correlation, as well as detailed knowledge about the required OTR and DO profiles as the batch progresses. This requires use of the "full data" scenario defined in Chapter 8.

Conceptual Overview

In the previous chapter, we learned how to minimize the sum of agitator plus compressor power when designing for peak OTR conditions. Because this was for the design phase of the project, we made a simplifying assumption that the piping,

Agitator Design for Gas–Liquid Fermenters and Bioreactors, First Edition. Gregory T. Benz.

sparge system, and filtration system would be designed based on a maximum permissible total pressure drop. Since this chapter is about maximizing operation, the actual pressure drop must be used to calculate compressor power. While detailed piping pressure drop calculations are very system-specific and beyond the scope of this book, we will make use of a simplifying principle: that at low absolute pressure (less than 5 bar absolute, typically), for a given piping system, the pressure drop is proportional to the square of airflow. We are also assuming in this book that the compressor efficiency is constant. The user is encouraged to do more detailed calculations for his particular set of conditions.

Typically, the pressure drop will be initially calculated for the maximum airflow.

With the above assumptions in mind, the actual pressure drop can be calculated as follows:

$$\Delta p = \Delta p(\max) * \left(Q_a / Q_a(\max)\right)^2 \tag{9.1}$$

Thus, the basic idea of this chapter is to use the principles of the previous chapter as delineated for full data, by breaking the process down into discreet steps and minimizing total power at each step, using Eq. (9.1) instead of a fixed pressure drop in the calculations. Then, we can sum the total energy consumed per segment and add up the total per batch.

This procedure works for batch and semibatch processes. It is not relevant for continuous flow systems operating at steady state. Those can be optimized using the procedure of Chapter 8.

Detailed Procedure

1) Divide the batch cycle into as many time increments as judged reasonable, based on incremental changes in OTR requirements
2) Calculate the system pressure drop for a reference condition. Usually, this will be at the maximum design airflow rate, so Eq. (9.1) can be used for other flow rates.
3) For each increment, perform steps 4–15 below to optimize operation for that condition
4) Based on OTR in that segment, calculate the minimum airflow required at 100% utilization. (That is, all of the oxygen in the feed gas must be transferred in order to meet the required oxygen transfer rate.)
5) Choose a beginning airflow (greater than the theoretical minimum calculated in step 1) for the first design calculation. I recommend beginning about 20% above the minimum.

6) Perform a mass balance and determine the exit gas flow rate and composition, including CO_2 respired.
7) Calculate the driving force for mass transfer. Because in small pilot-scale equipment, (typically <500l) there will not be significant variation in the DO value, it is OK to use a simple driving force to correlate data for the mass transfer relationship. However, for full-scale equipment there may be significant DO gradients, so use of a log mean driving force is recommended. The log mean driving force is as follows:

   $$((C_{sat}-C)_2-(C_{sat}-C)_1)/\ln((C_{sat}-C)_2/(C_{sat}-C)_1).$$

   The subscripts 1 and 2 refer to inlet and outlet conditions. Accurate saturation data are needed.
8) Calculate the required k_la. (k_la = OTR/driving force)
9) Calculate the actual volumetric airflow at the midpoint of the unaerated broth height, being sure to account for temperature, composition, back pressure, and liquid head. This is equivalent to averaging the flow rates from the inlet and outlet.
10) Calculate the superficial gas velocity at the midpoint of the unaerated broth height using the above gas flow rate.
11) Using the mass transfer relationship developed from pilot-scale data, solve the k_la equation for agitator power. (If the typical form of k_la correlation is used, this becomes $P = V^*(k_la/(A^*U_S{}^{C}))^{1/B}$). Add about 5% to account for gear drive and seal losses, to get motor power draw required.
12) Calculate compressor power. Include back pressure, liquid head pressure, losses from the sparge ring, piping, and filtration systems. Also include compressor efficiency (typically 50–80%) to calculate motor power draw required. For an existing plant, these should be calculated based on the actual piping system. For a new plant optimization study, a simplifying assumption can be made: The piping system will be sized based on a certain maximum pressure drop at the final airflow chosen. Since airflow is compressible, the procedure for calculating pressure drop and power is more complex than used for liquids, and will not be covered here. Air handling specialists should be consulted. However, once a pressure drop is established for the maximum design airflow, we can use a simplifying assumption to calculate pressure drop for low pressure systems at other flow rates, as stated in Eq. (9.1)
13) Calculate total power (agitator + compressor).
14) Repeat steps 5–13 for an increased airflow, typically incremented by 5–10% of the theoretical minimum.
15) Repeat step 14 until a minimum total power is found. Add safety factors to the results to account for data error and flexibility for future process changes. The factors depend on how much data scatter is found in the correlation and how

much confidence is needed. However, flexibility in airflow may also be provided. Future process changes are the most uncertain of all. There is no substitute for past experience with process changes as a guide for judgment. There are no fixed guidelines. If the process changes, it may be necessary to develop a new $k_l a$ correlation and redo both the initial optimization and this operational optimization with the new correlation.

16) For each segment, multiply the sum of agitator plus compressor power by the time interval, to get total energy per time interval. Add them all together to get total energy per batch. In other words:

$$E_{total} = \Sigma\left(P + P_c\right) * \theta \tag{9.2}$$

This can be compared to what the total would be if operated in a simpler way, such as fixed agitator speed and airflow. Then, the agitator speed or airflow can be set up with a control scheme to program them in, with a feedback on one or the other in response to DO changes.

Example Problem

Rather than go through each step in detail, we will primarily illustrate the results that can be had by using this technique.

The example we use here has the following attributes:

- Tank diameter = 6 m
- Working liquid volume = 350 m^3
- Ungassed liquid height = 12.38 m
- Operating temperature = 30°C
- Back pressure = 0.6 bar
- Liquid density = 1010 kg/m^3
- Barometric pressure = 0.99 bar
- C_{sat} at 1 atm, 21% oxygen = 7 mg/l
- Compressor efficiency = 70%
- Agitator mechanical efficiency = 95%
- DO at top of vessel = 1.5 mg/l
- DO at bottom of vessel = 3 mg/l
- $k_l a\ (s^{-1}) = 0.25(P/v,\ W/kg)^{0.55}(U_s,\ m/s)^{0.65}$

For simplicity, we will treat the fermentation cycle as batch, though many fermentations are fed batch. The same principles apply, but a few more calculations would be required, accounting for changes in liquid level. (Liquid level affects driving force, volume, and impeller coverage. We do not want to include power for impellers that are not submerged!) This model fermentation has a variable OTR requirement, as illustrated below in Table 9.1 and Figure 9.1.

Table 9.1 OTR profile.

Time (h)	OTR (mmol/l-h)
0	10
5	10
10	10
15	15
20	20
25	40
30	60
35	80
40	120
45	150
50	150
55	150
60	150
65	100
70	75
75	60
80	50

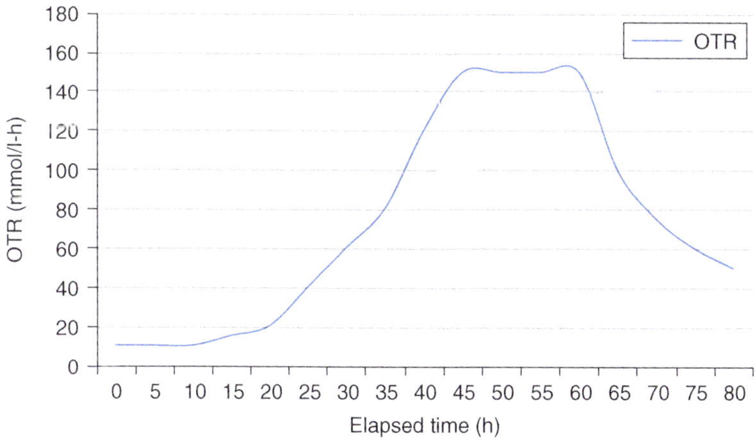

Figure 9.1 OTR profile graphed.

Minimizing Total Energy Usage

Using the techniques of reference 1, and the assumptions above, it has been determined that, at the peak OTR of 150 mmol/l-h, an airflow of 396 std. m^3/min (basis: 1 atm pressure and 0°C as standard conditions) is required (which requires a compressor brake power of 1780 kW) and an agitator brake power of 1222 kW is needed. If the agitator and compressor were operated at this capacity throughout the batch, the total energy usage of agitator plus compressor would be 240 000 kW-h per batch. But the fermenter need not be run this way. If the airflow were fixed but the agitator speed were varied, significant energy could be saved. Even more could be saved if both airflow and speed were varied to minimize total power input at each stage of the process, using the techniques described above. Table 9.2 compares three scenarios: running flat out (100% airflow and agitator speed throughout the batch), fixed airflow of 1 VVM (350 std. m^3/min) with variable agitator speed, and variable airflow and agitator speed. Figure 9.2 compares these scenarios graphically.

How do these three scenarios compare quantitatively and economically? As previously mentioned, the total energy running flat out is 240 000 kW-h per batch. Running a fixed airflow of 1 VVM but varying agitator speed would require 161 000 kW-h. Varying both airflow and agitator speed optimally at each different OTR in the process would require 128 000 kW-h. So, compared to running flat out, variable agitator speed saves about 33% of the energy and varying both agitator speed and airflow saves almost 47%. Assuming 90 batches could be made per year, and an electrical cost of $0.10/kW-h, the monetary savings would be $711 000/year for variable speed agitation only and $1 008 000/year for varying both compressor and agitator speed. This would be for one fermenter. Many facilities have more than one fermenter, so the savings would multiply.

Practical Design

It may be difficult to design a control system that senses DO and varies both agitator and compressor speeds optimally. A practical approach would be to calculate the optimum approach as carefully as possible for a fairly repeatable DO profile, then program in one of the variables as a fixed protocol, and then use feedback control on DO to vary the other. One example would be to create an airflow profile and then fine tune with agitation, though it could be vice versa.

Additional Considerations

As always, care must be taken to avoid unacceptable concentrations of CO_2 or other inhibitory metabolic by-products. Additional energy savings can be had by operating with low biomass, which will decrease the OUR. However, more fermenter volume would be required for the same net output.

Table 9.2 Power at different operating scenarios.

Time (h)	Airflow, std. m^3/min at optimum conditions	Airflow (VVM)	Airflow delta-P (bar)	Agitator brake P, kW, at optimum conditions	Compressor brake P, kW, at optimum conditions	Total brake P, kW, at optimum conditions	Brake P, running flat out (kW)	Brake P, 1 VVM, kW, variable agitator speed only
0	39.6	0.11	0.017	122	178	300	3002	1430
5	39.6	0.11	0.017	122	178	300	3002	1430
10	39.6	0.11	0.017	122	178	300	3002	1430
15	53.8	0.15	0.032	182	242	424	3002	1436
20	70.8	0.20	0.055	222	318	540	3002	1448
25	130	0.37	0.185	389	585	974	3002	1529
30	181	0.52	0.359	561	814	1375	3002	1660
35	232	0.66	0.590	713	1042	1755	3002	1843
40	326	0.93	1.166	1019	1462	2481	3002	2391
45	396	1.13	1.720	1222	1780	3002	3002	3000
50	396	1.13	1.720	1222	1780	3002	3002	3000
55	396	1.13	1.720	1222	1780	3002	3002	3000
60	396	1.13	1.720	1222	1780	3002	3002	3000
65	283	0.81	0.878	852	1271	2123	3002	2085
70	224	0.64	0.550	658	1004	1662	3002	1792
75	181	0.52	0.359	561	814	1375	3002	1660
80	156	0.45	0.267	477	699	1176	3002	1588

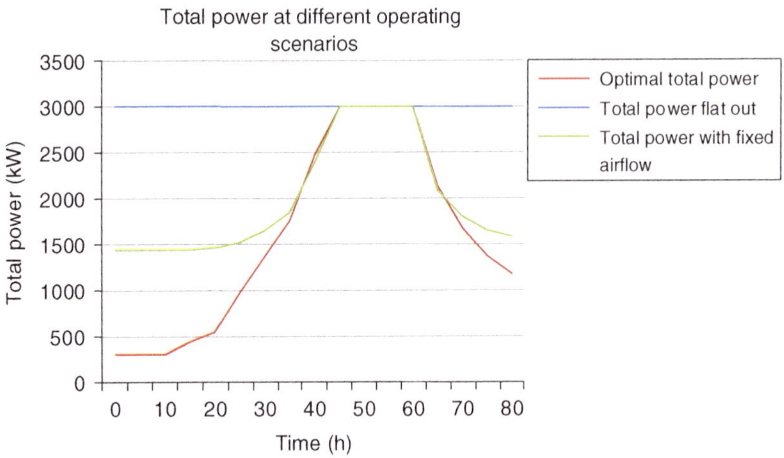

Figure 9.2 Total power graphically.

For gasses other than air, the economics are far more involved than just power cost. We will explore this more in Chapter 11.

Summary of Chapter

By judiciously applying the "full data" principles of Chapter 8 incrementally throughout a batch or semibatch process, we can minimize the total energy consumed. The savings can be considerable.

List of Symbols

C_L	concentration of dissolved species in liquid
C_{sat}	saturation concentration of gas in liquid
DO	dissolved oxygen concentration in liquid
E	energy used
k_la	overall mass transfer coefficient
OTR	oxygen transfer rate
p	pressure
P	agitator power
P_c	compressor power (usually in kW)
Q_a	airflow rate

U_S superficial gas velocity
VVM volume of gas per volume of liquid per minute
θ time increments

References

1 Benz, G. (May 2003). Optimize power consumption in aerobic fermenters. *Chemical Engineering Progress* 99: 100–103.
2 Fitzpatrick, J., Gloanec, F., Michel, E. et al. (2019). Application of mathematical modelling to reducing and minimising energy requirement for oxygen transfer in batch stirred tank bioreactors. *ChemEngineering* 3 (14): 1–21.

10

Heat Transfer Surfaces and Calculations

Purpose of This Chapter

Control of process temperature in a fermenter is important to assure that the organisms thrive, and, in some cases, it affects product distribution, yield, and product quality, as well as the rate of production. The use of heat transfer surfaces and equipment is necessary to assure good temperature control. In this chapter, we will explore some of the most common kinds of heat transfer surfaces used in fermenters, discuss pros and cons where applicable, and provide quantitative calculation methods where available.

Design Philosophy

Though it may not be immediately obvious, agitator design has a relatively small effect on heat transfer. As we will see later, most correlations have the process side convective heat transfer coefficient being proportional to the Reynolds number to the 2/3 power:

$$h_i \propto \left(N_{Re}\right)^{0.67} \tag{10.1}$$

Reynolds number is defined thusly:

$$N_{Re} = D^2 N\rho / \mu \tag{10.2}$$

Since agitator power is proportional to N^3D^5, it can be derived by substitution that:

$$h_i \propto P^{0.22-0.26} \tag{10.3}$$

Agitator Design for Gas–Liquid Fermenters and Bioreactors, First Edition. Gregory T. Benz.

The exponent depends on whether the shaft speed or impeller diameter is being manipulated.

Thus, the agitator design should not be on the basis of a targeted heat transfer coefficient. Rather, the agitator should be selected for the process requirements as previous chapters have indicated. Then the heat transfer area, type, and imposed temperature differential with cooling media should be selected based on the actual heat transfer coefficients created by the agitation system chosen. This is often an iterative process. A certain configuration is assumed, temperature differentials, coefficients, and area for heat transfer is calculated, and the results checked for sufficiency. If not sufficient, another design or set of conditions is tried until the results are satisfactory.

Overview of the Problem

Although aerobic fermentation is not strongly exothermic by chemical reactor standards, total heat generation is nonetheless significant. Excess heat must be removed to maintain a temperature conducive to organism growth and production of the desired product. Removing the heat is hampered by the fact that the operating temperature is quite mild. Most fermenters operate between 30 and 40 °C, with a few outliers near 50 °C. Future use of thermophiles, which can live at temperatures approaching 100 °C, would make the problem easier. Cooling tower water is generally in the range of 25–30 °C in the summer, so there is little temperature differential available if cooling tower water is used as the coolant. Often, chilled water must be used, at least in the summer.

Because of the above, it is often a challenge to get sufficient heat transfer area into the vessel. The problem gets more difficult as the vessel volume gets larger, as the area to volume ratio generally gets smaller and the heat transfer coefficients decrease as well. In addition to looking at various internal and external surfaces, as we will do in this chapter, increasing the aspect ratio can increase the area-to-volume ratio. This is discussed in more detail in Chapter 19.

Though pressure drop is an important design consideration, we will not be covering the calculations in this book. We prefer to leave that up to those responsible for pump and piping design.

Heat Sources

Essentially, at any given point in time, it is desired to operate the fermenter at a constant temperature. That means rate of heat transfer = rate of heat generation. So, we need to know what all the heat sources are, and how much heat is coming from each. There are three main heat sources: metabolic heat generation, agitator heat input, and compressor heat input. In some cases, there may be some pure chemical reaction going on also. All must be taken into account.

Usually, but not always, metabolic heat generation is the biggest contributor. In general, microbiologists can calculate the heat output per organism, times the organism cell count per volume, times total volume, to get the total metabolic heat generation. This will vary by organism, feedstock, and gas fed. Especially, it is apt to be different when the organisms are consuming other gasses besides oxygen, such as hydrogen, methane, etc., which will be discussed in Chapter 11.

When the gas being consumed is oxygen, the primary heat generation mechanism is oxidation of sugars. This means we can simplify the heat generation in terms of oxygen consumption [1]. In that reference, it was noted that, with very little error, the heat output from the consumption of oxygen equals 4.6E5 J/mol of oxygen consumed. So, if we have a working volume of 100 000 l and an OTR of 150 mmol/l-h, the metabolic heat generation would be about 100 000 l × 150 mmol/l-h × 1 mol/1000 mmol × 4.6E5 J/mol = 6.9E9 J/h, which equals 1917 kW.

Agitator heat input is also significant. Many times over the years, I have been asked, How much of the agitator power input goes into heat in the liquid? The answer is, all of it. This can be understood by imagining what happens if you suddenly removed the agitator from an agitated tank. Eventually, all the liquid motion stops. What happened to all that kinetic energy? Obviously, it must have been dissipated into heat. So, agitator heat input = agitator power input. For that same 100 000 l vessel, if we invested a specific power of 3 W/l, that would give us a total of 300 000 W, or 300 kW. That would be about 15% of the metabolic heat generation.

As a rough rule of thumb, compressor power is 2–5 times as high as agitator power at optimum conditions. While it must be calculated case-by-case, this rule of thumb serves to indicate that compressor power can be a significant contributor to the heat load. Compressor heat can sometimes be removed by an air-to-air heat exchanger before entering the tank, if the ambient air temperature is less than the fermenter temperature. This is usually less expensive than using cooling tower or chilled water to remove the compressor heat once it is already in the tank.

Cooling Sources

The primary cooling source will be cooled water or other thermal medium, circulating through heat exchange surfaces on, in or external to the vessel. These surfaces are the focus of this chapter. But there are two other potential cooling sources that can come into play.

One is the sparged gas. If it is chilled below process temperature prior to entry into the vessel, it will provide some cooling. However, as the heat capacity of gas is low on a volumetric basis, this is unlikely to be a major source of cooling.

Another is evaporative cooling. Whatever the humidity of the gas is exiting the compressor (or another source of gas), by the time it is dispersed and rises through the vessel, it is probably safe to say it is 100% saturated with moisture at the exit temperature and pressure of the fermenter. The enthalpy of evaporation can be calculated using standard psychometric methods. This can then be subtracted from the net heat duty. The author prefers not to take this into account, leaving it as a small safety factor.

Heat Exchange Surface Overview

There are many possible variations of heat transfer surfaces used to remove heat from a vessel. This overview is not exhaustive, as fabricators have exercised considerable creativity over the years with respect to heat transfer surface configurations. But these figures represent the most common types in use, based on the author's experience, and we will cover some of the pros and cons of each.

Broadly speaking, these surfaces may be divided into external (outside the main vessel wall) and internal (immersed in the process fluid). The external surfaces are basically vessel jackets. Many fermenter vessels have both jackets and internal surfaces. We will illustrate several types of each. We will not illustrate clamp-on surfaces, as these are not in common use for production bioreactors.

The combinations of jacket and coil in the illustrations can be mixed and matched. We chose to combine illustrations for the sake of brevity.

Figure 10.1a and b illustrates a vessel equipped with a simple jacket on the outside and a large-diameter helical coil on the inside.

The simple jacket has some of the following pros and cons.

Pros:

- Inexpensive, especially in small vessels.
- Low-pressure drop for a given flow rate.

Cons:

- Lower heat transfer coefficient than other jackets. This can be improved by adding spiral baffles, channeling the flow, and increasing velocity, though pressure drop will increase.
- Can require the main vessel to have thicker walls due to external crushing pressure. In some cases, this could negate any savings on the jacket itself, especially in larger vessels.

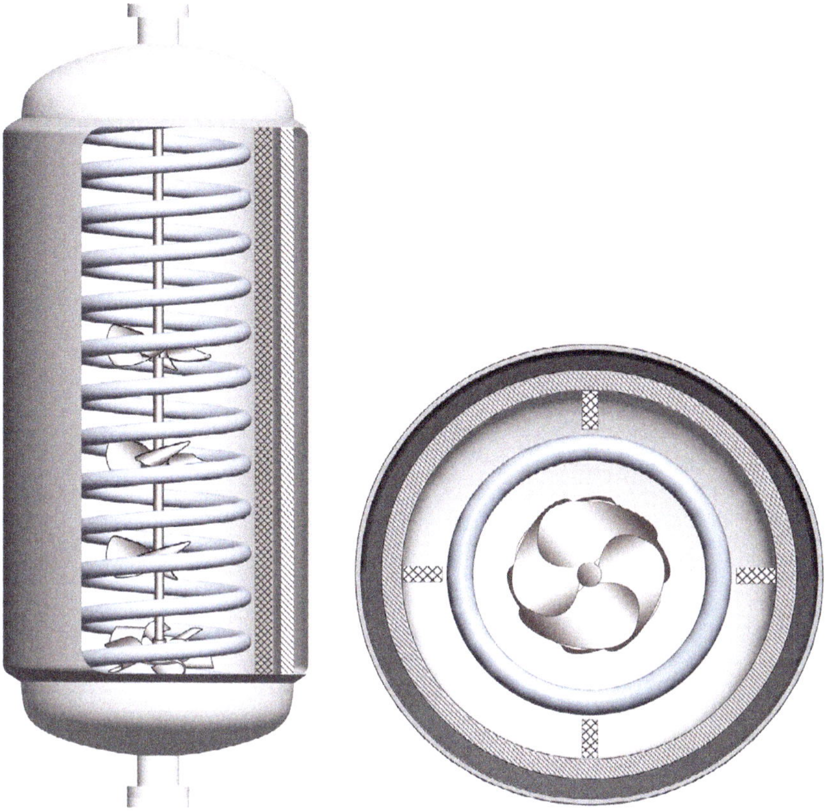

Figure 10.1 (a and b) Helical coil plus simple jacket, plan, and elevation views.

Large helical coils have some of the following pros and cons.

Pros:

- Relatively good heat transfer coefficients.
- Easy to fabricate.
- Mechanically strong.

Cons:

- Requires tank baffles; they do not create an antiswirl baffling effect.
- Possibly high-pressure drop, unless broken up into zones.
- May reduce agitation near vessel wall.
- If used in multiple banks, will negate the effect of wall baffles, requiring additional baffles inside the coil radius, which could create clearance problems with the agitator impellers.

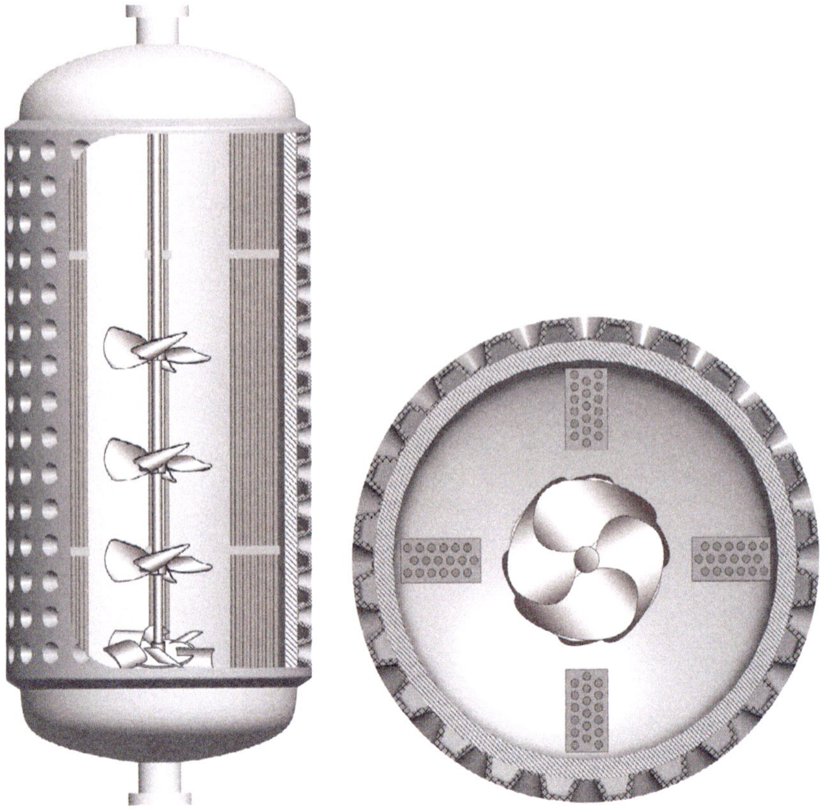

Figure 10.2 (a and b) Vertical tube bundles plus dimple jacket, plan, and elevation views.

- Length expansion due to temperature differential can be considerable. Support design must account for this.
- Limited area possible. More than three coil radii (one inside the other in the tank) not recommended.

Figure 10.2a and b shows a vessel equipped with a dimple jacket and vertical tube bundles. This is a fairly common combination in fermenters.

Dimple jackets are often considered to have the following pros and cons.

Pros:

- High external heat transfer coefficient.
- Mechanically strong.
- Generally do not require thicker vessel wall. May even reinforce the vessel.

- Easier to fabricate than half-pipe coils in some cases.
- Can easily be built in zones.

Cons:

- Relatively high pressure drop (This can be ameliorated by zoning, at the expense of reducing the heat transfer coefficient).
- More expensive than simple jacket in small vessel sizes.

Vertical tube bundles typically have the following pros and cons.
Pros:

- Mechanically strong.
- Act as tank baffles; no other baffles needed unless the bundles do not extend to full liquid height.
- Less total differential thermal expansion than large helical coils.
- Relatively high heat transfer coefficients.
- Versatile. 3–12 bundles can be used, with varying numbers of rows per bundle and tubes per row.
- Can be configured multiple ways to balance heat transfer coefficient vs pressure drop. For example, all tubes can be in series, or each bundle can be in series internally but bundles are piped in parallel, or rows in parallel then tubes in series, etc.

Cons:

- Manifolding complexity is greater than for helical coils, potentially adding to cost.

Figure 10.3a and b depicts a vessel equipped with a half-pipe coil jacket and small helical coil bundles. The small helical coils are an unusual geometry, but are shown as one more example of what is possible.

Half-pipe jackets are generally considered to have the following pros and cons.
Pros:

- Extremely strong. They actually reinforce the main vessel wall.
- Relatively high heat transfer coefficient.

Cons:

- Can be difficult to fabricate.
- Relatively high-pressure drop.
- Not as easy to zone as dimple jackets.

Small helical coils may have the following pros and cons.

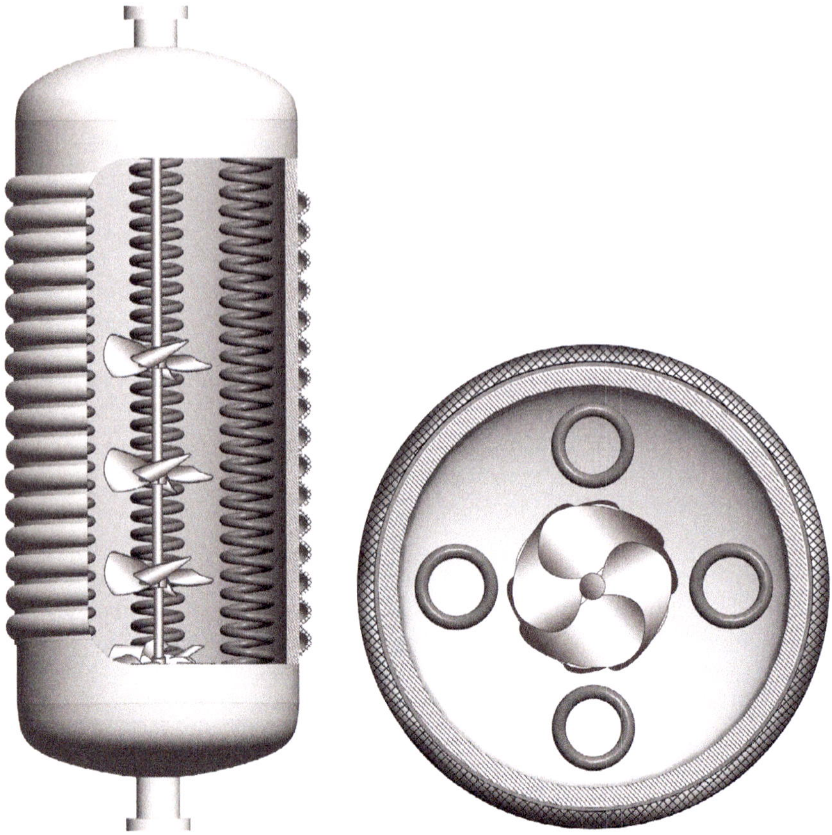

Figure 10.3 (a and b) Small helical coils plus half-pipe jacket, plan, and elevation views.

Pros:

- Act as tank baffles. No additional baffles needed as long as coils extend to liquid level.
- Relatively high heat transfer coefficient.
- Easy to fabricate.
- Less thermal expansion issue than large helical coils.

Cons:

- Not a common configuration. Not studied as much as other configurations.
- Support under agitated conditions needs careful consideration.
- Not as versatile as vertical tube bundles. Can only be configured a few ways, as each bundle has only one possible flow path.

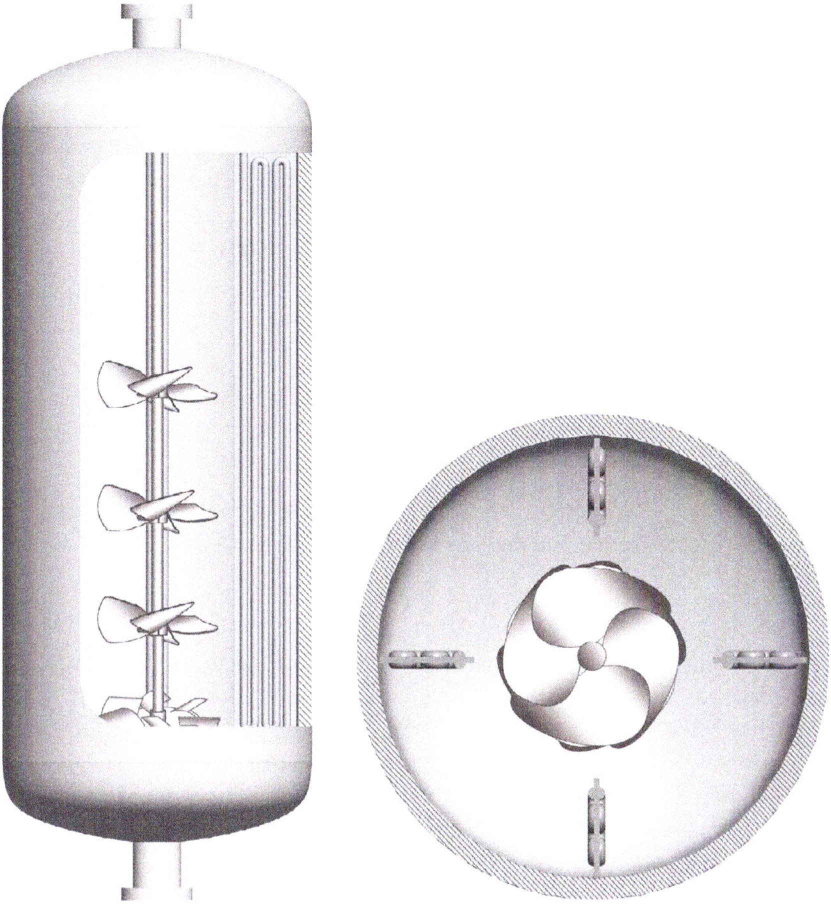

Figure 10.4 (a and b) Plate coils used as baffles, plan, and elevation views.

Figure 10.4a and b shows plate coil panels used as tank baffles, with no jacket. However, plate coil panels can be applied with or without a tank jacket.

Plate coils used as baffles have the following pros and cons.

Pros:

- Relatively high heat transfer coefficient.
- Easy to install.
- Inexpensive per unit area of heat transfer surface.
- Versatile. Multiple panels can be used. Some installations have used as many as 48 panels, but the author recommends no more than 18.

- Act as tank baffles. No other baffles are needed as long as the panels extend to the maximum liquid level.

Cons:

- Strength is an issue. Field experience in fermenters has included stress cracking of panels and their piping connections, resulting in coolant leaking into the broth and vice versa.
- Natural frequency may be a contributing factor to mechanical failure.

Principle of Heat Transfer Calculation

There are two basic equations that need to be solved simultaneously for each heat transfer surface used in the vessel:

$$Q_t = UA\Delta\theta_{LM} \tag{10.4}$$

where $\Delta\theta_{LM}$ is the log mean temperature difference, and

$$Q_m = FC_{Pm}\left(\theta_{out} - \theta_{in}\right) \tag{10.5}$$

We are using θ for temperature rather than T because we are already using T for tank diameter.

Equation (10.4) denotes heat transfer through the exchange surface. Equation (10.5) denotes the change of enthalpy in the heat transfer medium.

Note that under steady-state conditions, Q_t must equal Q_m. But, in order to calculate these values, we must know the value of θ_{out} for the heat transfer fluid. Thus, this is an iterative process. The steps involved are as follows:

1) Assume a value for θ_{out}.
2) If C_{Pm} varies with temperature, use a value at the mean temperature between inlet and outlet.
3) Calculate Q_m based on the above assumption.
4) Calculate U based on a mean estimate of the wall temperature's effect on wall viscosity.
5) Calculate $\Delta\theta_{LM}$, which equals $(\theta - \theta_{out}) - (\theta - \theta_{in})/\ln((\theta - \theta_{out})/(\theta - \theta_{in}))$.
6) Calculate Q_t based on the above.
7) If $Q_t \neq Q_m$, repeat starting with step 1, until the values of Q_t and Q_m match within reasonable accuracy. If the initial temperature guess is close, the next temperature estimate can be obtained by substituting the value of Q_t for Q_m in Eq. (10.5) and solving for θ_{out}. Usually, this method converges, but if the initial estimate is too far off, it sometimes does not. In such a case, one must try another temperature estimate until it does converge.

Table 10.1 Typical fouling resistances.

	Fouling resistances	
Substance	**[m^2-s-K/J]**	**[ft^2-h-F/BTU]**
Treated cooling tower water	1.80E-04	0.001
Untreated cooling tower water	5.30E-04	0.003
City or well water	5.30E-04	0.003
Brackish water	1.80E-04	0.001
Broth-typical	1.80E-04	0.001
Broth-sticky or clumping	5.30E-04	0.003
Engineered thermal fluids	0	0

Calculating U is obviously an important step. It is easier to do if we think in terms of resistances, or 1/U. The total resistance is the sum of the resistances on each side of the heat transfer surface, plus the fouling film resistances on both sides, plus the conductive resistance of the material barrier (usually stainless steel) between the broth and the coolant. Typical fouling resistances can be found in Table 10.1. In other words, for flat surfaces, such as plate coil baffles:

$$1/U = 1/h_i + f_i + t/k + f_o + 1/h_o \tag{10.6}$$

For round surfaces, such as coils or tube bundles use Eq. (10.7), based on the process side area.

$$1/U = d_t/\left(d_i * h_i\right) + f_o + d_t * \ln\left(d_t/d_i\right)/k + f_i + 1/h_o \tag{10.7}$$

No correction was made for area for the inside fouling resistance, as the effect is minor and fouling resistances are usually estimated to only one significant digit.

For vessel jackets, it is permissible to use Eq. (10.6), as the radius of the vessel is likely to be large enough that the diameter ratio (OD/ID) is about 1, which would make the two equations almost the same. For example, for a 3 m inside diameter vessel with a wall thickness of 6 mm, the OD is 3.012 m, the natural log of the diameter ratio is 1.998 E-3, and therefore, the middle term in Eq. (10.7) becomes 6.026E-3/k, whereas in Eq. (10.6), it is 6E-3/k. This is a difference of less than ½ of 1% on the conductivity resistance, and much less than that on the overall U value.

The next section will get into specific correlations for h values for various heat exchange surfaces. The general form of such equations is:

$$N_{Nu} = K * N_{Re}{}^{a} * N_{Pr}{}^{b} * \left(\text{geometric factors}\right) * \left(\text{wall viscosity factor}\right) \tag{10.8}$$

Keep in mind that the correlations have about a ± 30% error band, so safety factors should be applied. Meanwhile, several tables have been supplied with useful data. We have not included data for unusual vessel materials that are not common in fermenters. However, these are easily found in the literature.

For data on other thermal fluids, other metals, and other cool stuff (pun intended), an excellent resource is The Engineering Toolbox, www.engineeringtoolbox.com.

Calculations By Type of Surface

Vessel Jacket, Agitated Side

Because the flow pattern is different for different impellers and is different on the sidewall from the bottom, we use separate correlations for the sidewall and bottom, and modifications for different impeller types. The jacket type has no influence on the agitated side coefficients. For the vessel bottom:

$$N_{Nu} = K\left(N_{Re}\right)^{2/3}\left(N_{Pr}\right)^{1/3}\left(\mu/\mu_w\right)^{0.14}\left(3D/T\right)^{0.13} \tag{10.9}$$

where $N_{Nu} = h_o T/k$ and $N_{Re} = D^2 N\rho/\mu$ and K depends on impeller type. Estimates are:

K = 0.5 for Rushton and concave radial impellers

K = 1.08 for 45° pitched blade turbines

K = 0.9 for low solidity hydrofoils

K = 1.0 for high solidity hydrofoils (author's estimate)

For the sidewall:

$$N_{Nu} = K\left(N_{Re}\right)^{2/3}\left(N_{Pr}\right)^{1/3}\left(\mu/\mu_w\right)^{0.14}\left(T/Z\right)^{0.15}\left(3D/T\right)^{0.13} \tag{10.10}$$

K varies by impeller type, as before

K = 0.74 for Rushton and concave radial turbines

K = 0.31 for low solidity hydrofoils

K = 0.45 for 45° pitched blade turbines

K = 0.4 for high solidity hydrofoils (author's estimate)

Similar equations can be found in Ref. [2], though these equations came from Ref. [3].

Some might observe that the coefficient is higher for Rushton impellers than for hydrofoil impellers, even though hydrofoils are much better at producing flow than Rushtons. However, note that the power number of a Rushton is about 5.5, whereas the power number of a low solidity hydrofoil is about 0.3. Therefore, at equal power input, the hydrofoil will be $(5.5/0.3)^{0.2}$, or 1.79 times as large in diameter. That would make the Reynolds number term 2.17 times as large, roughly making up for the lower K value on the sidewall. For the bottom, the heat transfer coefficient is substantially higher for hydrofoils than for radial impellers. This is because of the direction of flow: axial flow benefits the bottom coefficient, whereas radial flow benefits the sidewall coefficient.

It is common to have multiple impellers in a fermenter. Sometimes they are of unequal type and/or diameter. This does not affect the vessel bottom coefficient; just use the data for the bottom impeller. The sidewall can be divided into zones for each impeller. The h_o value can be computed for each zone and the value averaged. If the jacket is zoned, the calculation can be done separately for each jacket zone.

Simple Unbaffled Jacket, Jacket Side

For most enclosed exchanger surfaces, we use the Sieder-Tate expression, modified to use equivalent diameter, d_e, for the reference diameter [4]. We wrote out the Nusselt number below to make this clear. It will be assumed that the flow is turbulent ($N_{Re} > 2100$) on the media side. If it is not, heat transfer will be poor. Then the design should be changed to increase liquid velocity.

$$h_j.d_e/k = 0.027 * \left(N_{Re}\right)^{0.8} * \left(N_{Pr}\right)^{0.33} * \left(\mu/\mu_w\right)^{0.14} \tag{10.11}$$

So what is this equivalent diameter? For any noncircular cross section, equivalent diameter is defined as:

$$d_e = 4\left(\text{flow area}\right) / \left(\text{wetted perimeter}\right) \tag{10.12}$$

For a simple unbaffled jacket, this becomes:

$$d_e = 4\left(T_J - T_o\right) * L / \left(2 * \left(T_J - T_o\right) + 2L\right) \tag{10.13}$$

For a baffled jacket, the flow area and equivalent diameter would have to be determined based on the actual baffle configuration and flow path.

The above methodology does not take into account the effect of the Dean number, which can be important for flow along a circular path. However, the effects are expected to be small, as the radius of curvature is large compared to d_e. Any errors are likely to be conservative.

Dimple Jacket, Jacket Side

The methodology presented here comes from Ref. [5]. Though we will not address it here, the same reference covers calculation of pressure drop through a dimple jacket.

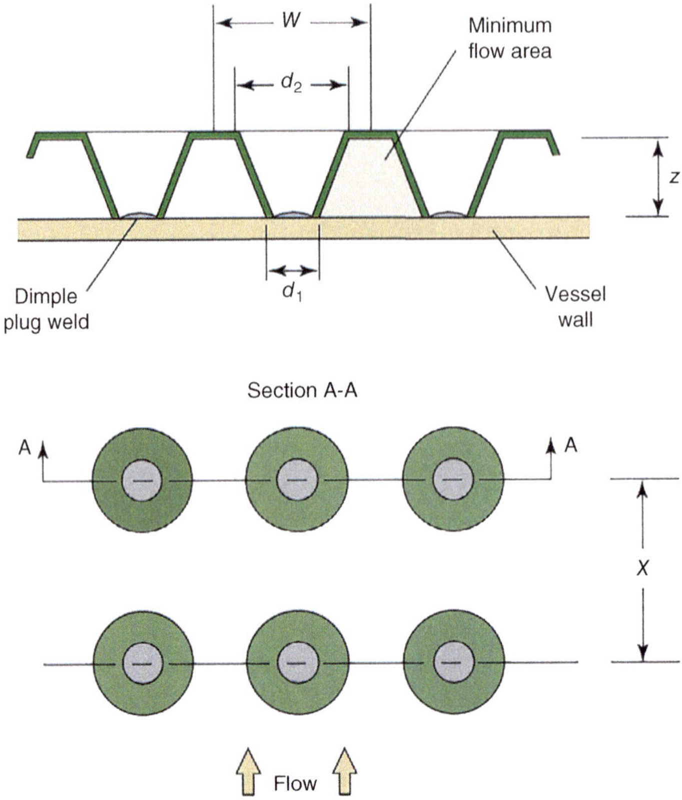

Figure 10.5 Dimple jacket cross section.

Several concepts need to be addressed before getting into the equations in detail. These are aided by referring to Figure 10.5, which is a cross section and plan view of a simplified dimple shape. We show the dimple diameter at the wall and the dimple jacket shell as d_1 and d_2, respectively. The mean dimple diameter, d_m, is simply $(d_1 + d_2)/2$. The Reynolds number is based on d_m. The maximum flow area per dimple is the jacket height times the dimple pitch, or:

$$A_{max} = W * z \tag{10.14}$$

The minimum area is the area approximated by a trapezoid, equal to the mean diameter times the internal height:

$$A_{min} = d_m * z \tag{10.15}$$

Now we can introduce the Coburn heat transfer factor, J, which corrects for the dimple pitch and geometry, and uses d_m in the Reynolds number, and the velocity in the Reynolds number is V_{max}.

$$J = 0.0845(W/X)^{0.368}(A_{min}/A_{max})^{-0.383}(N_{Re})^{-0.305} \tag{10.16}$$

$$N_{Re} = d_m * V_{max} * \rho/\mu \tag{10.17}$$

$$V_{max} = F_d / A_{min} \tag{10.18}$$

Now the Nusselt number ($h_i d_m/k$) can be expressed as a function of the J factor, the Reynolds number, and the Prandtl number:

$$N_{Nu} = J * N_{Re} * N_{Pr} \tag{10.19}$$

Of course, we can combine Eqs. (10.19) and (10.16):

$$N_{Nu} = 0.0845(W / X)^{0.368}(A_{min} / A_{max})^{-0.383}(N_{Re})^{0.695}(N_{Pr}) \tag{10.20}$$

These correlations are reported to have good accuracy at velocities of 1.5 m/s and higher, but may result in values 15–20% high when the velocities' are as low as 0.5 m/s. (presumably, we are talking about V_{max}, but the author does not say). The Reynolds number range is 1000–50 000.

The linearity with the Prandtl number strikes this author as strange. Most other correlations have the Prandtl number to about a 1/3 exponent. I don't have an explanation for this, but it is reported to fit the data, so we will go with it.

One comment on area used in the total heat transfer calculation may be needed. Though the actual contact area of the fluid on the vessel wall is less than 100%, the fin effect of the metal surrounding the dimples makes it acceptable to use 100% of the area in the calculations.

Half-Pipe Coil, Jacket Side

The term "half-pipe" implies that a pipe is cut in half and then formed to wrap around the vessel shell. More commonly, metal plate is formed into a curve and then bent to fit the vessel shell. The radius of the pipe may or may not conform to standard pipe dimensions. The tube geometry may not actually be a half-cylinder either. It could be made of any cross-sectional angle of 180° or less, though the flow area would be highly restricted if the angle were less than 90°. In practice, the majority of coils use either a 180° angle or a 120° angle.

The method of calculation is basically a modified Sieder-Tate method, using the equivalent hydraulic diameter, d_{eq}, in the Reynolds and Nusselt numbers. This diameter can be derived for any coil angle, but we will present the results here only for 180 and 120° angles. Reference [6] has other details. We will assume the coil has a circular segment cross section, with an inside diameter of d_i.

For 180°:

$$d_{eq} = d_i * \pi/2 \tag{10.21}$$

$$A = (\pi/8) * d_i^2 \tag{10.22}$$

For 120°:

$$d_{eq} = 0.708 d_i \tag{10.23}$$

$$A = 0.154 d_i^2 \tag{10.24}$$

The Sieder-Tate correlation, modified for the Dean number effect due to the curved flow path, may be written as

$$N_{Nu} = 0.027 (N_{Re})^{0.8} (N_{Pr})^{0.33} (\mu/\mu_w)^{0.14} (1 + 3.5 d_i/d_c) \tag{10.25}$$

where d_c is the mean diameter of the coil. To calculate that, we need to know the coil outside diameter, or d_o. That is, for a 180° coil:

$$d_o = T + d_i + 2t \tag{10.26}$$

where t refers to the vessel shell thickness

For a 120° coil:

$$d_o = T + 0.5 d_i + 2t \tag{10.27}$$

The coil diameter, d_c, can now be calculated for either 120 or 180° coils:

$$d_c = (d_o + T)/2 \tag{10.28}$$

Note that the Dean number correction, the last term in Eq. (10.25), increases the Nusselt number only slightly for larger vessels. The Nusselt and Reynolds numbers for the inside coefficient for the half-pipe coils are defined as follows:

$$N_{Nu} = h_i d_{eq}/k \tag{10.29}$$

$$N_{Re} = d_{eq} V \rho/\mu \tag{10.30}$$

where $V = F_t/A$

Equation (10.25) is intended for use when the Reynolds number is at least 10 000, which should be the case most of the time. For laminar flow ($N_{Re} < 2100$), the following is proposed from the same reference:

$$N_{Nu} = 1.86 (N_{Re} N_{Pr} d_{eq}/L)^{0.33} (\mu/\mu_w)^{0.14} \tag{10.31}$$

For transition flow ($2100 \le N_{Re} < 10\,000$), calculate the Nusselt numbers in both turbulent flow, based on an artificial Reynolds number of 10 000, using Eq. (10.25), and laminar flow based on a Reynolds number of 2100 using Eq. (10.31). Call these N_{NuT} and N_{NuL}, respectively. Then calculate the actual Reynolds number. Then do a linear interpolation as follows:

$$N_{Nu} = N_{NuL} + \left(N_{NuT} - N_{NuL}\right) * \left(N_{Re} - 2100\right) / 7900 \qquad (10.32)$$

$$(7900 = 10000 - 2100)$$

Helical Coil, Inside

The relationship for helical coil inside Nusselt number is basically the Sieder-Tate correlation in turbulent flow, similar to Eq. (10.25) with the Dean number modification for coil curvature:

$$N_{Nu} = 0.027\left(N_{Re}\right)^{0.8}\left(N_{Pr}\right)^{0.33}\left(\mu/\mu_w\right)^{0.14}\left(1 + 3.5 d_i / d_c\right) \qquad (10.33)$$

Here, the Nusselt number and Reynolds numbers are based on the tube inside diameter, d_i. The coil diameter, d_c, is at the centerline of the flow channel. The flow velocity is just the coolant flow rate divided by the cross-sectional area of the tube, if only one flow zone is used. If there are multiple coils or multiple zones, the actual flow per coil or zone must be used.

Equation (10.33) is for fully developed turbulent flow. It may take a number of pipe diameter lengths for turbulent flow to be established in the entrance zone, reducing heat transfer. To "encourage" turbulent flow to develop, one can add a flow disruptor to the entrance. This could be a right-angle bend, a slight washer-like intrusion into the pipe, or a Static Mixer® element, as just a few examples.

Correlations for laminar flow are problematic, as they all involve a d_i/L term. Some correlations have the Nusselt number approaching zero as d_i/L approaches zero (very long coils). One that does not come from Ref. [7]:

$$N_{Nu} = 3.66 + (0.0668\left(d_i / L\right) N_{Re} N_{Pr} / \left(1 + 0.04\left(\left(d_i / L\right) N_{Re} N_{Pr}\right)^{2/3}\right) \qquad (10.34)$$

Laminar flow in the heat transfer medium should be avoided, of course. If the fermentation medium is low viscosity (<5 cP, for example), it is not unusual for the limiting resistance to be the medium side.

Helical Coil, Process Side

Reference [2] uses a popular correlation based on the work of Oldshue and Gretton in 1954. In this chapter, we have updated it for some modern impellers, as the original paper was based on Rushton turbines. However, 45° pitched blade turbines seem to have the same coefficient, probably due to their mixed flow pattern. The Nusselt number correlation is:

$$N_{Nu} = K\left(N_{Re}\right)^{0.67}\left(N_{Pr}\right)^{0.37}\left(D/T\right)^{0.1}\left(d_t/T\right)^{0.5} \quad (10.35)$$

In the above expression, the Reynolds number is the impeller Reynolds number, and the Nusselt number is based on the tube OD, d_t:

$$N_{Re} = D^2 N\rho/\mu \quad (10.36)$$

$$N_{Nu} = h_o d_t/k \quad (10.37)$$

where k is the broth thermal conductivity

The value for K depends on impeller type. Table 10.2 has suggested values of K. The value does not seem to be strongly dependent on impeller type.

It is not known to the author what the Reynolds number range of validity for Eq. (10.35), but other correlations by the same authors are reported to have validity for $N_{Re} > 100$ [4].

There is no viscosity ratio correction, possibly because the flow over the tubes is strong in an agitated tank (or maybe because the original authors did not study this).

The viscosity to be used in the case of non-Newtonian fluids is the apparent viscosity at the impeller. The methods of Chapter 3 can be used to determine this.

Equation (10.35) is for a single helical coil, with reasonable coil pitch. Good practice would be to have a coil pitch of at least two tube diameters. If more than one coil is used concentrically (this does not apply to coil bundles), apply an effectiveness factor to Eq. (10.35) on the Nusselt number, as follows:

$$E = 0.92^{(n-1)} \text{ for } n > 1 \quad (10.38)$$

More than three rows of coils are not recommended. They may under baffle the tank and create dead zones.

Table 10.2 K values for helical coils.

Impeller description	K	Comments
Rushton	0.17	—
Concave radial	0.16	Author's estimate
45° pitched blade, four blades	0.17	—
Low solidity hydrofoils	0.14	—
High solidity hydrofoils	0.15	Author's estimate

To avoid swirl and loss of both heat transfer and gas dispersion capability when large helical coils are used, it is recommended that vertical baffles be placed both between the helical coil and the wall and between the helical coil and the impellers. The inner section of the baffles should be at least 5% of the coil average diameter in width, or $0.05d_c$. No baffles are needed for small helical coil bundles unless they do not extend to the maximum liquid level, as they act as baffles. It may not be necessary to place baffles inside the coils if the coil pitch is $3d_o$ or more.

Equation (10.35) was developed for a square batch. If multiple impellers are used, one can imagine a Z value per impeller. If this does not correspond to a Z/T of 1 per impeller, we can borrow a geometry term from Eq. (10.10) by multiplying the Nusselt number by $(T/Z)^{0.15}$ where the Z referenced is the Z per impeller, i.e., Z/n_I.

Vertical Tube Bundle, Inside

Since vertical tubes are merely straight pipe, the standard Sieder-Tate relation may normally be used:

$$N_{Nu} = 0.027\left(N_{Re}\right)^{0.8}\left(N_{Pr}\right)^{0.33}\left(\mu/\mu_w\right)^{0.14} \tag{10.39}$$

The Reynolds number and Nusselt number are both based on the I.D. of the tube, d_t. However, the velocity term must be figured based on how the flow is divided up. If all of the tubes in the tank are connected in series, the velocity is merely the medium flow rate divided by the cross-sectional area of the tube. However, it is more common to manifold the bundles in such a way that some tubes are in parallel and some in series. The velocity depends on the actual flow rate per tube. For example, it is normal to pipe each bundle in parallel with the others. So, if there are six bundles, each bundle gets 1/6 of the total flow. Within each bundle, there are a multitude of tubes. If they are in a single row, they can be in parallel or in series. If there are several rows, each row can be in parallel or series with the others. For example, if we have six bundles with three rows of six tubes each per bundle, if we pipe each bundle in parallel with the others but all the tubes within a bundle are in series, the flow rate per tube is 1/6 of the total flow rate. If we pipe each row in series with the others, but the tubes in each row are in parallel, the flow rate per tube is 1/36 of the total flow rate. If we pipe all of the tubes within a bundle in parallel, the flow rate per tube is 1/108 of the total flow. Other configurations are possible. Using a manifold at the first inlet will tend to eliminate entrance effects.

The manifold design must be a balance between heat transfer and pressure drop. Higher velocities create a higher heat transfer coefficient, but also a higher pressure drop. Usually, practical designs have a flow velocity inside the tube between 0.9 and 3.5 m/s.

Table 10.3 Recommended minimum tubing diameters for tube bundles.

Vessel diameter, m	Minimum tubing OD, mm
1.5–3	50
3–5	75
5–10	100
10+	150

Choice of tube diameter is influenced not only by the need for area, but mechanical strength. See the section on mechanical loads on baffles later in this chapter. But as a guide based on practical experience in the field, I suggest Table 10.3 for minimum tubing sizes. With more bracing, it may be possible to use smaller tubes. This needs to be carefully analyzed by the vessel fabricator.

Sometimes, it is difficult to get the velocity high enough and also have sufficient area for heat transfer, especially when complying with the recommendations in Table 10.3. In some cases, this can be helped by using an insert inside the pipe. It may be necessary to go up a pipe size to do this. As an example, if we have an outer tube of 100 mm OD, 90 mm ID, we might insert another tube inside it of 75 mm OD. This tube would be sealed off, and only serves to increase the velocity in the annulus. For this particular case, the annular area would be 30.5% as much as the open area of the outer tube, resulting in 3.27 times as high a flow velocity. It is that velocity that would be used in the Reynolds number calculation.

The diameter to be used in the Nusselt and Reynolds numbers would be d_{eq}, based on four times the flow area divided by the wetted perimeter. That would be 15 mm for the above example.

Vertical Tube Bundle, Process Side

The Dunlap–Rushton correlation referenced in [2] is most commonly used. It has been modified for modern impellers.

$$N_{Nu} = K\left(N_{Re}\right)^{0.65}\left(N_{Pr}\right)^{0.3}\left(D/T\right)^{0.33}\left(2/n_b\right)^{0.2}\left(\mu/\mu_w\right)^{0.14} \quad (10.40)$$

In this case, the Nusselt number uses the tube OD, d_o, and the Reynolds number is as defined for agitator impellers, $D^2N\rho/\mu$. The "K" value for various impellers is listed in Table 10.4.

As in the case of multiple helical coils, there should be an effectiveness factor, E, applied to multiple rows of tubes per tube bundle. We repeat the same equation here.

Table 10.4 K values for tube bundles.

Impeller description	K	Comments
Rushton	0.09	—
Concave radial	0.08	Author's estimate
45° pitched blade, four blades	0.09	—
Low solidity hydrofoils	0.074	Author's estimate
High solidity hydrofoils	0.08	Author's estimate

$$E = 0.92^{(n-1)} \text{ for } n > 1 \tag{10.38}$$

Minimum spacing of vertical tubes is about $1.75d_o$, center to center. A value of 2.0 or more is preferred. This applies within a row and between rows.

Equation (10.40) was developed for a square batch ($Z/T = 1$). If there is one impeller per square batch section, no correction need be applied. Though not studied specifically, the $(T/Z)^{0.15}$ factor from Eq. (10.10) may be applied judiciously. Here, Z is understood to be per impeller, or Z/n_I.

Since it is common practice to use more than one impeller in a fermenter, and these impellers may differ in both diameter and type, one might ask how to handle such a case. We can do the calculations by zone and average them for an overall Nusselt number.

Plate Coil, Inside

There are several different manufacturers of plate coil or panel coil products that can be used both as baffles and heat transfer surfaces. Their internal configuration varies from manufacturer to manufacturer, or even within different model series at the same manufacturer. Some have a serpentine series of tubes welded to a flat plate. Some are essentially dimple jackets on a flat plate. Some are dimple jackets welded back to back. Some are made of dual flat plates with flow-directing baffles on the inside. Probably, the best way to estimate the Nusselt number is to use the Sieder-Tate correlation, with no correction for the Dean number (as there is no curvature), and using the equivalent hydraulic diameter, d_{eq}, in the Nusselt and Reynolds numbers. The velocity will depend on the exact configuration as well as how the baffles are connected to each other (series, parallel, or a combination).

I was unable to get correlations directly from the manufacturers.

Though as many as 48 such baffles have been used in production fermenters, more than about 18 of them is not recommended, as they will create a "tank

Table 10.5 K values for plate coil baffles.

Impeller description	K	Comments
Rushton	0.031	—
Concave radial	0.028	Author's estimate
45° pitched blade, four blades	0.031	—
Low solidity hydrofoils	0.025	Author's estimate
High solidity hydrofoils	0.028	Author's estimate

within a tank" effect, resulting in substantially reduced power draw and poor movement at the vessel walls. They can be applied wider than the customary T/12 for baffles, but probably should not exceed T/8 for the same reason as avoiding too many of them.

Plate Coil, Process Side

Fortunately, the agitated side coefficients have been studied and published [4].

$$N_{Nu} = K\left(N_{Re}\right)^{0.66}\left(N_{Pr}\right)^{0.33}\left(T/Z\right)^{0.15}\left(\mu/\mu_w\right)^{0.14} \tag{10.41}$$

Where $N_{Nu} = hW_{PC}/4k$

Though only a Rushton turbine was studied, by analogy to tube baffles, we have created Table 10.5 for this geometry.

Additionally, this relationship was normalized for four baffles. For other numbers of baffles, by analogy with tube bundles, the Nusselt number should probably be multiplied by a factor of $(4/n_b)^{0.2}$.

Presumably, the baffle width used in the study was around T/12 to T/10. It may be that there should be another correction for extremely wide baffles, beyond that built into the Nusselt number. We do not have any information on this.

Example Problem: Vertical Tube Bundle

Problem Statement

Our production fermenter has an inside diameter of 4 m, a straight side of 12 m, and ASME dished heads. It will contain a working broth volume of 109 m^3 and have a gassed liquid level of 10 m. The broth properties are a density of 1050 kg/m^3, a viscosity of 10 cP or 0.01 Pa-s, a thermal conductivity of 600 mW/m-K or 0.6 W/m-K and an isobaric heat capacity, C_P, of 4.1 kJ/kg-K. Operating temperature is 30 °C.

To remove the process heat, chilled cooling water is supplied at a temperature of 10 °C, at a flow rate of 0.06 m^3/s.

The cooling system consists of [6] vertical tube bundles. Each bundle has three rows of six tubes each (18 tubes per bundle), and the tubes are 8 m long. Each tube is 75 mm OD, 67 mm ID (4 mm wall thickness). The system is piped such that the total coolant flow is evenly divided among bundles (connected in parallel), and the tubes within each bundle are all in series. The tubes are stainless steel.

The agitation system consists of a lower deep concave turbine and two upper up-pumping high solidity impellers. All are 1.6 m diameter. The shaft speed is 106 rpm or 1.77/s. Its power draw is 218 kW.

What is the cooling capacity of this system? And, assuming the compressed air feed is cooled to 30 °C before sparging (so there is no heat input from the compressor), and the agitator inputs 218 kW of power as stated above, what OUR can this system support, based on a metabolic heat generation of 460 kJ/mole of oxygen consumed?

Problem Solution

The prime basis for solution is to first solve Eq. (10.7), which is repeated here, and then solve for Q_t, which must be equal to Q_m. As we shall see, this is an iterative process. We solve Eqs. (10.4) and (10.5) after solving 10.6. These, too, are repeated here.

$$1/U = d_t/(d_i * h_i) + f_o + d_t * \ln(d_t/d_i)/k + f_i + 1/h_o \tag{10.7}$$

$$Q_t = UA\Delta\theta_{LM} \tag{10.4}$$

where $\Delta\theta_{LM}$ is the log mean temperature difference, and

$$Q_m = FC_{Pm}(\theta_{out} - \theta_{in}) \tag{10.5}$$

We will work on Eq. (10.6) piecewise. First, we will assign a value for the fouling resistance of 1.8E-4 m^2-s-K/J for both f_o and f_i. Next, we will work on the agitated or broth side convective coefficient, or h_o, since this part is not iterative, but is needed for the later iterations. We use Eq. (10.40) plus Table 10.4. Equation (10.40) is repeated here.

$$N_{Nu} = K(N_{Re})^{0.65}(N_{Pr})^{0.3}(D/T)^{0.33}(2/n_b)^{0.2}(\mu/\mu_w)^{0.14} \tag{10.40}$$

The lower impeller has a K value of 0.08, and so do the upper impellers. So, each impeller will produce the same Nusselt number. We will apply a T/Z and effectiveness factor later. We will simply assume the wall viscosity is twice the bulk viscosity. We will cover the implications of that in a later section. Solving Eq. (10.40) term by term, we get:

$$N_{Nu} = 0.08\left(1.6m^2 * 1.77s^{-1} * 1050\,kg/m^3/0.01\,Pa\text{-}s\right)^{0.65}$$
$$\left(4100\,kJ/kg\text{-}K * 0.01\,Pa\text{-}s/0.6\,W/m\text{-}s\right)^{0.3}$$
$$*\left(1.6\,m/4\,m\right)^{0.33}\left(2/6\right)^{0.2}\left(10/20\right)^{0.14} = 566$$

Now applying a T/Z factor for a Z/T of 0.8333 per impeller or $(1.2)^{0.15} = 1.03$, and an effectiveness factor from Eq. (10.38) of 0.92^2, or 0.85, we get a revised Nusselt number of 566 * 1.03 * 0.85, or 495.

Now we can solve for h_o using the definition of Nusselt number:

$$N_{Nu} = h_o d_t/k, \text{ or } h_o = N_{Nu} k/d_t = 495 * 0.6\ W/m\text{-}K/0.075m$$
$$= 3962\ W/m^2\text{-}K$$

The bulk process thermal properties are fixed because the broth temperature is fixed. But the cooling water enters at one temperature and leaves at a higher one. For this reason, and to solve Eqs. (10.4) and (10.5), we need to make an initial guess of the outlet temperature of the cooling water, and iterate until $Q_t = Q_m$. Let's make an initial guess of 20 °C. We will use the average thermal properties over that temperature interval, looking them up in Table 10.6 for a temperature of 15 °C. Therefore, density = 998.2 kg/m^3, C_P = 4.19 kJ/kg-K, μ = 1.14E-3 Pa-s (we will half that for wall viscosity, so the ratio will be 2), and k = 0.589 W/m-K. Before we can solve Eq. (10.39) (repeated below), we need to determine coolant velocity. Our total flow is 0.06 m^3/s. This is divided into six because the bundles are in parallel. However, within each bundle, all tubes are in series. Therefore, the flow rate in each tube is 0.06/6, or 0.01 m^3/s. The inside diameter of the tubes is 67 mm or 0.067 m. Therefore, the flow area in each tube is 3.526E-3 m^2. The velocity is thus 0.01 m^3/s/3.526E-3 m^2 = 2.836 m/s. Now we can use Eq. (10.39):

$$N_{Nu} = 0.027\left(N_{Re}\right)^{0.8}\left(N_{Pr}\right)^{0.33}\left(\mu/\mu_w\right)^{0.14} \tag{10.39}$$

Solving,

$$N_{Nu} = 0.027\left(0.067\ m * 2.836\ m/s * 998.2\ kg/m^3/1.14E\text{-}3\ Pa\text{-}s\right)^{0.8}$$
$$\times\left(4190\ J/kg\text{-}K * 1.14E\text{-}3\ Pa\text{-}s/0.589\ W/m\text{-}k\right)^{0.33}\left(2.0\right)^{0.14} = 891.9$$

Now we can solve for h_i using the definition of Nusselt number:

$$N_{Nu} = h_i d_i/k; h_i = N_{Nu} k/d_i = 891.9 * 0.589\ W/m\text{-}k/0.067\ m$$
$$= 7841\ W/m^2\text{-}K.$$

Table 10.6 Thermal properties of water.

Temperature	Density		Thermal conductivity		Viscosity		Heat capacity, C_P	
[°C]	[kg/m^3]	[lb$_m$/ft^3]	[mW/m-K]	[BTU/h-ft-F]	[Pa-s]	[Lb$_m$/ft-h]	[kJ/kg-K]	[BTU/lb-R]
0.1	999.85	62.4186	556	0.321	1.79E-03	4.32	4.22	1.01
1	999.9	62.4218	560	0.324	1.73E-03	4.19	4.22	1.01
4	999.97	62.4264	570	0.329	1.57E+00	3.8	4.21	1.01
10	999.7	62.4094	579	0.335	1.31E+00	2.75	4.2	1
15	999.1	62.3719	589	0.340	1.14E-03	2.75	4.19	1
20	998.21	62.316	598	0.346	1.00E-03	2.42	4.18	1
25	997.05	62.2436	607	0.351	8.89E-04	2.15	4.18	1
30	995.65	62.1563	615	0.355	7.97E-04	1.93	4.18	1
35	994.03	62.0554	622	0.359	7.20E-04	1.74	4.18	1
40	992.22	61.942	629	0.363	6.54E-04	1.58	4.18	1
45	990.21	61.8168	635	0.367	5.97E-04	1.44	4.18	1
50	988.04	61.6813	641	0.370	5.47E-04	1.32	4.18	1

Table 10.7 Thermal conductivities of common vessel materials.

	Thermal conductivity	
Material	**[W/m-K]**	**[BTU/h-ft-F]**
Alloy C	8.7	5.03
Alloy 400	26	15
Alloy 600	15	8.67
Carbon steel	54	31.2
Nickel	85	49
304 SS	14.4	8.32

Now we can go back to solve the overall coefficient, U, by using Eq. (10.7). We use the conductivity k for stainless of 14.4 W/m-K from Table 10.7:

$$\begin{aligned}1/U &= d_t/\left(d_i * h_i\right) + f_o + d_t * \ln\left(d_t/d_i\right)/k + f_i + 1/h_o \\ &= 0.075\,m/\left(0.067\,m * 7841\,W/m^2\text{-}K\right) + 1.8E\text{-}4\,m^2\text{-}K/W \\ &\quad + 0.075\,m * \ln\left(0.075/0.067\right)/14.4\,W/m\text{-}K + 1.8E\text{-}4\,m^2\text{-}K/W \\ &\quad + 1/3962\,W/m^2\text{-}K = 1.34E\text{-}3\,m^2\text{-}K/W.\end{aligned}$$

Inverting this, we get U = 744.8 W/m^2-K. (For those more used to British units, this number equates to 236.5 BTU/h-ft^2)

Now we want to solve Eq. (10.4), repeated below, to see how much heat is transferred based on the temperatures assumed and the U value:

$$Q_t = UA\Delta\theta_{LM} \tag{10.4}$$

To do so, we have to calculate the area and the log mean temperature difference. The area equals the number of tubes times the length of tubes times the outer perimeter of the tubes, or (6 * 3 * 6) * 8m * π * 0.075m = 203.6 m^2. The log mean temperature difference, based on the assumed outlet temperature of 20 °C for the cooling water, is ((30–20) – (30–10))/ln((30–20)/(30–10)) = 14.43 °C. Thus, solving Eq. (10.4), we get Q_t = 744.8 W/m^2-K * 203.6 m^2 * 14.43 °C (C is equivalent to K, in terms of size of the degree), or 2.189E6 W (2189 kW).

But our job is not done. We have to check this against Eq. (10.5), repeated below.

$$Q_m = FC_{Pm}\left(\theta_{out} - \theta_{in}\right) \tag{10.5}$$

The easiest way to check it is to plug in Q_t for Q_m and see what value of θ_{out} comes back. In other words,

$$\theta_{out} = Q_i / \left(FC_P\right) + \theta_{in} \tag{10.42}$$

Plugging in, we get θ_{out} = 2.189E6 W/(998.2 kg/m^3 * 0.06m^3/s * 4190 W/kg-K) + 10C = 18.72C. (Note that we converted C_P from kJ/kg-K by using the fact that 1 W = 1 J/s).

We had assumed that our outlet temperature would be 20 °C, but we get 18.72 °C based on heat transferred. This was a reasonable first guess but we must iterate until we get acceptable agreement between Q_t and Q_m. We would then go through the entire above procedure with a new temperature guess and repeat until they agree sufficiently. Rather than take up too much book space, we will present the results. Our second guess was an outlet temperature of 19 °C, and adjusting for the slight changes in the thermal properties of water, we solved Eq. (10.42) to get an outlet temperature of 19.1 °C, and a Q value of 2.28E6 W or 2280 kW. We could go through more iterations, but considering the accuracy of heat transfer correlations (typically ± 30%), a difference of 0.1 °C is judged to be "good enough."

Interestingly, the chief heat transfer resistance is the tube wall conductivity resistance. It is about triple the other resistances, which are roughly equal in magnitude to each other. Fouling factors are very similar to the convective resistances.

Part two of the problem is to determine the OUR this heat exchange rate could support. From the problem statement, the agitator heat input (which equals its power draw) is 218 kW. Since we are transferring 2280 kW, that leaves a net cooling rate from the process of (2280–218), or 2062 kW. Each mole of oxygen consumed releases 460 kJ of energy. 2062 kW equals 2062 kJ/s or 7.42E6 kJ/h. If we divide 7.42E6 kJ/h by 460 kJ/mol of oxygen consumed, we get a gross oxygen uptake of 16135 mol/h. Since OUR is usually stated as mmol/l-h, we would get:

$$\text{OUR} = 16137\,\text{mol}/\text{h} * 1000\,\text{mmol}/\text{mol} / \left(109\,\text{m}^3 * 1000\,\text{l}/\text{m}^3\right) = 148\,\text{mmol}/\text{l}-\text{h}.$$

With a 218 kW agitator power input into 109 m^3 liquid volume, the P/V is 2W/l. If a typical airflow of 0.7VVM were used, the agitation system plus the airflow would probably support an OTR of 100–150 mmol/l-h. Thus, the heat transfer design for this problem appears reasonable. More area may be needed as a safety factor if the desired OTR is near the top of the range. However, there will often be cases where the fermentation is heat transfer limited. Possible solutions to a heat transfer limited problem (not an exhaustive list) could include, using this example problem as a guide:

- Add bundles (up to maybe 12).
- If above is done, cooling water flow can be increased also.
- Increase cooling water flow to existing bundle design. But it cannot increase by much, as the velocity is already 2.8 m/s. (Usually a velocity much greater than 3m/s is not advisable due to pressure drop).

- Change the manifolding to put each row in parallel, and increase the coolant flow. The flow could be tripled and the velocity would still be reasonable. But this is probably not enough to solve the problem.
- Consider smaller tube diameter, with more tubes/row to increase area. But be careful about using tubes smaller than the minimum diameters recommended in Table 10.3.
- Angle the bundles 45° into the impeller flow and add about 20% more tubes.
- Add a jacket to the vessel, with additional coolant to the jacket.
- Use a lower coolant inlet temperature. 5 °C is generally the lower limit.
- Use an external heat exchanger loop. We will discuss this briefly in a later section of this chapter.
- If it won't harm the organisms, allow operation at a higher temperature to increase the temperature driving force. Mass transfer capacity will be reduced somewhat due to decreased gas solubility.
- Use a higher aspect ratio vessel to get more area per volume

Additional Consideration: Effect on Power Draw

Tank baffles affect flow pattern and power draw. Heat exchange surfaces that act as baffles do likewise. The effect of baffles depends on impeller type, D/T ratio, the angle of the baffles, width of the baffles, the number of baffles, the Reynolds number and, to some extent, on the shape of the baffles. Reference [8] studied the effect of baffles on various impeller types and D/T ratios. It demonstrated the rationale behind why, in the USA at least, the term "standard baffling" refers to 4 baffles at 90°, 1/12 of the tank diameter. With this kind of baffling, the power draw is at or near the maximum for all common impellers, and that is true across a spectrum of D/T ratios. But nonflat baffles were not studied. Although more recent papers have investigated nonflat baffles, such as wedges, semicircles, or circles, they did not study such baffles at greater than normal baffle quantity or width. In [8], the authors introduced the concept of the baffle number, N_b:

$$N_b = n_b w_b / T \tag{10.43}$$

where n_b is the number of baffles and w_b is the projected width of the baffles.

Standard baffling thus corresponds to a baffle number of 0.333 or 1/3. Figure 10.6 illustrates the effect of baffle number on power draw relative to standard baffling, using a radial turbine at a D/T of 1/3.

This chart can be used as an initial guide to estimating the power draw effects of heat exchange surfaces in a tank. For example, the example problem would have a baffle width of 6 * 75 mm or 450 mm (we don't need to add each row when they are in line with each other), and the number of bundles was six, so we would

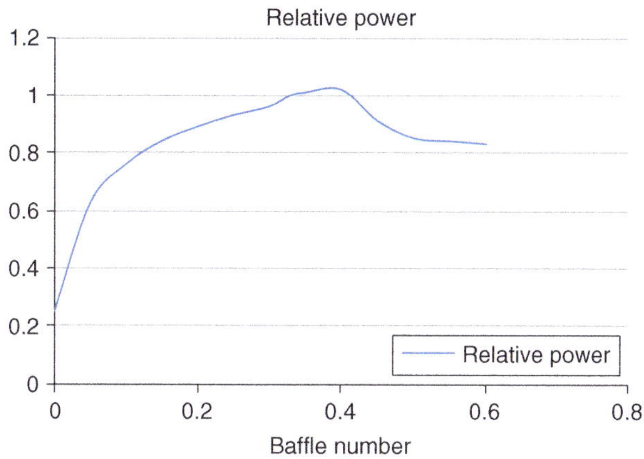

Figure 10.6 Effect of baffle number on Power Draw.

have a baffle number of 6 * 450/4000 = 0.675. Though Figure 10.6 stops at a baffle number of 0.6, we can easily see that the power draw might be reduced by 15–20% due to the tube bundles. Since mass transfer depends actual power draw, this power draw reduction would have a deleterious effect on mass transfer. To correct for that, the shaft speed would need to be increased by 5–8%, or the impeller diameters would need to be increased 3–5%.

There are more extreme cases than that. The author has seen systems with 12 tube bundles with a 30–40% reduction in power and plate coil baffles numbering 48 with a width of T/4 (baffle number of 12!) with more than a 50% power draw reduction.

Because of the lack of published studies of such systems, the author recommends that an actual physical model be built of the proposed system when the baffle number exceeds about 0.4, and lab/pilot testing be conducted to verify the gassed and ungassed power draw characteristics. Typically, such tests are conducted at equal Reynolds number (unless fully turbulent in both scales), equal Froude number, and equal aeration number. Most of the larger agitator companies are capable and willing to build such models and do such tests, especially for the sake of getting an order.

Additional Consideration: Forces on Heat Exchange Surfaces Used as Baffles

Considering the previous section, up to 70% of the agitator power can be attributed to drag on the baffling surfaces. This drag can be expressed as torque, in SI units, as:

$$T_q = P / 2\pi N \tag{10.44}$$

For example, in our tube bundle sample problem, the invested power was 218 kW at a shaft speed of 106 rpm or 1.77/s. Our torque is thus 218 000 W/(2 * π * 1.77/s) = 19 600 n-m.

For design purposes, recognizing that the load fluctuates, we can apply a 0.7 factor for the % going to the baffles, and a factor of 2 for fluctuations, resulting in multiplying Eq. (10.44) by 1.4:

$$\text{Design}\, T_q = 1.4P / 2\pi N \tag{10.45}$$

For our sample problem, that means we would use a value of 27 440 N-m.

This torque must be applied at a radius and divided among the baffles:

$$F_b = \text{Design}\, T_q / (n_b r) \tag{10.46}$$

For our sample problem, the radius of the inner edge of the tube bundles is about 1.07 m. That would result in a design load on the baffles of 27 440/(6 * 1.07) or 4274 Newtons. In English units, that would be about 940 lb. The total force can be divided up locally in proportion to the impeller power in each zone. It is spread out along the length of each impeller zone. The lower impeller usually draws more power, so the bottom section may need more reinforcement.

Besides the design value of the forces applied, the frequency of those forces must be considered. The agitator produces frequencies at a value of N; N times the number of blades ($N * N_{bl}$), and higher-order frequencies due to turbulence. In addition, Von Karman vortex frequencies created by flow across tubes should be considered. The natural frequency of the heat exchange surfaces and their supports should be well above any frequencies generated in the tank. This level of design may need a combination of finite element analysis and computational fluid dynamics to solve, and these topics are beyond this author's expertise. But is it probably a good idea to make sure the natural frequency of heat transfer internals it at least 10 times the agitator shaft speed.

Additional Consideration: Wall Viscosity

As you may have noticed, most Nusselt number correlations have a μ/μ_W term. Specifically, most have that term raised to the 0.14 exponent. Logically, we need some way to estimate the wall viscosity to calculate the Nusselt number. For Newtonian liquids, this basically means estimating the mean wall temperature, and then looking up or calculating the viscosity corresponding to that temperature. This is an iterative process, as the heat transfer calculations depend on an estimate of μ_W, but the wall viscosity depends on the wall temperature. All of the individual heat fluxes must be equal. That is, the heat flux from the bulk liquid must equal the flux across each fouling film, the conductive flux across the tube wall and the flux within the heat transfer liquid. This requirement would lead to solving for wall temperature on both inside and outside heat transfer surfaces in order to determine the wall viscosity. The procedure would be something like this:

1) Assume a wall viscosity for both coolant and process fluid.
2) Do the heat transfer calculations as described above.
3) Solve for wall temperatures based on equal heat flux across all resistances.
4) Determine mean wall temperatures on both sides.
5) Determine wall viscosities corresponding to the temperatures in step 4.
6) Go back to step 1 with the new viscosities and repeat until the values match reasonably well.

In this author's opinion, that is a lot of work for a term that is only there to the 0.14 exponent. Moreover, the original correlations were developed at a time when computational fluid dynamics was not available, so the determination of wall viscosity was probably not that accurate. My suggestion would be to make a reasonable guess and be done with it. I usually use a value of about 2 for cooling, and 0.5 for heating.

In the case of non-Newtonian liquids, the situation is more complicated. Most non-Newtonian fermentation broths are shear thinning. This means the wall viscosity must be evaluated at the temperature and shear rate at the wall. There is little published on the shear rates at the vessel wall or other heat transfer surfaces. To make matters worse, internal heat transfer surfaces can interfere with the flow pattern for highly shear-thinning liquids, creating dead spots. Often, it is better to use a jacket plus an external loop if needed. Reference [9] describes a method for predicting jacket coefficients with pseudoplastic fluids. It uses a modified Reynolds number in the correlation. The procedure is rather involved. Rather than repeat it here, the reader is encouraged to obtain the reference.

Additional Consideration: Effect of Gas

All of the foregoing heat transfer correlations were developed based on ungassed liquids. However, aerobic or other gas–liquid fermenters always have gas present during the heat-generating stages of operation. The question naturally arises, What effect does the presence of gas have on heat transfer coefficients? Unfortunately, there seems to be very little published on the subject. Mechanistically, there are two possible effects gas may have on heat transfer. One is that, if the aerated liquid is treated as a homogeneous fluid, the average thermal properties will be affected. Specifically, the heat capacity and thermal conductivity will be less than in the unaerated liquid, and this will tend to decrease heat transfer. On the other hand, bubbles near the actual heat transfer surface will provide additional flow and turbulence, reducing the wall viscosity effect and perhaps increasing the coefficient enough to offset any reduction in thermal properties. My own opinion is that the effects cancel out for low viscosity liquids and a gas holdup fraction of less than 0.2. Use caution for higher gas holdups; perhaps reducing the process side coefficient calculated by about 10%.

For viscous fluids, especially shear-thinning ones, I believe the gas bubble effect will be a net benefit by breaking up the film near stationary surfaces. But, until someone studies this in more detail, this remains just my opinion.

External Heat Exchange Loops

Besides vessel jackets and internal heat exchange surfaces, it is possible to remove heat using a pump-around loop and an external heat exchanger. This can be a viable choice for some fermentations. As usual, there are pros and cons to this approach.

Pros

- No limit to heat transfer capacity, as the exchanger is not limited by space available in the tank.
- Possibly less expensive per unit of cooling capacity than tank internals.
- Less subject to mechanical failure, as it is isolated from agitation forces.

Cons

- Requires additional space outside the vessel.
- May need periodic maintenance, depending on style.
- Requires a fairly large pump to circulate the broth. This is an additional capital, power, and maintenance expense.
- Because flow produced by such pumps is at least an order of magnitude smaller than that produced by an agitator, the retention time in the loop can be several seconds or more, resulting in the organisms being exposed to chilled conditions potentially long enough to shock them into a reduced metabolic state. The contact time at chilled surfaces within the tank, by contrast, is likely in the millisecond range except for viscous liquids.
- Related to the above, the recirculating loop may be oxygen deprived, which could also shock the organisms.
- Passing the broth through a pump and heat exchanger is a possible source of additional shear, particularly if a low retention time, high-pressure drop design is used.

Summary of Chapter

In this chapter, we identified the fact that heat exchange is a very important consideration in the design of gas–liquid fermenters. We described the most common heat exchange systems used in fermenters (based on the author's experience), and calculation methods where possible. We advised caution regarding the mechanical design of internal heat exchange surfaces, and some simplified load formulae. We brought up the issue of the effect of tank internals on agitator power draw. In summary, there is a lot to the design of heat exchange systems for fermenters, and the subject could easily merit a book by itself. We hope this introduction to the issues proves valuable to the reader, and that it will be sufficient for many applications.

List of Symbols

A	Flow area of half-pipe or simple jacket
A_{max}	Maximum flow area of dimple jacket, per dimple
A_{min}	Minimum flow area of dimple jacket, per dimple
a	Exponent
b	Exponent
C_P	Heat capacity at constant pressure
C_{Pb}	Heat capacity at constant pressure, broth
C_{Pm}	Heat capacity at constant pressure, cooling medium
D	Impeller diameter
D_{eq}	Equivalent impeller diameter for heat transfer
d_1	dimple diameter at weld to vessel wall
d_2	Dimple diameter at outer wall of jacket
d_c	Mean diameter of coil
d_{eq}	Equivalent hydraulic diameter
d_i	Tube inside diameter
d_m	Mean diameter of dimple
d_o	Outside diameter of tube coil
d_t	Tube outside diameter
E	Tube or coil effectiveness factor
F	Flow rate of cooling medium
F_b	Force applied per baffle
F_d	Flow rate of cooling medium per dimple
F_t	Flow rate of cooling medium per tube or coil
f_i	Inside fouling resistance
f_o	Outside fouling resistance

h_i	Inside or media side convective heat transfer coefficient
h_o	Outside or broth side convective heat transfer coefficient
J	Coburn heat transfer factor
k	Thermal conductivity
K	Coefficient on Nusselt number correlation
L	Coil length along the vessel
N	Shaft speed
n	Number of banks of coils or rows of tubes
n_b	Number of tube baffles or bundles
n_I	Number of impellers
N_b	Baffle number, $n_b w_b/T$
N_{bl}	Number of blades on an impeller
N_{Nu}	Nusselt number
N_{NuL}	Laminar Nusselt number
N_{NuT}	Turbulent Nusselt number
N_P	Power number
N_{Pr}	Prandtl number = $C_P\mu/k$
N_{Re}	Reynolds number
OUR	Oxygen Uptake Rate
Q_m	Rate of heat transferred to or from medium
Q_t	Rate of heat transferred through a heat exchange surface
r	Radius of inner edge of tank baffles
t	Vessel or tube wall thickness for heat transfer
T	Tank diameter
T_J	Jacket inside diameter
T_o	Tank shell outside diameter, i.e., T plus two times wall thickness
T_q	Torque
U	Overall heat transfer coefficient
V	Velocity inside pipe or tube
V_{max}	Maximum velocity in conduit (minimum flow area)
W	Dimple jacket pitch across flow direction
W_{PC}	Width of plate coil baffles.
w_b	Projected width of baffles, normal to tank tangent
X	Dimple jacket pitch in flow direction
Z	Ungassed liquid level
z	Dimple jacket inside height
ρ	Liquid density
μ	Viscosity, bulk
μ_w	Viscosity on vessel wall or tube surface
θ	Temperature, usually of broth
θ_{in}	Inlet temperature of heat transfer medium
θ_{out}	Exit temperature of heat transfer medium

References

1 V'ant Riet, K. and Tramper, J. (1991). *Basic Bioreactor Design*. Marcel Dekker.
2 Dickey, D. and Hicks, R. (1976). Fundamentals of agitation. *Chemical Engineering Magazine*: 14–21.
3 Fasano, J. et al. (1994). *Advanced Impeller Geometry Boosts Liquid Agitation*, 7. Chemical Engineering.
4 Paul, E. et al. (eds.) (2004). *Handbook of Industrial Mixing*. Wiley Interscience.
5 Garvin, J. (2001). Estimate heat transfer and friction in dimple jackets. *Chemical Engineering Progress*: 73–75.
6 Dream, R. (1999). Heat transfer in agitated jacketed vessels. *Chemical Engineering Magazine*.
7 Holman, J. (ed.) (1972). *Heat Transfer*, 3rde. McGraw-Hill.
8 Bates, R. et al. (1963). An examination of some geometric parameters of impeller power. *I & EC Process Design and Development* 2 (4).
9 Ali, A. et al. (2014). Heat transfer to Newtonian and non-Newtonian fluids in mechanically agitated vessel. *International Journal of Scientific Engineering and Technology Research* 3 (14): 3031–3035.

Further Readings

https://www.checalc.com/solved/jacketHeat.html (simple or spiral baffled jacket).
https://cheguide.com/jacket_vessel.html (half-pipe jacket).

11

Gasses Other Than Air and Liquids Other Than Water

This chapter presents an overview of some of the issues involved in fermentations utilizing fed gasses other than air. Though not very common now, it is expected that more such fermentations will be developed over the years as new strains of organisms are found or developed that can metabolize such gasses. Some of the organisms are extremophiles, using not only unusual gasses but unusual operating temperatures and/or pressures. Others are variants of existing organisms, with modified metabolic pathways or a natural ability to metabolize different gasses into different products from those they would make with air as the feed gas.

General Principle

The general design principle is the same for all gasses. The approach is still to design for a given mass transfer rate (MTR):

$$MTR = k_l a * (\text{driving force}) \tag{11.1}$$

The main differences between gasses are solubility, which affects driving force, diffusivity, which affects the k_l term, and various economic, disposal, and safety concerns.

In general, the design procedure is to identify which gas is limiting and design for mass transfer for that gas.

Comments on Some Specific Gasses

Ammonia

Ammonia can be an excellent nitrogen source. Its solubility in water is so high that it generally needs no help from an agitator.

Agitator Design for Gas–Liquid Fermenters and Bioreactors, First Edition. Gregory T. Benz.

Carbon Dioxide

Usually, this would be introduced as part of a mixture with other gasses, such as in a Syngas fermentation. The other gasses may include carbon monoxide, nitrogen, and hydrogen, for example. Carbon dioxide is usually not the limiting gas, because it is highly soluble compared to the other gasses.

Carbon Monoxide

Due to lower solubility, the mass transfer rate for CO is more likely to be limiting than it is for CO_2. There are various safety issues which will be discussed later.

Hydrogen

Although hydrogen diffuses easily, its low solubility often makes it the limiting gas. It is normally part of a Syngas mixture. In nonbiological systems, such as oil hydrogenators, it may be fed as a pure gas. In such cases, it is not vented; it is expected to be fully absorbed. This calls for different agitator designs from those presented here. Safety is an issue.

Methane

This is also usually part of a mixture. It may or may not be the limiting gas. Safety is an issue.

Oxygen

In this chapter, we are mainly talking about oxygen as a supplemental gas, beyond the 21% concentration found in air. There are different economics for adding pure oxygen from those for air, and there are safety issues as well. It may be viable to supplement with extra oxygen during high OTR conditions if they last only part of the production cycle, as this may save on capital costs for larger agitators and compressors used to capacity for only a fraction of the batch cycle. Since most fermentations using oxygen also discharge carbon dioxide, if the carbon dioxide is to be recovered for other uses, using oxygen can increase the concentration of CO_2 in the off gas, decreasing recovery cost. This, combined with the reduced agitator power needed, can help offset the cost of oxygen.

Economic Factors

As we saw in Chapters 8 and 9, when air is the gas, the primary cost is the power needed to deliver it to the fermenter (Another cost is the air filtration system and its filter changes, but we did not get into that, as it should be a minor cost compared to the power cost). Thus, it is possible to optimize the production cost with respect to minimizing power cost. Often, the optimum point for air is in the range of 50% oxygen utilization. With gases other than air, the economic model is not so simple. Virtually all other gases used in fermenters have material costs beyond the cost of power and transportation. The economic optimum must balance the cost of gas with the cost of agitator power and other costs. With expensive gases such as hydrogen and methane, for example, it may be that the overall economic optimum is at or above 90% utilization. However, the agitator power needed approaches infinity as the utilization approaches 100%. There is no simple step-by-step procedure for optimizing such gases. The only way is to look in great detail at an array of scenarios to determine overall CAPEX and OPEX. This analysis will not only involve the cost of the gas, the equipment, and the power, but also the disposal cost of unused gas, as described in the next section.

Disposal Factors

Unlike air, oxygen-supplemented air and carbon dioxide, which can be discharged directly to the atmosphere after filtration to remove broth spray and pathogens, the rest of the common gases must be treated or destroyed prior to discharge. In some cases, it may be possible to separate and recycle them. In most cases, it will be necessary to destroy them, usually by combustion. Depending on the exit gas composition, it may be possible to burn the gases directly, or it may be necessary to add an additional fuel source to assure complete combustion. Carbon monoxide, methane, and hydrogen burn quite easily, of course. In the case of hydrogen, it may be necessary to dilute it to have a safe mixture to burn. Chemical absorption may be an alternative method to capture gases in some cases. Whatever method is used, it is a part of the overall cost calculation.

Effects of Different Gasses on $k_l a$

As stated in Eq. (11.1), the mass transfer rate is equal to the product of $k_l a$ and driving force. We will deal with these factors separately. *Caution! The methods described to estimate $k_l a$ below are to be used only for feasibility studies in the absence of data taken with the actual gases. It is always best practice to develop a broth-and-gas-specific $k_l a$ correlation before designing a fermenter.*

$k_l a$ itself is the product of the liquid film coefficient, k_l, and the interfacial area per volume. These two factors may respond somewhat differently to changes in gas behavior. A key influence on the k_l term is the diffusivity, D_{ab}, which varies with the gas, the liquid, the temperature, and the viscosity. It can also vary with absolute pressure and other things such as ionic strength, surface tension, etc. References

Table 11.1 Diffusion coefficients of different gases in water.

	Diffusion coefficients ($cm^2/s*10^5$)					
	Temperature (°C)					
Gas	10	15	20	25	30	35
Ammonia		1.3	1.5	1.64		
Carbon dioxide	1.26	1.45	1.67	1.91	2.17	2.47
Carbon monoxide				2.03	2.43	
Hydrogen	3.62	4.08	4.58	5.11	5.69	6.31
Methane	1.24	1.43	1.62	1.84	2.08	2.35
Oxygen		1.67	2.01	2.42		

[1–4] are example of possible sources of diffusivity data. Table 11.1 lists some of these data, compiled from several sources.

Though other factors may influence k_l besides the diffusion coefficient, the effect of the diffusion coefficient is a square root function, according to reference [5]. That is,

$$k_l \propto \sqrt{D_{ab}} \tag{11.2}$$

If we assume that the hydrodynamics are the same as for an air–water system, then the change in k_la is the same as the change in k_l. That would lead us to approximately the ratios of actual k_la compared to an air–water k_la as listed in Table 11.2. It should be noted that reference [6] concludes that the exponent on D_{ab} is 0.37 rather than the value of 0.5 used in Eq. (11.2). Experimental verification is recommended in any case for a particular system.

Though there are some gaps in the reported data, the Stokes–Einstein equation can be used to interpolate. It states that D_{ab} is directly proportional to absolute temperature and inversely proportional to the liquid film viscosity at those temperatures. In other words:

$$D_{ab1} / D_{ab2} = (\theta_1 / \theta_2) * (\mu_2 / \mu_1) \tag{11.3}$$

Because so many things influence k_la, I must emphasize again that this method may not always be correct, even in terms of ratios. I have personal knowledge where the k_la ratio experienced with methane was actually lower compared to air than predicted by this method. Only a broth-and-gas-specific k_la correlation will yield a reasonably safe design calculation.

Table 11.2 Relative k_la values.

Temperature (°C)	20		25		30		35	
Gas	D_{ab}	Relative k_la to oxygen at 20 °C	D_{ab}	Relative k_la to oxygen at 20 °C	D_{ab}	Relative k_la to oxygen at 20 °C	D_{ab}	Relative k_la to oxygen at 20 °C
Ammonia	1.5	0.86	1.64	0.90				
Carbon dioxide	1.67	0.91	1.91	0.97	2.17	1.04	2.47	1.11
Carbon monoxide			2.03	1.00	2.43	1.10		
Hydrogen	4.58	1.51	5.11	1.59	5.69	1.68	6.31	1.77
Methane	1.62	0.90	1.84	0.96	2.08	1.02	2.35	1.08
Oxygen	2.01	1.00	2.42	1.10				

Effects of Different Gasses on Driving Force

This part is fairly straightforward. Solubilities of gasses in water have been published for most gasses, as in references [1–4]. Table 11.3 lists absolute and relative solubilities for a few gases at 30 °C. Note that the absolute solubilities are based on 100% gas concentration at 1 atm. pressure. The absolute solubility will be proportional to the partial pressure of the gas. For example, with air at 21% oxygen, the solubility will be 8.4E-3 g/kg, or 21% of 0.04.

The actual solubility will also depend on the presence of dissolved species. For this reason, solubility tests should be conducted to determine the Henry's law constant for the actual broth at operating temperature, or at least at a reference concentration and pressure, which can then be factored by the partial pressure ratio to get the C_{sat} term in the driving force. The rest of the driving force is the same as previously discussed.

From Table 11.3, it is easy to see that the dispersion of ammonia and carbon dioxide are almost trivial problems, due to their extreme solubilities. Methane, carbon monoxide, and hydrogen are all less soluble than oxygen. The high diffusion coefficient of hydrogen makes up for that to some extent.

Operating Condition Effects

As you might have surmised, most of the calculations presented thus far for air were based on the ideal gas law. That should provide reasonable accuracy for the conditions typical of aerobic fermenters (<1 bar-g pressure, 25–40 °C). As more organisms

Table 11.3 Solubility of some gases in water.

Gas	Solubility at 30 °C (g/kg)	Relative to oxygen
Ammonia	405	10125.00
Oxygen	0.04	1.00
Hydrogen	0.00155	0.04
Carbon monoxide	0.023	0.58
Carbon dioxide	1.4	35.00
Methane	0.018	0.45

are pressed into service for fermentation, including some extremophiles, the operating temperatures and pressures may be higher. Higher temperatures should pose no problems as long as the reduced solubility is taken into account. But I have seen some studies where the operating pressures being considered were in the range of 5–10 bar. For hydrogen, this probably makes little difference. But it may create measurable error for larger molecules such as methane or carbon dioxide, or molecules that have attraction to each other such as ammonia or hydrazine.

Constraints on Outlet Concentration

While target mass transfer rates can be achieved over a range of gas flows and inlet concentrations by varying agitator power input, there can be constraints on outlet gas concentration. Some may be based on gas toxicity to the organism. This can be the case with normal aerobic fermentations, with CO_2 being the gas of concern. This is more likely to be a problem with oxygen-enhanced fermentation, as the total gas flow is less, allowing the CO_2 concentration in the discharge to be higher.

The constraints can be purely economic. Aside from wasting gases that cost money, there may be disposal costs to get rid of any unused gas, or to concentrate and recycle it.

The constraints can be safety-related. We will talk about that more in the next section.

Safety

There are a number of safety issues that can arise in the operation of fermenters with various feed gases. An entire book could probably be written on the subject. This section is just an overview of some of the safety issues which have occurred to the author, who is not an expert in this field. A thorough safety review is a must to establish proper equipment design, operating conditions, protocol, and operator training to create an accident-free workplace.

Toxicity can be a concern, not only in the case of tank entry, but in case of leakage. This is especially true for carbon monoxide, in part because it is odorless. This differentiates it from ammonia, for example, which, although toxic, is very easily detected by the human nose. Careful purging procedures must be used prior to tank entry, and CO detection equipment must be installed in any area where it is in use. Nitrogen, while not toxic per se, can be a dangerous gas when used to purge a vessel, as there may be insufficient oxygen in a purged vessel to sustain human life, and it is odorless and undetectable by human senses, so oxygen levels must be checked before tank entry.

Flammability is an issue with several gases being used in fermenters. Gases such as methane and other alkanes are very flammable, but Class 1 Group D Division 1 motors will not ignite these gases. If a VFD is used, the motor must still carry its XP rating under such operation. However, the area must be monitored for leakage to avoid a potential buildup of flammable gases. Ammonia and hydrazine are also potential flammable gases that may be used in fermenters. Their solubility makes it unlikely they will have much leakage, and they are easy to detect. Standard explosion-proof motors should work for these gases also.

Explosion potential within the tank is a real potential hazard, and not easy to solve. Some fermentations have a mixture of air, methane, and hydrogen, or other mixtures which are potentially flammable/explosive. Within the gassed liquid, this is not a problem. But in the headspace, it is desirable to avoid an explosive mixture. To do so, the mixture must be either above the upper explosive limit, in terms of reducing gases, or below the lower explosive limit. The first case is fairly easy to handle if the process needs only a small amount of oxygen. Then the combustible gases can be at a high concentration, above the upper explosive limit. The second case can be difficult. Gases such as methane and hydrogen can burn at fairly low concentrations of both the gas and oxygen. It may be that under normal operation, the organisms in the broth absorb enough of the reactive gas to keep the vapor space below the lower explosive limit. But under upset conditions, this may not be so. Therefore, the gas composition in the vapor space must be monitored. Options for a dangerous condition include shutdown and nitrogen purging. The first is not very desirable, for obvious reasons. The second is better, but when nitrogen is added to the vapor space, some of it is incorporated into the liquid below and that reduces the overall mass transfer driving force. It may be necessary to increase agitator power input to compensate for this.

Some might ask, Why is this a concern if there is no ignition source? The answer is, there is a potential ignition source. If there is a failure in the mechanical seal area, temperatures due to friction can get high enough to cause ignition.

Oxygen supplementation also poses special risks. It increases the danger of ignition and lowers the lower explosive limit. If allowed to leak, it can increase risk of fires in the general area.

Liquids Other Than Water

Water is essential to life as we know it. Therefore, virtually all fermenters use water as the primary constituent of the fermentation medium. There are some that have an oil phase present, and this has strong effects on mass transfer and holdup. However, almost all use water as the continuous phase. Nonetheless, it is possible in principle to have a nonaqueous continuous phase. In such cases, the living organisms may be suspended in aqueous droplets dispersed within the oil phase. Or there may be organisms that can actually live in the oil phase; the author is not an expert on this.

Nonetheless, the same mass transfer principles apply. Gas solubility in the continuous phase will be used to determine the mass transfer driving force, and it may be very different in an oil phase from what it is in water. Likewise, gas diffusivity may be very different. Overall, there are enough differences that trying to extrapolate from $k_l a$ correlations based on water is extremely risky. The researcher is strongly encouraged to develop a specific correlation in the actual broth using the principles outlined in Chapter 7.

Summary of Chapter

In this chapter, we demonstrated that all gases obey the same fundamental relationships with respect to mass transfer. They primarily differ in magnitudes such as diffusion and solubility. From an economic standpoint, there are costs beyond power to consider, such as the cost of the gas and potentially the cost of disposal. Safety is often an important issue that must be addressed. Nonaqueous media need a good experimental protocol to come up with design relationships.

List of Symbols

a interfacial area per volume
C_{sat} saturation concentration of gas in liquid
D_{ab} diffusion coefficient of species a in species b
k_l liquid film coefficient

$k_l a$ overall mass transfer coefficient
MTR mass transfer rate per volume
OTR oxygen transfer rate per volume

References

1 The Engineering Toolbox. www.engineeringtoolbox.com (accessed 10 October 2020).
2 Wikipedia. www.wikipedia.com (accessed 10 October 2020)
3 *Perry's Chemical Engineering Handbook*, various editions. McGraw-Hill.
4 *CRC Handbook of Chemistry and Physics*, various editions. CRC Press.
5 Kawase, Y. and Moo-Young, M. (July 1990). The effect of antifoam on mass transfer in bioreactors. *Bioprocess and Biosystems Engineering* 5 (4): 169–173.
6 Ho, C., Stalker, M., and Baddour, R. (1987). The oxygen transfer coefficient in aerated stirred reactors and its correlation with oxygen diffusion coefficients. In: *AIChE Biotechnology Processes* (eds. C. Ho and J. Oldshue), 85–95. AIChE.

12

Viscous Fermentation

General Background

Most fermenter broths are low viscosity (<50 cP) and Newtonian. A few rise into the range of several hundred cP (say, up to 500 at a shear rate of 1/s). Some of these are Newtonian and some are not. Most that are non-Newtonian are shear thinning, following the power law model described in Chapter 3. Above 500 cP, almost all are non-Newtonian. A former mixing expert at a major chemical company that recently merged with another major chemical company, forming one of the largest chemical companies in the world, once said "any stated viscosity above 200 cP is wrong." While not completely true (there are Newtonian fluids that are highly viscous, such as corn syrups, glycerin, and silicone oils), at least a great many viscous fluids are non-Newtonian, meaning their viscosity cannot be described by a single number. In this chapter, we will explore how to disperse gas into viscous liquids of several rheologies. We will also discuss why the process result limitation may have more to do with blending and motion control and avoidance of dead spots than mass transfer per se.

Sources of Viscosity

There are several possible sources of viscosity in a fermentation broth. The main ones are soluble species (which may be part of the medium or secreted by cells), interactions with suspended solids (generally part of the medium, such as starches, and have little effect until above 15% by weight), and interactions with the structure of the cells themselves, such as in filamentous/hyphal/fungal fermentations.

Solution viscosities can be Newtonian or non-Newtonian. For example, sugar solutions are generally Newtonian, even up to very high dissolved solids levels. Examples mainly include honey and corn syrups, which can be thousands of cP at

Agitator Design for Gas–Liquid Fermenters and Bioreactors, First Edition. Gregory T. Benz.

room temperature. On the other hand, soluble gums and polymers tend to be shear thinning and may even have a yield stress. Examples include xanthan and gellan gums, Carbopol, CMC (carboxymethyl cellulose), and similar substances.

High-solids suspensions are usually shear thinning and may have a yield stress. Some, such as corn starch, exhibit dilatancy (shear thickening behavior) at very high-solids concentrations. However, such high concentrations are not used in fermenters, except maybe in the solid-state variety, which is not addressed in this book.

Filamentous broths exhibit strongly non-Newtonian behavior. They almost always have a yield stress and may be shear thinning above the yield point. The liquid between the filaments may be of low viscosity, but the structure created by the filaments results in a bulk viscosity effect.

Viscosity Models for Broths

At the risk of seeming repetitively redundant again, I am repeating the viscosity models from Chapter 3 so you don't have to go back there to look them up. However, I am not including the graphical figures here, so if you want to see those, you still have to go back to Chapter 3.

Newtonian fluids are those which have a shear stress that is directly proportional to shear rate (also called velocity gradient). Mathematically, this can be represented as:

$$\sigma_s = \mu * \gamma \tag{12.1}$$

where μ, the coefficient on shear rate, is also known as viscosity. This can be arranged to:

$$\mu = \sigma_s / \gamma \tag{12.2}$$

Next are shear-thinning or pseudoplastic fluids. These fluids are called shear thinning because the slope of the shear stress versus shear rate curve decreases as the shear rate increases. The behavior may often be described by a power law model:

$$\sigma_s = \mu * \gamma^n \tag{12.3}$$

The exponent n is called the power law exponent. When it has a value of 1, the equation reverts to Newtonian. When it is less than 1, it is shear thinning. The apparent viscosity can be expressed by re-arranging the above:

$$\mu_a = M * \gamma^{(n-1)} \tag{12.4}$$

M is the viscosity coefficient or the viscosity at a shear rate of 1 (usually in units of 1/s).

Bingham plastic fluids exhibit a yield stress. That is, until the fluid is exposed to a certain minimum shear stress (called the yield stress), it will not move. Above the yield stress, Bingham fluids exhibit a constant slope.

Mathematically:

$$\sigma_s = \sigma_y + M * \gamma \tag{12.5}$$

and

$$\mu_a = \sigma_y / \gamma + M \tag{12.6}$$

Herschel–Bulkley fluids combine a yield stress with shear-thinning behavior above the yield stress:

$$\sigma_s = \sigma_y + M * \gamma^n \tag{12.7}$$

and

$$\mu_a = \sigma_y / \gamma + M * \gamma^{(n-1)} \tag{12.8}$$

One may recognize that all three previous models are special cases of the Herschel–Bulkley model, making this the most versatile model commonly in use.

Note that any fluid that has a yield stress will exhibit some behaviors that are more commonly associated with solids than liquids. For example, when poured into a container or onto a table top, the fluid surface will not fully flatten out. It will form what is called in the solids handling industry an "angle of repose." If in a container and the surface is disturbed, say, by poking a finger into it, the fluid will retain at least some of the shape caused by the disturbance indefinitely. If it does eventually flatten out, that is proof positive that it does not have a significant yield stress.

Effect of Viscosity on Power Draw

Viscosity affects both ungassed and gassed power draw. The ungassed effect is relatively straightforward. The impeller Reynolds number can be calculated based on the viscosity and density, then the power number can be calculated, and then the power draw can be calculated. In the case of non-Newtonian fluids, the apparent viscosity is used. This is based on the Metzner–Otto shear relationship:

$$dv / dx = K_M * N \tag{12.9}$$

Table 12.1 lists the constant, K_M, repeated from Chapter 3.

Table 12.1 Metzner–Otto constants for impellers.

Impeller type	K_M
Propeller	10
Low-solidity hydrofoil	10
High-solidity hydrofoil	11
Pitched blade turbine	11
Radial turbines	11.5
Chaos (reversing pitch) turbines	40

The viscosity effect on gassed power is not that straightforward. The presence of gas may alter the shear rate field around the impeller, affecting the apparent viscosity the impeller sees. Increasing viscosity above about 100 cP also reduces gas holdup (due to larger bubble size), thus increasing the gassing factor. Unfortunately, there is little published on the subject of viscous, gassed power draw, and what there is published is mostly confined to Rushton turbines and Newtonian fluids, whereas we are more interested in axial/radial impeller combinations with deep concave turbines, and non-Newtonian fluids. Some manufacturers of agitation equipment have proprietary software to do this, but its accuracy is unknown. Until more is published, I suggest that the best way is to do a "wet" laboratory test under conditions most likely to represent the true gassed power draw conditions. Since gassed impeller power will depend on Reynolds number, Aeration number, and Froude number, a good test would consist of scaling down while holding all three of these constant, and as close to geometric similarity as possible. This will require a different fluid to be used in the laboratory scale from that used in the production scale, as we will see in the example problem below.

Example Problem

Suppose we have a proposed production fermenter that has a 4m tank diameter. For simplicity, we will describe just one of the impellers: a deep concave turbine of 1.65 m diameter, rotating at 100 rpm. Gas flows into the impeller at an actual flow rate of 1.0 m^3/s. The fluid density is 1050 kg/m^3 and the viscosity is 5000 cP. A laboratory tank and agitator are available. The tank is 0.5 m diameter. A 203 mm impeller is available. At what speed and gas flow rate should we operate the impeller, and with what fluid viscosity, assuming a density of 1000 kg/m^3 for the laboratory fluid?

Example Problem Answer

To get exact geometric similarity, the laboratory impeller would need to be 1.65 m * (0.5/4), or 0.206 m. Our 203 mm impeller is close to that, so we should be OK if we calculate based on it, since we are not interested in duplicating agitation, just gassed power number. Let us define a scale ratio, R, based on impeller diameter:

$$R = D_2 / D_1 = 0.203 / 1.65 = 0.123 \quad (12.10)$$

Now, since Froude number is N^2D/g, we may calculate the following:

$$N_2 / N_1 = (1/R)^{0.5} \text{ or } N_2 = N_1 (1/R)^{0.5} = 100(1/0.123)^{0.5} = 285.1\,\text{rpm}. \quad (12.11)$$

Next, we must determine airflow. We want to keep aeration number constant. Aeration number is Q_g/ND^3. Our production aeration number thus equals $(1\,m^3/s)/([100/60\,s][1.65\,m])^3 = 0.134$. Solving for gas flow at equal aeration number on the laboratory unit, we get the following:

$$Q_g = 0.134(ND^3) = 0.134(285.1/60\,s)(0.203\,m)^3$$
$$= 5.33E-3\,m^3/s, \text{or more conveniently}, 320\,\text{actual l/min}. \quad (12.12)$$

Reynolds number equals $D^2N\rho/\mu$. To keep it dimensionless, we need to use viscosity in Pa-s. Our production unit has a Reynolds number of $(1.65\,m)^2(100/60\,s)(1005\,kg/m^3)/(5000 * 0.001\,kg/m\text{-}s) = 1505$. This is in transition flow, but mostly turbulent. Re-arranging the Reynolds number definition to calculate viscosity for the laboratory unit, we have the following:

$$\mu = D^2N\rho / N_{Re} = (0.203\,m)^2 (285.1/60\,s)(1000\,kg/m^3)/1505$$
$$= 0.1301\,Pa-s \text{ or } 130.1\,cP. \quad (12.13)$$

As one can readily see, this scale-down method requires a much lower viscosity in the laboratory scale than in the production scale. Ideally, the rheology would be similar as well, just with the M value multiplied by a factor. However, the production fluid is likely non-Newtonian, but most 130 cP fluids are fairly Newtonian. We cannot always tailor-make our fluids for an exact replica of the flow phenomena, adjusted for scale. So, this simulation represents the best we can do with the given set of conditions.

Effect of Viscosity on $k_l a$

Up to about 50 cP, viscosity seems to have little or no effect on $k_l a$ for a mechanically agitated gas–liquid system (Ref. [1]). Above 50 cP, various authors report exponents ranging from –0.5 to –0.9 on viscosity. In other words,

$$k_l a \propto (\mu/50)^{(-0.5\,\text{to}\,-0.9)} \text{ for } \mu \geq 50\,cP \quad (12.14)$$

Given this large degree of uncertainty on the exponents, it should be obvious that testing on the actual broth is needed, rather than just trying to adjust a water-based correlation. A viscosity term can be added to the correlation. Moreover, the above equation is based on Newtonian liquids. For non-Newtonian liquids, the viscosity depends on shear rate. This leads to the question, What shear rate should we use?

Though we usually treat k_la as a single parameter, it actually consists of two variables: k_l and a. These may be affected differently by apparent viscosity and at different shear rates. For example, k_l might mostly be affected by the apparent viscosity at the gas–liquid interface. But the shear rate on the surface of a bubble is difficult to determine and is likely fairly high. On the other hand, the "a" term is probably controlled by bulk hydrodynamics, which may be at a lower shear rate than at the bubble surface. Some would argue that the apparent viscosity at the agitator impeller should be used. Though that seems reasonable, in my experience it is hard to get a data fit that works well that way. And such a procedure would predict a lower k_la in the production scale than in the laboratory scale, because the actual shaft speed is lower in the production scale, leading to lower shear rates and higher apparent viscosities at the impeller for pseudoplastic fluids. What I have found to work empirically (and I do not have a theoretical explanation for it) is to fit the viscosity effect at a fixed shear rate for pseudoplastic fluids. I generally use 1/s, which means the viscosity to be used is the M value in Eq. (12.4). In other words, the form of correlation is as follows:

$$k_la = A(P/V)^B (U_S)^C (M)^{D'}; \text{ for } \mu \geq 50\,\text{cP} \tag{12.15}$$

This form of correlation has resulted in data fits with an R^2 value of 0.9 and higher for pseudoplastic fluids. It is not necessarily recommended for use with yield stress fluids. Reference [2] reports a value for D' of about −0.82, with some dependence on D/T ratio.

Most yield stress fermenter broths are fungal in nature. Yield stress probably has little effect on k_l, as it is mostly the soluble species that will affect the film properties. However, yield stress has a huge effect on overall hydrodynamics and, therefore, the "a" term. Reference [1] found decent correlation (±30% typically) using the apparent viscosity, in the form:

$$k_la = A(P/V)^B (U_S)^C (\mu_a)^{D'}; \text{ for } \mu \geq 50\,\text{cP} \tag{12.16}$$

However, the rheological data were modeled as a pseudoplastic (power law) fluid, even though fungal fermentations are known to have a yield stress. The exponent D′ they calculated for their experiments was −0.39, which was a lower effect than most other authors reported.

Reference [3] fitted the rheological data to the Herschel–Bulkley model, which is a reasonable approach. However, no explicit k_la correlation was derived. Probably, a fit similar to Eq. (12.16) would have worked.

In my own experience, when the biomass concentration gets high for a fungal fermentation (15–20% by weight), the best fit is a Bingham plastic model. If the nutrient medium is low viscosity, it almost has no effect on the overall apparent viscosity. The viscosity correlation takes the form

$$\mu_a = \sigma_y / \gamma + M; \sim \sigma_y / \gamma \text{ if } M \text{ is small} \tag{12.17}$$

In one case I worked on, M had a value of 0.022 Pa-s (22 cP). Yet the apparent viscosity at scale was about 1000 cP. Not enough data were available to do a k_la fit. It could be that using yield stress instead of apparent viscosity would give a decent fit. If apparent viscosity is used, in effect the shaft speed is implicitly being curve-fitted with a negative exponent, though agitator power has a positive exponent.

The work done in Refs. [3, 4] was done under conditions of full motion. If there are dead spots (nonmoving areas outside the "caverns"), these relations will not hold. We will cover caverns in a subsequent section. Numerical methods such as used in Ref. [5] can assist in many aspects of evaluation.

We will briefly cover the effects of agitation on mycelial morphology under a separate heading.

Effect of Viscosity on Holdup

Not much has been published on the effect of viscosity on gas holdup. However, the main effect viscosity has is to increase bubble size and decrease holdup. In some cases, it can decrease it as much as 50%. That might mean a change from 20–30% holdup to 10–15%. In practice, most viscous fermentations start out at low viscosity, so the vessel must allow room for the initial condition. In the rare event that significant viscosity exists from the start, it may be possible to reduce the allowance for gas holdup and save a little money on the vessel. On the other hand, plant production personnel will always increase the batch size until broth comes out the vent, so a little more vessel volume may translate into an incremental productivity increase.

Effect of Viscosity on Blend Time

This topic is somewhat complex, as there are a lot of influences on blend time. Recalling Chapter 3, Figure 3.6, the dimensionless blend time is a constant in turbulent flow. This means viscosity has no effect on the blend time as long as the

Reynolds number is above 10 000, and it increases only about 10% for a Reynolds number of 5000. Going back to our example problem above for power draw simulation, we found that the 1.65 m impeller diameter operating at 100 rpm would have a Reynolds number of 1505 at a viscosity of 5000 cP. That means that the viscosity corresponding to a Reynolds number of 5000 would be 1505 cP. A very few fermentations have a viscosity that high, with the exceptions being gum fermentations and various mycelial/fungal fermentations. However, those that are highly viscous are also highly non-Newtonian. With insufficient agitation, they can form "caverns," or reasonably well-mixed zones, surrounded by dead or slowly moving areas that do not contribute to production. This is a more serious problem than just mixing. We will discuss it further in the next section.

Studies of agitated blend time have largely been confined to a single impeller in a liquid-only system. There are a few studies involving multiple impellers of equal size and type. A very few have studied unequal diameters and types, which is commonly the situation in a modern fermenter. Almost none have attempted to account for the effect of gas. My own experience is that the presence of gas increases the blend time when down-pumping impellers are used and decreases the blend time when up-pumping impellers are used. This effect is more pronounced as the viscosity increases. As a rough guide, an up-pumping system halves the gassed blend time compared to down-pumping in a water-like fluid and reduces it to about 1/3 as long in viscous fluids such as Xanthan gum. Nonetheless, quantitative prediction of gassed blend time by simple calculations has not been developed at this time. It is probably possible to do so using numerical methods (CFD), but this approach has not been validated. It may also be possible to establish a dimensionless blend time in the laboratory by operating the laboratory impeller at equal Reynolds number. If the same fluid is used in both scales, the laboratory impeller would have to turn much faster. Using the data from our sample problem above, we would need to rotate the impeller at 6610 rpm if it were in a 5000 cP Newtonian fluid. If it were in a shear-thinning fluid, the speed could be lower, as the apparent viscosity would decrease with increasing shaft speed. However, such high speeds are impractical in the laboratory scale. A larger test vessel would reduce the speed requirement. A lower viscosity fluid with the same power law exponent would help, but might not be possible to find. How to scale down the gas flow effect on blend time is uncertain. Running at several gas flow rates between equal superficial gas velocity and equal VVM would give an idea of the sensitivity to gas effects.

Ultimately, however, we care less about blend time per se than about dead zones and DO gradients. The dead zone issue we will discuss in the next section and illustrate more fully in our Xanthan gum discussion. DO gradients are hard to calculate and measure, particularly throughout the vessel. Some CFD simulations have been done (Ref. [6]), and the models used are getting more sophisticated. But they are quite costly to run, and the results have not been validated, as there is no good way to measure DO in any location except near the vessel wall.

Effect of Viscosity on Flooding

As viscosity goes up, gas handling capacity decreases. However, I was unable to find good quantitative relationships in the literature. Though flooding is rare in viscous fermenters, the best way to be sure is to model the proposed production design at equal Reynolds number, Froude number, and gas flow number in the laboratory scale. Up-pumping impellers are more resistant to flooding than down-pumping. Some would even argue that up-pumping impellers cannot flood, but this is incorrect. Would an impeller turning at zero rpm not be flooded, whether up or down-pumping? There would always be some speed below which there is no mass transfer enhancement.

Caverns

When a fluid has a yield stress, if enough torque is not applied, the moving zone, or "cavern," will not extend to the walls of the tank, resulting in dead zones that contribute nothing to productivity. P. Arratia and others used visualization techniques with dyes and photography (Ref. [6]) to show this phenomenon. Others have used different techniques, such as conductivity probes and salt solutions.

Figure 12.1 comes directly from Ref. [6]. The left image, "a," shows a cavern which has a diameter less than the tank diameter. The "b" image shows a cavern that touches the wall. We would definitely prefer the latter condition! The height of the cavern at the vessel wall is about 40–50% of the tank diameter, with a Rushton turbine. Increasing agitation once the cavern reaches the vessel wall results in a cavern height proportional to the shaft speed, up to an aspect ratio of about 1 for the cavern. For axial impellers, the cavern height is closer to 75% of the tank diameter when the cavern just touches the tank wall.

If one looks carefully, one can also see the torus-shaped areas above and below the plane of the impeller, as were conceptually sketched in Chapter 5. These do not blend quickly even within the cavern. However, gas bubbles will disrupt these areas.

Obviously, for a typical tall fermenter vessel, multiple impellers are needed. But multiple radial impellers are not the best choice. The same article used for Figure 12.1 used different dyes for a multi-impeller system. The result, Figure 12.2, showed that each impeller created a zone, or cavern, that had almost no intermixing with its neighbors. This is why axial impellers are normally used as upper impellers in almost all modern fermenters, but, especially, in viscous fluids. Unfortunately, I do not have a similar image with axial impellers, but other laboratory simulations do show top to bottom mixing with them.

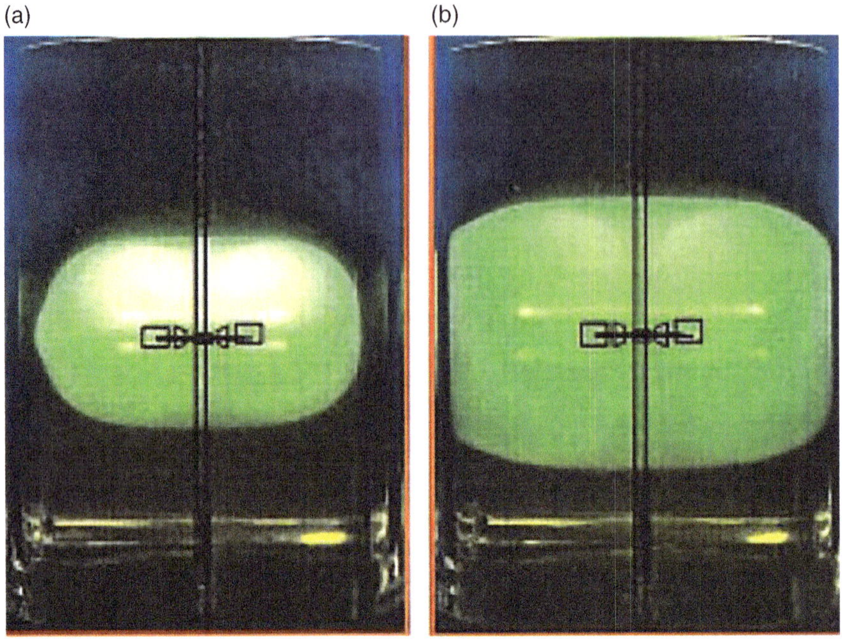

Figure 12.1 Cavern visualization, permission of Wiley Interscience. *Source:* Arratia et al. [6].

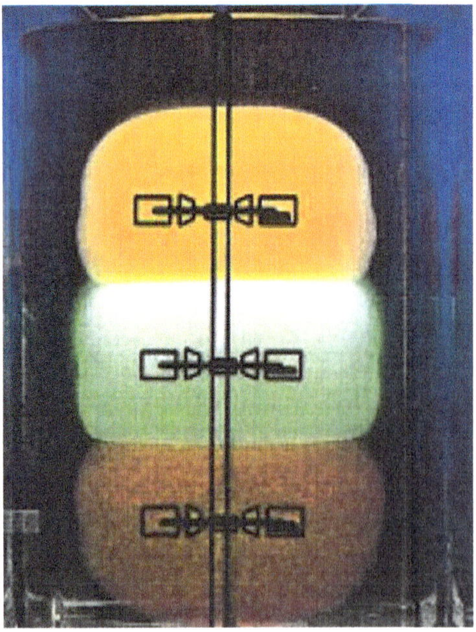

Figure 12.2 Cavern segregation, permission of Wiley Interscience. *Source:* Arratia et al. [6].

Estimating Cavern Size

In Ref. [7], the authors derived an equation for the cavern diameter, based on a cavern shape of a cylinder with an aspect ratio of 0.4. Their equation is below.

$$(D_C / D)^3 = (1.36 / \pi^2) * (N_P \rho N^2 D^2 / \sigma_y) \tag{12.18}$$

If we want to have the cavern reach the wall, this means that $D_C = T$.

With a few derivations, the minimum torque per impeller for full motion with the number of impellers set to (Z/0.4T) becomes:

$$\text{Minimum torque / impeller}(N - m) = 0.1156T(m)^3 * \sigma_y(\text{dynes / cm}^2) \tag{12.19}$$

Since zones do not fully intermesh, the actual minimum torque is probably 20–30% higher than this for multiple impellers. Also, per Ref. [8], getting full movement at the bottom requires more torque, as the tank bottom is another shear surface that must be overcome.

The authors of Refs. [7, 9] did not note any difference in torque requirement as a function of D/T for minimum movement for pure Bingham plastic fluids. Of course, less power is required at a larger D/T due to the lower shaft speed required to invest the same torque (Power = torque times shaft speed).

However, few fermenter broths are purely Bingham fluids. Most are shear thinning, whether they have a yield stress or not. Shear-thinning fluids have a more rapid velocity decay away from the impeller than those that are not shear thinning. Hence, larger D/T ratios are beneficial for such fluids.

In the strictest sense, fluids without a yield stress cannot form caverns. This is because any finite stress imposed on the fluid will result in a finite velocity gradient greater than zero. In practice, however, highly shear-thinning fluids can form pseudocaverns, in which the apparent boundaries of the cavern are moving so slowly that there is little intermixing between the impeller, or "cavern" zone and the fluid near the vessel wall. Some authors have sought to describe such boundaries using a low-velocity criterion, either in absolute terms or relative to the average velocity. The latter approach has the odd feature that the cavern boundaries remain the same independent of shaft speed, even though the overall velocity increases with speed. Though equations estimating cavern size for pseudoplastic fluids exist (Ref. [10]), they require impeller parameters, such as the axial force number, that are not generally available or published. The concept is that the cavern is bounded by a low-velocity virtual surface, with the velocity chosen by the user. Several models and equations from this reference can also be found in Chapter 18 of Ref. [11]. Converted into the terms used in this book, we have, for a torus-shaped cavern and a pseudoplastic fluid without a yield stress:

$$\left(D_C/4\right)^{(1-2/n)} = v_o\left[\left(2/n-1\right)\left(4\pi^2K/F\right)^{1/n}\right] + \left(T/4\right)^{(1-2/n)} \quad (12.20)$$

F represents the total force applied by the impeller to the liquid, as defined below. The value of v_o is set to some arbitrary low figure, such as 0.001 m/s.

$$F = \rho N^2D^4\left[N_T{}^2 + \left(4N_P/3\pi\right)^2\right]^{0.5} \quad (12.21)$$

N_T is the axial impeller thrust number, previously defined in Chapter 3. The definition is repeated here:

$$N_T = F_T/\rho N^2D^4 \quad (12.22)$$

Not much has been published on this subject. Reference [12] has some data. For laminar flow, for low-solidity axial flow impellers with a turbulent power number of about 0.3, they give thrust numbers of about 0.3 in turbulent flow. For pitched blade turbines with a turbulent power number of about 1.3, they give thrust numbers of about 0.5. In an unpublished student report, those numbers increase to about 0.9 for both impellers in laminar flow, and drop to less than 0.1 when the Reynolds numbers are between 30 and 100. Though unpublished, it is expected that high-solidity hydrofoils with a turbulent power number of about 1 will have similar thrust numbers to a pitched blade turbine. For best results, measure the thrust number curve for the impellers in question, or, better still, model the cavern if a fluid can be found to match rheology. This is not always feasible.

All of the above discussions and equations were based on agitating an unaerated liquid. But fermenters will always have a significant amount of gas introduced. As long as the gas is dispersed and the impellers are not flooded, the gas should actually help the overall circulation and reduce dead spots (increase cavern size), especially if the axial impellers are up-pumping.

Xanthan and Gellan Gums

Xanthan and Gellan gums are two of the most common biopolymers produced by fermentation. They are found in a wide range of uses as thickeners, binders, lubricants, and texturizers. Examples include additives for oil well drilling mud fluid; binders in toothpaste and cosmetics; humectants in breads; thickeners in ice cream, sauces, and medicines such as cough syrups, gelling agents in gels, and many other uses. The two gums are fairly similar, but Gellan produces its effects at a much lower concentration than Xanthan.

Though considered safe (Xanthan gum is GRAS as a food additive), there is one hazard to be aware of: It is slippery! As I once told the president of a previous employer, one can think of it as synthetic snot. Once, I was supervising testing of a Xanthan gum solution in a fermenter originally designed to produce amino acids, which is a low-viscosity process. While we were setting up for testing, powdered Xanthan gum was spilled on the smooth concrete floor of the mezzanine. The client decided to clean it up by hosing it down with hot water. This made the floor extremely slippery and hazardous. Trying to cross it with normal work boots was like ice skating with flat deck shoes! Luckily, I did not fall down and crack my skull! Use of a vacuum cleaner to pick up the powder would have been better and safer!

Commercially, Xanthan gum is normally produced as an excretion by the bacterium Xanthomonas campestris in aerobic fermentation. Likewise, Gellan gum is produced by fermentation of Sphingomonas elodea. Both fermentations start out in a water-like medium, and the viscosity rises as the product is excreted. This means the fermenter must have baffles to disperse the gas that is introduced, even though baffles may hinder the results in the later stages of fermentation, when the gum concentration is highest.

Viscosity Models for Gums

There seems to be some disagreement in the literature whether Xanthan gum solutions have a yield stress or not. References [6, 10, 13, 14] report a yield stress, with shear-thinning behavior. In other words, they tended to use the Herschel–Bulkley model. Reference [2] did not detect a yield stress and used a power law model. The overall consensus seems to be that there is a yield stress. However, the power law model and the Herschel–Bulkley model often fit the data equally well. For purposes of mass transfer, heat transfer, and impeller power draw, it probably makes little difference which model is used, as long as the caverns touch the vessel wall. A yield stress makes cavern formation more likely.

In this author's opinion, pure Xanthan gum solutions behave somewhat differently from Xanthan gum broths, which have cells, dissolved salts, and sugars, etc., which affect the rheology. My own observation is that the broth does not appear to have a yield stress, based on visual observation by disturbing the surface. It seems to return to a flat surface fairly quickly, at concentrations up to 6%. Most authors of papers about Xanthan gum rheology (Ref. [2] excepted) use pure solutions, and many stop at 1% concentration, with some going as high as 2%. However, most commercial Xanthan gum fermenters are run until the production rate slows down too much to be practical. This usually ends up between 3 and 6%, depending on agitator design, though it is possible to go higher.

Xanthan gum comes in different grades and formulations as well. I recommend testing the actual broth for fermenter design purposes. Sophisticated viscometers can be used, though a simple Brookfield viscometer can be sufficient for agitator design, if the data are treated according to the methods of Ref. [15].

Though not definitive, Table 12.2 gives an idea of rheology for both pure Xanthan gum and broth at a range of concentrations, from two different sources. The temperature was 29–30 °C. The first few entries were from Ref. [14], included a yield stress, and were fitted to the Herschel–Bulkley model. The rest of the entries, where "NR" (Not Recorded) was entered for yield stress, did not attempt to measure yield stress, but instead, fitted the data to a power law model. Later, we will use the first few points and the higher viscosity broth data for visualization of the progress of fermentation.

I was unable to obtain such detailed data on Gellan gum broths. Some manufacturers provide data for pure Gellan solutions at 2% in various grades. This can be used for general guidance, but testing of the actual broth is recommended. In general, Gellan is qualitatively similar to Xanthan from a rheology standpoint, albeit at a lower concentration. I have curve-fitted data for two Gellan gum 2% solutions in Table 12.3, based on shear rates ranging from 0.1 to 50/s. Though Gellan probably does have a yield stress, the curve fits are just as accurate using both Herschel–Bulkley and Power Law models. I would suggest using the Power law model for impeller power and the Herschel–Bulkley model for cavern size.

Installation Survey

I have compiled a list of installations I am aware of, in terms of % Xanthan gum in the final broth, vessel working volume, power, D/T, and impeller type. These are shown in Table 12.4. The term "IDT refers to "Invisible Draft Tube," which we will describe later.

One can see that there is little consistency between various installations. Different impeller systems, different D/T ratios, and different P/V levels were used. These data can be condensed into a figure, as done in Figure 12.3.

What this figure shows is that most Xanthan gum fermenters invest between 2 and 4 kW/m^3, regardless of vessel size or % gum, except for one outlier, which was in a small vessel. One might get the impression that the designs were based on rules of thumb, rather than first principles. But it is expected not only by this author but all who have studied the matter that larger D/T ratios at constant power input will improve productivity by enlarging the caverns, and hence the active fermenter volume, until they at least touch the vessel wall and overlap from top to bottom. The next section will help visualize this.

Table 12.2 Rheological properties of Xanthan Gum.

Data source	Material	% by wt.	Temp, C	Yield stress, Pa	Viscosity Coefficient, M	Power law exponent (n)	R^2
[a]	Pure XG	0.4	20	4.66	0.87	0.64	0.966
[a]	Pure XG	0.6	20	7.35	3	0.47	0.974
[a]	Pure XG	1	20	14.53	13.08	0.3	0.958
Private	Pure XG	1	NR	NR	7.97	0.2833	0.79
Private	Pure XG	2.5	NR	NR	1.79	0.5417	0.93
Private	Pure XG	4	NR	NR	44.99	0.2016	0.98
Private	Pure XG	5	NR	NR	114.8	0.2396	0.97
Private	XG in broth	0.28	NR	NR	0.662	0.3925	0.92
Private	XG in broth	0.333	NR	NR	0.935	0.385	0.96
Private	XG in broth	0.667	NR	NR	4.302	0.2279	0.96
Private	XG in broth	1	NR	NR	8.717	0.1912	0.97
Private	XG in broth	4	NR	NR	40.06	0.2016	0.94
Private	XG in broth	6	NR	NR	44.4	0.3246	0.98

[a] Szymon Wozwodski and Lubomira Broniarz, 14th Eurpean Conference on Mixing.

Table 12.3 Rheological properties of 2% Gellan Gum solutions.

Gellan sample number	Viscosity model	Yield stress, Pa	Viscosity coefficient (M)	Power law exponent (n)	R^2
1	Power Law	N/A	229	0.36	0.999
1	Herschel–Bulkley	10.62	218	0.387	0.999
2	Power Law	N/A	544	0.0371	0.999
2	Herschel–Bulkley	487	64.6	0.722	0.999

Table 12.4 Xanthan Gum installation data.

Facility	% XG	Motor Kw	Working Volume, m^3	Nominal P/V, kW/M^3	D/T	Impellers	Comments
"Austrian"	~3	310	93	3.33	~0.6	Intermig	
"American"	2.5–3.5	750	230	3.26	0.45/0.33	HE-3/ CD-6	
"American"	<2.5	94	230	0.40	0.45/0.33	HE-3/ CD-6	
"American"	~5–6	1125	600	1.87	~0.55	Maxflo (up)	IDT
"American"	Unknown	7.5	5.7	1.31	0.45	Maxflo or A-315	
"American"	~2	6.7	0.57	11.75	0.44	Maxflo	
"American"	~3–4	188	45	4.17	0.41	Maxflo	
"American"	4+	188	45	4.17	0.45	Maxflo	
"American"	2.5	188	132	1.42	0.4	S-4	
"American"	3–4	263	132	1.99	0.45	D-6	
"American"	4+	450	132	3.40	0.5	A/315/ Rushton	
"American"	4+	450	151	2.98	0.47	Maxflo	
"American"	4+	450	144	3.12	0.47	Maxflo	
"Chinese"	~5–6	355	115	3.08	0.6	Smith. HS axial	
"German"	~2–3?	355	225	1.57	0.57	Intermig	

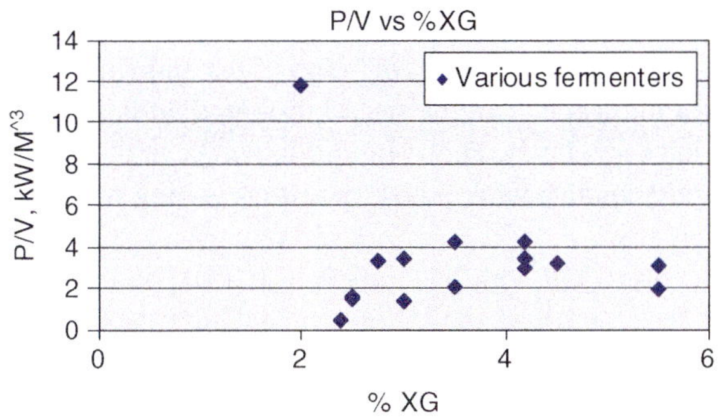

Figure 12.3 Xanthan Gum installations.

Effect of D/T and No. and Type of Impellers on Results in Xanthan Gum

As many of the papers listed as references for this chapter attest, a larger D/T ratio will improve overall blending and productivity. We will try to illustrate that with figures based on equations presented in the "Estimating Cavern Size" section in this chapter, using some artistic imagination to depict the shapes and sizes of dead zones when the caverns overlap due to multiple impellers. In addition to the D/T effect, we simultaneously look at the effect of adding more impellers, since the cavern height is generally considerably less than 1 tank diameter. The simulated fermenters have an aspect ratio (Z/T) of 2.5, which is a fairly common ratio. The simulations are based on a 4 m tank diameter, with 122 m^3 working volume. For a typical low-viscosity fermentation, 3 impellers would be used: a lower deeply concave radial turbine, and two upper up-pumping high-solidity axial impellers. The D/T ratio is usually between 0.25 and 0.4 for low-viscosity fermentations. We chose to use 0.3 as typical for an agitator designed for low-viscosity fermentation. We compare this to a system with five impellers, set to a D/T of 0.6. In all cases, the gassed power is the same, set to a value of 2 kW/m^3, which is in the range of what is found in industrial scale Xanthan gum fermenters. For power draw purposes, the gas flow was set at 0.5 VVM.

For the simulations of 1% and below, we used the Herschel–Bulkley model of Ref. [14], with yield stress accounted for. For 4–6%, we used a power law model for broth from the private data in Table 12.2. The cavern size did not seem to change at the higher concentrations, so we combined them into one figure for each D/T ratio.

Figure 12.4a–e track the cavern development at water-like conditions, 0.4% Xanthan gum, 0.6, 1, and 4–6%, respectively, at a D/T ratio of 0.3 and with three impellers. One can see that, at higher concentrations, a substantial portion of the tank volume is dead or moving very slowly. This is bound to reduce productivity, as continued production is only possible where the organism is getting sufficient nutrients and oxygen.

Figure 12.5a–e do the same thing, but with five impellers at a D/T ratio of 0.6. The differences are massive. The small dead zone at the top of Figure 12.5e would probably be broken up by the rising gas bubbles, so nearly 100% of the fermenter volume would be active. It should be noted that, although all of these runs use the same power input, the large D/T design invests over four times as much torque as the small D/T design. So, the operating cost is the same, but the capital cost is probably 2–3 times as high for the larger impeller design. Is it worth it? If making more product is worth anything, yes!

Production Curve

Based on a little data and a lot of assumptions, we have prepared a hypothetical set of curves of Xanthan gum concentration versus time at different D/T ratios, at equal invested power and airflow. Although quantitative values are shown, these curves may not be accurate for any specific strain or process, but are intended to qualitatively illustrate both the trend and the possibilities. It is reasonable to expect substantially higher gum concentrations for a given batch time by using higher D/T and more impellers, or a shorter batch time. These curves are shown in Figure 12.6.

Heat Transfer

It is not easy to get good heat transfer in a viscous fluid, and little is published on the subject of heat transfer in a non-Newtonian fluid. The techniques described in Chapter 10 should still work using the apparent viscosity at the turbine (the power law model works best for this), *provided that the heat transfer surface is not in a dead zone.* Unfortunately, the equations in this chapter for estimating cavern size do not include the effects of heat transfer surfaces such as vertical tube bundles or coils. In fact, their use can create larger dead zones. CFD modeling may prove valuable or a "wet" test simulating flow patterns. An external heat exchange loop may be the best option, particularly at high concentrations and large vessel sizes.

All-Axial Impeller Design

In Chapter 6, we discussed various impeller systems used in fermenters. The most common system in use in new installations today consists of a lower deeply concave radial turbine with a plurality of up-pumping, high-solidity axial flow impellers.

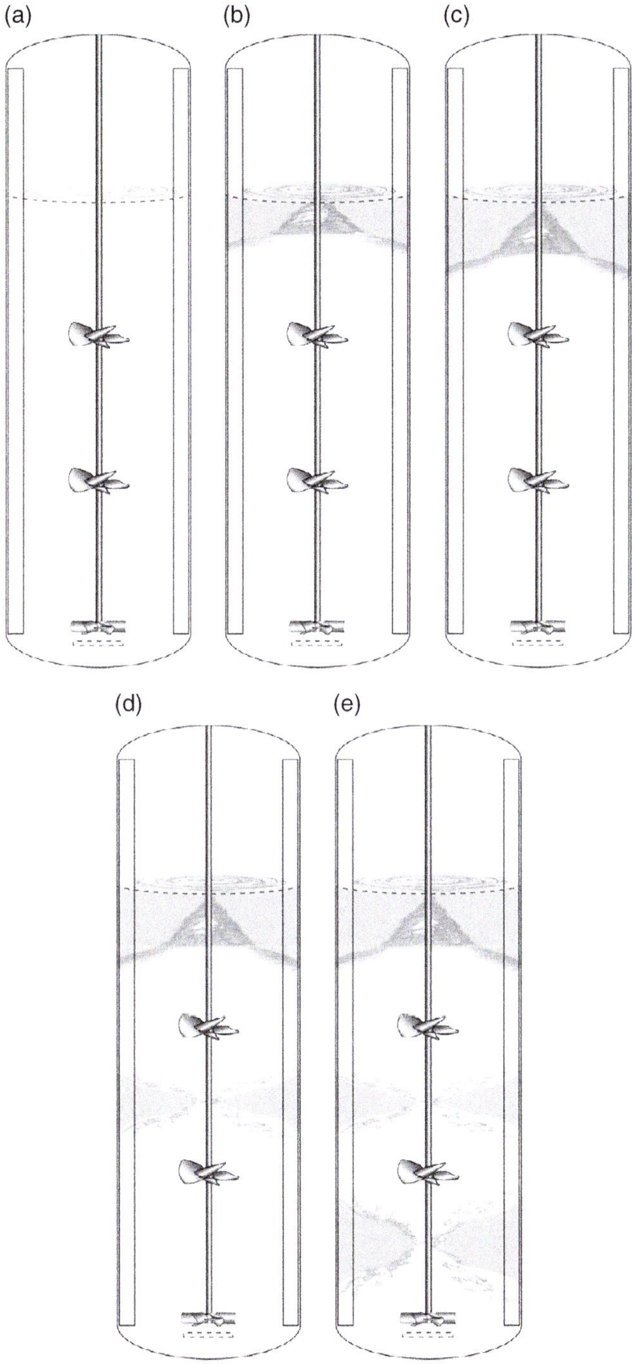

Figure 12.4 (a–e) Low D/T results.

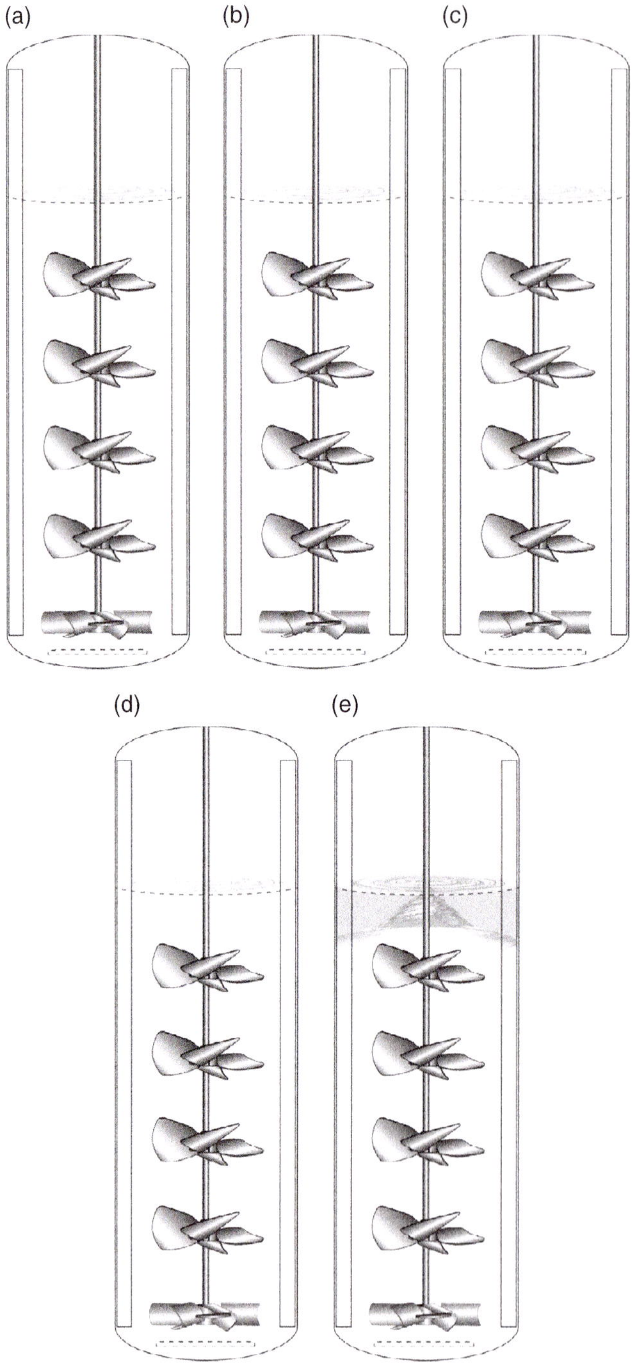

Figure 12.5 (a–e) Large D/T results.

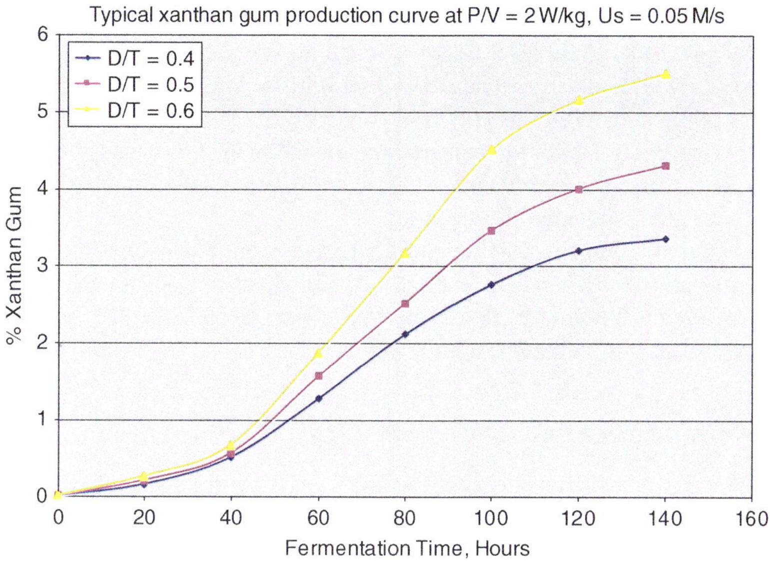

Figure 12.6 Hypothetical Xanthan Gum production curve.

This is true both for low-viscosity and high-viscosity fermentation. However, some experiments conducted by Xanthan gum producers in the 1980 and 1990s used all-axial systems. These, whether up-pumping or down-pumping, outperformed systems of multiple Rushton turbines. Most of these were retrofits, so they invested the same power at the same shaft speed as the original Rushton turbines. Investing the same power means the k_la was probably not higher; it may have even been less, as some authors have reported lower mass transfer coefficients for axial impellers than radial. Since the torque was the same, the cavern size theoretically would be equal based on yield stress, though it may have been larger for shear-thinning fluids, as the D/T ratio is higher with axial turbines than with radial. Moreover, as Figure 12.2 shows, multiple Rushton turbines create zones around each impeller with very little intermixing between them. Thus, one could envision a scenario in which the bottom zones are oxygen-rich, but the upper zones are being fed oxygen-depleted air from the lower zones. Due to such zoning, the high DO broth from the bottom of the tank does not intermix with the broth in the top of the tank, resulting in very low DO there. This renders the upper zones of the vessel unproductive or less productive. With multiple axial impellers, the tank is closer to being well mixed, i.e., has a less severe DO gradient, so the active tank volume is greater.

Up-pumping impellers work better than down-pumping in viscous fermentation for several reasons. One is that the gassed blend time is much shorter (less than 50% as long as down-pumping), so the DO gradient should be smaller. Another is that the gas passes through the impellers, resulting in smaller mean bubble size and likely a higher k_la value. A third is that foaming should be reduced, particularly when the liquid level passes through an impeller, as in a fed-batch fermentation. (This results in more usable vessel volume, as well as less antifoam usage. Less antifoam addition usually results in a higher k_la value.) Finally, the vibration velocity is cut 50% or more, increasing the life span of the steady bearing and seals, increasing the mean time between mechanical failures. That is an important source of productivity increase.

My own opinion is that an axial–radial combination is still best overall, but with increased D/T and more impellers, about 0.5T spacing between each one. I usually recommend a D/T of 0.6–0.7 for viscous gums. But even 0.5 is better than the "traditional" value of 1/3 found in some fermenters.

Invisible Draft Tube vs. Axial/Radial Combination

A variation of the all-axial impeller concept, known as the "invisible draft tube," or IDT for short, was introduced in the 1990s. The concept was to create a pure axial flow pattern in the tank, similar to what would be created if an actual draft tube were installed. In other words, the flow would rise up a virtual, pipe-like conduit to the surface, then travel back down in the annular space between the virtual conduit and the vessel wall. The natural diameter of this virtual conduit would be $\frac{\sqrt{2}}{2}$, or 0.707, times the tank diameter, as that would make the cross-sectional area equal inside and outside, resulting in equal average velocities in each direction.

If no gas were present, this could be done either up- or down-pumping. However, with gas present, down-pumping introduces radial flow as the rising gas fights the impeller flow, and this breaks up the intended IDT flow pattern. Therefore, all IDT designs for gas dispersion must be up-pumping.

Based on experiments done by the author, there are very specific geometry constraints to make such a system work as intended. These include, for example:

- D/T between 0.6 and 0.7.
- High-solidity impellers are preferred to maintain gas dispersion.
- Up-pumping impeller orientation.
- Use of shallower chord angle than normal to minimize radial flow. Most high-solidity impellers have a chord angle between 25 and 40°. In an IDT, it should be less than 25°. 15–20° is an ideal compromise, as this range gives power numbers between 0.3 and 0.6, as opposed to the 0.75–1.0 power number range of typical high-solidity impellers, allowing use of higher shaft speeds and therefore less torque for a given power input.
- Impeller spacing of no more than 0.5D. This means, for example, for a Z/T of 2.5, a typical low-viscosity fermenter would have three impellers, a recommended high-viscosity Xanthan gum design would have five, and an IDT would have 8–9 impellers, depending on D/T ratio.

The end result of adhering to the above geometric constraints is a much more expensive agitator than even the recommended impeller system design in this book, because the torque is higher even with the use of low chord angle impellers (because there are more of them), and of course, additional impellers add considerable expense. As a rough guide, if a low-viscosity design has a relative cost of 1, my recommended high-viscosity design will be about 1.5–2.5 times that price, and the IDT would be 3–4 times that price, or 2+ times as expensive as the recommended high-viscosity design. The question, then, is whether the user gets additional productivity from an IDT over this author's recommended design. Unfortunately, I do not have any operating data comparing the two systems in the same process, same tank size, and same power input. My suspicion is that there is little or no additional benefit to be gained over the recommended system, as the IDT is hardly ever specified or quoted any more.

Yet another variant is an IDT with a concave radial impeller at the bottom. In my own testing of such a system, the gas appeared better dispersed than with a pure IDT, and the holdup was greater. However, gassed blend time was about 20% longer, so there might be a slightly larger DO gradient. To my knowledge, there are no full-scale operating data with such a system, so it would be hard to say which would work better. My suspicion is that an IDT, with or without a lower radial turbine, would not work better than the recommended system except in very large fermenters (say, >500 m^3 working volume), where blending and DO gradient may be more important than mass transfer rate per se. I would be eager to see this tested.

Mycelial Broths

Unlike gum fermentations, wherein the main contributor to viscosity is the soluble gum excreted by the cells as a product of their metabolism, the main reason for high viscosity in a mycelial or fungal broth is mechanical entanglement due to the fibrous structure or morphology of the cells themselves. The viscosity in the free liquid may be quite low, and therefore, the viscosity at the bubble film may be quite low. One might expect, therefore, that the viscosity has little effect on the k_l value and that may indeed be true. However, the bulk apparent viscosity still has a profound effect on the interfacial area per volume, due to both its effect on bubble size and holdup, and therefore, a profound effect on the overall k_la. Reference [4] found an exponent of −0.39 on apparent viscosity in their k_la correlation, which is significant, but not as large an effect as some have reported for viscosity. They used a power law model to fit viscosity, and this should work well enough for apparent viscosity near the impeller. It may not work as well as other models for predicting cavern size.

Typical Viscosity Model

Due to the fibrous structure, mycelial broths have a yield stress. Likely, the best of the common rheology models to use is the Herschel–Bulkley model. However, the yield stress may dominate rheology at low shear rates. One sample I tested had a yield stress of about 10 Pa, but the slope of the yield stress vs shear rate curve above the yield stress was nearly constant, corresponding to about 20 mPa-s. The high yield stress has a major effect on cavern formation, though the material flows pretty well once the yield stress has been overcome. So the yield stress must be taken into account to avoid dead spots, but a power law model may be sufficient for prediction of $k_l a$ and heat transfer.

Morphology Effects

This is where things get a bit messy. Although bacterial, yeast and organ cell fermentations seem to have their shear effects limited to those caused by collapsing bubbles this does not seem to be the case with fungal fermentations. References [3, 4] report definite effects of agitation on shear damage to the structure, or morphology, of the fungi. This, in turn, affects the rheology, and hence the mass transfer, heat transfer, etc. With time and shear, the morphology itself changes. But what aspects of agitation can be used to predict shear damage?

Although the maximum shear rate is associated with impeller tip speed, regardless of impeller type [16], shear damage is cumulative, depending not only on maximum shear rate, but frequency of exposure to various shear rates. Some have suggested that P/V and time may be used to track such damage. However, a small impeller may have the same P/V based on overall liquid volume as a large one, but is dissipating it in a smaller volume around the impeller. Thus, some suggest using power per impeller swept volume. Still others suggest that the impeller circulation rate affects the frequency of exposure.

One result of all this discussion is a concept called the "Energy Dissipation/Circulation Function," or EDCF for short. Reference [3] discusses several versions of this function. The simplest forms to use are as shown below.

$$EDCF_{max} = \varepsilon_{max} / \theta_c \tag{12.23}$$

where ε_{max} is the maximum local energy dissipation rate and θ_c is the circulation time for one circuit.

The circulation time can be estimated as follows:

$$\theta_c \cong V / \left(N_Q N D^3\right) \tag{12.24}$$

The maximum energy dissipation rate can be approximated by:

$$\varepsilon_{max} \cong 1.04 \rho \left(N_P\right)^{0.75} N^3 D^2 \tag{12.25}$$

The empirical evidence suggests that the EDCF is better at correlating damage to the hyphal structure, as evidenced by "holes" in the structure, than either tip speed or P/V. This is good news for scale-up, as both P/V and EDCF decrease when scaling up at equal mass transfer rate, and circulation time becomes longer as well. These equations also suggest that the use of lower power number impellers would be beneficial, even though they have a higher tip speed, because they have a lower power number and a larger diameter for a given power input. The equations also suggest that larger D/T ratios would be beneficial for a given power input. We can illustrate this by an example. To keep it simple, we will compare a Rushton turbine of 1.2 m diameter and a deeply concave radial turbine of 2 m diameter in a 4 m diameter tank, with a volume of 50 m^3 and a density of 1000 kg/m^3. We will assume the Rushton has a gassed power number of 2.75 and a pumping number of 0.72 and the concave impeller has a gassed power number of 1.9 and a pumping number of 0.7. The Rushton impeller will be operated at 80 rpm, which would result in a power draw of 16 220 W. To draw the same power, the concave turbine would rotate at 38.6 rpm or 2.07 times the torque due to the low speed. Using Eq. (12.24), we get a circulation time of about 30 s for the Rushton, and about 14.1 s for the concave impeller. Using Eq. (12.25), we get an ε_{max} value of 7581 w/m^3 for the Rushton, and a value of 1793 for the large concave turbine. (Note that we are using SI units in the equations). Thus, using Eq. (12.23), the maximum value of the EDCF is 253 w/m^3-s for the Rushton and 127 for the large deeply concave impeller. Obviously, this is a significant difference. The expectation would be that the large concave impeller would produce less hyphal/morphological damage than the more conventional Rushton design.

Multiple impellers can probably be calculated in zones. However, the lower radial impeller will have the most power draw and the highest power number, so this is where most of the damage will occur. Multiple axial impellers may be a reasonable option, though larger D/T axial/radial combinations may have better mass transfer and still a low level of damage.

Recommendations

By both the conclusions of a plurality of authors, some of whom have been referenced in this chapter, and observation of many viscous broths in the laboratory, combined with a survey of some operating facilities, better results will be obtained in viscous fermentations by using larger impellers and more of them than would typically be found in low-viscosity gas–liquid systems. Table 12.5 gives recommended minimum D/T values as a function of the pseudoplastic power law exponent, n. Table 12.6 gives guidelines for the maximum aspect ratio, or Z/T, per impeller, for gas–liquid systems. For yield stress fluids, try to fit them to a power law model; if the yield stress is very high, use the recommendations for an n value

Table 12.5 D/T guidelines for pseudoplastic fluids.

N	D/T minimum
<0.05	0.70
0.05	0.68
0.10	0.65
0.15	0.63
0.20	0.60
0.25	0.58
0.30	0.55
0.35	0.52
0.40	0.50
0.45	0.47
0.50	0.44
0.55	0.41
0.60	0.38
0.65	0.35
0.70	0.31
0.75	0.28
0.80	0.24
0.85	0.20
0.90	0.15
0.95	0.15
1.00	0.15

Table 12.6 Z/T per impeller guidelines in gas/liquid systems.

	Z/T per impeller (maximum)	
n	Radial turbines	Axial turbines
<0.05	0.4	0.5
0.05	0.45	0.5
0.10	0.5	0.55
0.15	0.5	0.55
0.20	0.5	0.55

0.25	0.55	0.6
0.30	0.55	0.65
0.35	0.6	0.7
0.40	0.6	0.7
0.45	0.6	0.75
0.50	0.65	0.75
0.55	0.65	0.8
0.60	0.65	0.85
0.65	0.7	0.85
0.70	0.7	0.9
0.75	0.75	0.95
0.80	0.8	1
0.85	0.8	1.05
0.90	0.8	1.1
0.95	0.85	1.15
1.00	0.85	1.2

of less than 0.05. These recommendations are a combination of the author's own experience and private communications between the author and his esteemed former colleague, Dr. Julian B. Fasano.

It must be stressed that these are guidelines and may be stretched a little, particularly at high P/V levels, such as 3 kw/m^3 or higher.

Summary of Chapter

We have identified the fact that increasing viscosity poses challenges for mass transfer, power draw, heat transfer, and blending. We have discussed dead spots and cavern formation, and related this and also potential morphology damage in mycelial fermentation to impeller D/T ratio. All indicators point to the benefits of using an increased quantity of larger impellers for viscous fermentations. The only downside is that such systems are more expensive. But higher productivity should actually increase the return on investment for the agitator.

List of Symbols

A Constant in $k_l a$ correlation
a Interfacial area per volume

B	Exponent in $k_l a$ correlation
C	Exponent in $k_l a$ correlation
D	Impeller diameter (may have subscripts)
D_C	Cavern diameter
D'	Exponent in $k_l a$ correlation
EDCF	Energy dissipation/circulation function
F	Impeller force
F_T	Axial thrust
g	Acceleration due to gravity
K_M	Metzner–Otto constant
k_l	Liquid film coefficient
$k_l a$	Overall mass transfer coefficient
M	Viscosity coefficient
N	Shaft speed (may have subscripts)
N_P	Power number
N_Q	Pumping or flow number
N_{Re}	Reynolds number
n	Power law exponent
Q_a	Aeration number
R	scale ratio
T	Tank diameter
v_0	Velocity at cavern boundary
V	Liquid volume; volume per impeller in Eq. (12.24)
VVM	Gas flow expressed as volume of gas per volume of liquid per minute, at a reference condition
Z	Liquid height
	Greek letters
μ	Viscosity
μ_a	Apparent viscosity
σ_s	Shear stress
σ_y	Yield stress
ρ	Liquid density
ε_{max}	Maximum local energy dissipation rate

References

1 van't Riet, K. and Tramper, J. (1991). *Basic Bioreactor Design*. Marcel Dekker.

2 Xueming, Z. et al. (1994). Rheological characteristics, power consumption, mass and heat transfer during Xanthan Gum fermentation. *Chinese Journal of Chemical Engineering* 2 (4): 198–209.

3 Hardy, N. et al. (2017). Scale-up agitation criteria for Trichoderma reesei fermentation. *Chemical Engineering Science* 172: 158–168.
4 Albaek, M. et al. (2011). *Modeling Enzyme Production With Aspergillus oryzae in Pilot Scale Vessels With Different Agitation, Aeration and Agitator Types*. Wiley Online Library.
5 Benz, G. et al. (2014). Computing dissolved oxygen profiles in aerobic fermenters. Presented at North American Mixing Forum NAMF (June).
6 Arratia, P., Kukura, J., Lacombe, J., and Muzzio, F.J. (2006). Mixing of shear-thinning fluids with yield stress in stirred tanks. *AIChE J.* 52 (7): 2310–2322. https://doi.org/10.1002/aic.10847.
7 Elso, T. et al. (1986). X-ray studies of cavern sizes and mixing performance with fluids possessing a yield stress. *Chemical Engineering Science* 41 (10): 2555–2562.
8 Benz, G. (2012). Determining torque split for multiple impellers in slurry mixing. *Chemical Engineering Progress* 108: 45–48.
9 Etchells, A. et al. (1987). Mixing bingham plastics on an industrial scale. *Fluid Mixing III, European Federation of Chemical Engineering Publication Series, No. 63*, from a conference held at the University of Bradford, Bradford, England (8–10 September 1987).
10 Amanulla, A., Hjorth, S., and Nienow, A. (1998). A new mathematical model to predict Cavern diameter in highly shear thinning power law liquids using axial flow impellers. *Chemical Engineering Science* 53 (3): 455–469.
11 Paul, E., Atiemo-Obeng, V., and Kresta, S. (2004). *Handbook of Industrial Mixing*. Wiley Interscience.
12 McCabe, W., Smith, J., and Harriott, P. (eds.) (2001). *Unit Operations of Chemical Engineering*, 4the. McGraw-Hill.
13 Whitcomb, P., Ek, B., and Macosko, C. (1977). *Rheology of Xanthan Gum Solutions*. Washington, DC: Sandford and Lake Extracellular Polysaccharides, ACS Symposium Series, American Chemical Society.
14 Woziwodzki, S. and Broniarz-Press, L. (2012). Mixing of shear-thinning fluid with yield stress in a vessel with unsteadily rotating impeller. *14th European Conference on Mixing*, Warzawa, Poland (10–13 September 2012).
15 Mitscha, P. (1982). Simple conversion of Brookfield R. V. T. readings into viscosity functions. *Rheologica Acta* 21: 207–209.
16 Fondy, P. and Bates, R. (1963). Agitation of liquid systems requiring a high shear characteristic. *AIChE Journal* 9: 338–342.

13

Three Phase Fermentation

In a sense, all fermenters operate with at least three phases: liquid, gas, and suspended solids. However, the suspended solids generally have such a low settling rate that the broth essentially just acts as a liquid, so the net behavior is that of a two phase system. What we are specifically discussing in this chapter is a three phase system consisting of the gas phase plus two liquid phases: an aqueous phase and an oil phase.

General Problem

Although it is theoretically possible for a fermentation to be run as an oil-based system or as oil-continuous system with a dispersed aqueous phase, the author is unaware of any such systems in operation. If there are, I would be interested in learning more about them. What we are talking about in this chapter are systems where the continuous phase is aqueous, and the oil phase is a dispersed additive. The oil may be added as a viscosity modifier, an antifoam agent, or a nutrient that is consumed. The effects on the process performance are complex and cannot be fully generalized. What we will do in this chapter is discuss some of the issues and some ideas on how to deal with them.

Effect on Mass Transfer

To put it simply, the effect of an oil phase on mass transfer is complex. One of the more common uses of an oil phase is foam control or antifoam. In Ref. [1], the authors studied three different antifoam additives: polypropylene glycol, silicone, and soybean oil. As expected, the additions in small concentrations reduced k_la considerably. This is reasonable, because the oil phase coats the

Agitator Design for Gas–Liquid Fermenters and Bioreactors, First Edition. Gregory T. Benz.

bubbles, creating an additional film that the gas must pass through, and with a material in which oxygen has a lower solubility than it does in water. But, in the case of silicone and soybean oils (but not for the glycol), at higher concentrations, k_la actually improves.

Prins [2] reports a decrease in k_la with addition of antifoam.

Sometimes, oil can be added as a nutrient or even to enhance k_la. Though it is far from obvious why adding oil can increase k_la, several authors have reported it. For example, in Ref. [3], the authors report that silicone oil hinders mass transfer, whereas decane, hexadecane, and perfluorocarbon PFC 40 have no noticeable effect. In Ref. [4], it has been reported that n-decane, dodecane, and n-heptane significantly improve mass transfer, whereas toluene, anisole, and 2-ethyl-1-hexanol retard mass transfer.

Xanthan gum is a particular case where the addition of a nutrient, palm oil, can have two effects. It serves as an antifoam as well as a k_la enhancer, but only up to about a 10% concentration. Above that, it retards mass transfer. In an open-access paper [5], the authors examine this at different agitation intensities and air flow rates. This is illustrated in Figure 13.1 for one case (one particular rpm).

So, the empirical evidence is that an oil phase can increase or decrease mass transfer, depending on the organic compound used, its concentration and possibly the agitation and airflow conditions. But why is this so? At this point, explanations are speculative. The oil film may reduce bubble coalescence, in addition to causing a resistive film. These two phenomena affect mass transfer in opposite ways, and one or the other may dominate for a particular case. The presence of organics may have an effect on bulk viscosity, which may help or hurt mass transfer. In the case of an organic gas such as methane which is highly soluble in oils, droplets of oil may help carry the dissolved gas through the vessel, effectively increasing the area available for mass transfer. But that explanation would not work for oxygen.

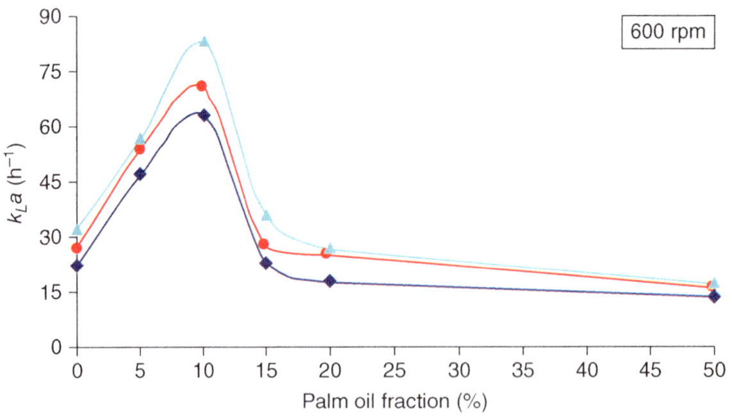

Figure 13.1 Effect of palm oil on xanthan k_la. *Source:* Sauid et al. [5]. Licenced under CC by 3.0.

The practical result of all this is that specific mass transfer correlations are needed for the specific system employed. Optimization of the oil concentration is an obvious area of research.

Effect on Foam

Most oil phases have an inhibitory effect on foam formation and even help break up foam that has already formed, so oil addition is a common way to control foam. At typical foam control concentrations, mass transfer is often inhibited, more so with silicone oils than with carbon-based organics. A question arises, What becomes of the oil phase? Presumably, if it cannot be used as a nutrient source by the microorganism, it must be removed from the final product.

Emulsion vs. Suspension

Most of the time, the oil phase will form a suspension, rather than an emulsion. This is especially true if the oil is being consumed as a nutrient. As a suspension, the oil will have effects on the mass transfer coefficient both by affecting the liquid film coefficient, k_l, and the interfacial area per volume, a. It may also indirectly affect the overall mass transfer coefficient, k_la, by affecting the apparent viscosity.

In the case of the oil phase forming a true emulsion (unlikely unless surfactants are deliberately added to facilitate this), the emulsion essentially acts as a single phase liquid. This means the effect on k_la will be governed by viscosity effects and diffusivity and therefore can be treated by the techniques described in Chapters 11 and 12. Typically, an emulsion will be more viscous than either the water phase or the oil phase alone, and the diffusivity will be lower also. Gas species solubility can be affected as well.

Empirical testing is advised whether a suspension or emulsion is formed.

Complexity: How to Optimize Operation

Gas–liquid contacting is obviously a complex subject. The addition of a second liquid phase makes it considerably more so. In addition to the inherent complexity introduced by the second liquid phase (changes to viscosity, diffusivity, solubility, interfacial area, hold-up, foaming, etc.), there is the matter that desired mass transfer rate likely changes over the course of the fermentation. The amount of oil phase present may also change over the course of the fermentation. This might be due to varying amounts being added to control foam, to aid in mass transfer in some cases (such as xanthan gum), or as part of the nutrient feeding strategy.

To optimize the overall operation, one would need to optimize the process stepwise. This involves not only agitator power and airflow, but the amount and cost of the oil phase. Indeed, I have been told by many that antifoam is often the second highest material cost of the broth, second only to the nutrient cost.

The amount of data required to do the above tasks is quite large and may take many months or even several years to acquire. That may not be economically practical, as there is an opportunity cost associated with delaying the construction of a plant and getting the product to market. It may be a better strategy to focus on the most difficult part of the process and make sure the agitation and airflow are adequate for that condition and then go ahead and build the production facility. Concurrently with the operation of the facility, it may be possible to continue laboratory studies or even production scale studies to gradually improve the efficiency of operation so that subsequent plant design and operation can be optimized.

Summary of Chapter

We have identified most of the issues surrounding the presence of a second liquid phase in a fermentation process. Because of the inherent complexity of such systems, there is a paucity of design information available. Users of such processes will find they must do a considerable amount of experimental work to predict performance, and optimization may take years. It may be best to optimize concurrent with operation of the actual plant to avoid serious delays getting product capacity to market. The basic methodology for predicting required gas flow and agitator power remains unchanged, but the exact data require experimental work, as always. Optimization will likely involve more than just power minimization.

List of Symbols

a	Interfacial area per liquid volume
$k_l a$	Overall mass transfer coefficient
k_l	Liquid film coefficient.

References

1 Moral, A. et al. (1999). Effect of antifoam addition on gas-liquid mass transfer in stirred fermenters. *Bioprocess Engineering* 20: 165–172.

2 Prins, A. and v'ant Riet, K. (1987). Proteins and surface effects in fermentation: foam, antifoam and mass transfer. *Trends in Biotechnology* 5 (11): 296–301.

3 Dumont, E., Andres, Y., and Le Cloirec, P. (2006). Effect of organic solvents on oxygen mass transfer in multiphase systems: application to bioreactors. *Biochemical Engineering Journal* 30 (3): 245–252.

4 Kundu, A. et al. (2008). Mass transfer characteristics in gas-liquid-liquid system. *The Canadian Journal of Chemical Engineering* 81: 640–646.

5 Sauid, S. et al. (2013). Enhancement of oxygen mass transfer and gas holdup using palm oil in stirred tank bioreactors with Xanthan solutions as simulated viscous fermentation broths. *Hindawi Publishing Corporation, BioMed Research International* 409675: 9. pages, an open access article distributed under the Creative Commons Attribution License, which permits unrestricted use, distribution, and reproduction in any medium, provided the original work is properly cited.

14

Use of CFD in Fermenter Design

Purpose of This Chapter

This chapter describes the use of CFD (computational fluid dynamics) in fermenter agitator design, told from the viewpoint of one who is not a CFD expert. As such, theoretical discussion will be very brief. The focus of this chapter is on what CFD can do to help the designer of agitation equipment predict its performance and the owner of fermenters to get an idea of the possibilities.

CFD can do much more than described herein. A more complete set of examples and descriptions comes as a CD included with Ref. [1]. Our focus is on CFD simulations that are most useful for fermenter design, though some basic uses are also illustrated.

Since CFD results are presented visually and generally coded by color, the images presented in the print edition of this book may not show up well, as they are in black and white. I recommend purchasing the digital version to get the color images.

The main sources of the images used in this chapter are www.Bakker.org, private images from Tim Laska, an independent CFD consultant, Tridiagonal Solutions (in collaboration with Gregory Benz), Chemical Engineering Progress Magazine (by blanket permission), and Ref. [2]. All are used with permission and will be attributed as appropriate.

Basic Theory

In the purest form, CFD numerically solves the Navier–Stokes equations and equations of continuity. This can be done by setting up a series of grids or nodes at all surfaces, whether moving or stationary, and setting a boundary condition that a differential layer of fluid is attached, or moves with, the surface. Examples of

Agitator Design for Gas–Liquid Fermenters and Bioreactors, First Edition. Gregory T. Benz.

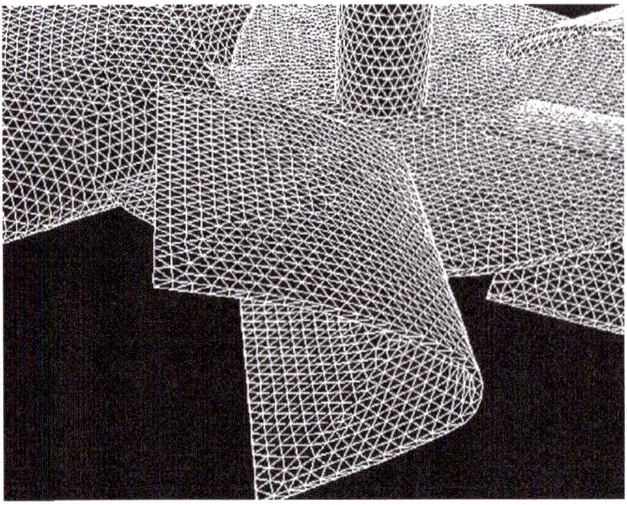

Figure 14.1 BT-6 grid. *Source:* Courtesy of Bakker.org.

grids are shown for a deeply concave impeller (Chemineer BT-6) and the tank wall in Figures 14.1 and 14.2.

The most basic output from these calculations is a velocity distribution. Additional processing can be done to determine pressure distribution, power dissipation, impeller power draw, shear distribution, transient blending progress, etc. With additional functions added in, such things as solids suspension profile, bubble size distribution, holdup distribution, mass transfer coefficient distribution, dissolved species distribution, reaction progress, product distribution, local heat transfer coefficients, and many other things can be calculated. Not all of what can be calculated has been validated experimentally. We will explore some of the more useful calculations in subsequent sections, focused on fermentation needs. However, we will cover some basic uses first.

Not all simulations start with an impeller grid. Some start with experimentally determined impeller boundary conditions and then solve for velocity distributions throughout the tank. This method saves time setting up the CFD model and may be more accurate. However, it requires "wet" testing of impellers under conditions not too far from the model being simulated in order to create the boundary conditions. It is the hope of CFD modelers to come up with accurate enough CFD codes so that the wet testing can be skipped. In some cases, they may have done so well enough to skip the "wet" experimental program.

We will comment on whether the CFD models have been validated in subsequent sections.

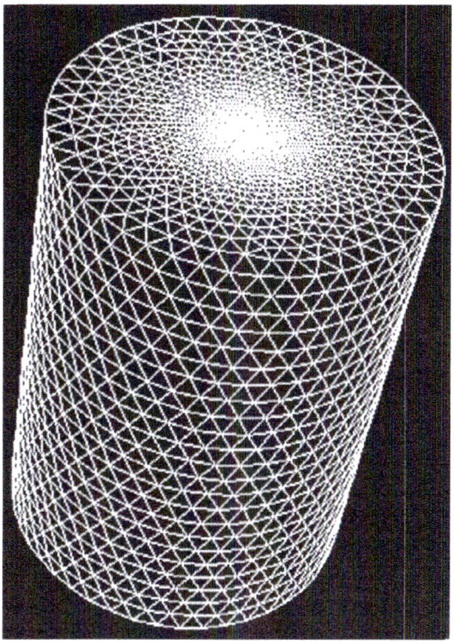

Figure 14.2 Tank grid. *Source:* Courtesy of Bakker.org.

Methods of Presenting Data

Some data, such as velocity distribution, have both magnitude and direction. Thus, they can be presented as vector plots. Typically, the direction of the vector indicates the direction of flow (it can be instantaneous or time-averaged), the length of the vector is proportional to velocity, and the color of the vector is coded to velocity.

If only the magnitude of the variables is to be presented, a scalar plot is used. This is often in the form of a raster plot, in which various regions are color coded based on magnitude. For velocity raster plots, for example, the color is coded to velocity. By convention, the color red represents a high magnitude, with the magnitude reducing as the color goes through orange, yellow, and various shades of green and blue. Purple or indigo are not normally used, so we do not have the full "Roy G. Biv" rainbow color spectrum.

Simulated particle tracks can be shown, as well as stressed virtual surfaces. Results can be presented in two dimensions or simulated three dimensions. Animations or movies can be created. New methods of presenting data visually are constantly being developed.

While some of these simulations are visually amazing and fun to watch, they can take up a lot of personnel time and expense to create and a lot of computer time to execute, with no certainty that the results are valid. Just as important, many such simulations do not actually help us design fermenters more effectively. Most information that is useful for fermenter design can be presented as a quasi-steady-state simulation in 2D using a raster plot. We will show a few of these in subsequent sections. The reader may think of specific needs not addressed here, and there are CFD modelers eager to help.

We will start with the simplest and most commonplace CFD simulation.

Velocity Distribution

Though it is possible to derive a velocity distribution from basic theory and surface-based boundary conditions, the most common simulations use boundary conditions around the impeller and then solve numerically for the rest of the tank. The impeller performance data usually come from either Laser-Doppler Velocimetry (LDV) or Digital Particle Image Velocimetry (DPIV), though other methods could be used. These techniques can also be used to validate the results. Commonly, both vector and raster plots are presented. These can be either 2D or simulated 3D. Generally, for an axisymmetric tank as is typical of fermenters, a 2D presentation will suffice, and it is an easier model to set up as well as one requiring less computation time. Most agitator vendors will prepare such plots in two dimensions for free, as part of their equipment proposal.

Figures 14.3 and 14.4 show vector and raster plots, respectively, of a dual impeller system. These are reproduced from an article I wrote in Chemical Engineering Progress (CEP) magazine and are used by blanket permission of CEP.

The output of these kinds of plots has been compared to the measured velocity fields using both DPIV and LDV, and the error is generally less than 5% for Newtonian fluids. It can rise to near 10% for some classes of non-Newtonian fluids. This is still less error than the error band of many phenomena of interest in fermenter design, such as $k_l a$ prediction. Thus, we can consider this type of information to be validated. However, these plots do not include the effect of gas. While plots can be created that include the effects of gas on the flow pattern, I am unaware of any experiments that have validated the results. Most people do not attempt this.

Cavern Formation

In highly shear-thinning fluids or those with a yield stress, it is possible to form "caverns" or moving zones surrounded by relatively stagnant zones. This was a major topic of Chapter 12. In addition to the methods described in Chapter 12, cavern formation can be predicted using CFD techniques. Velocity can be used as a direct way of visualizing caverns. However, a CFD consultant friend of mine, Tim Laska, timmy2x@gmail.com, devised another way to visualize caverns. Given that shear-thinning fluids have a higher viscosity as shear rate gets lower and that

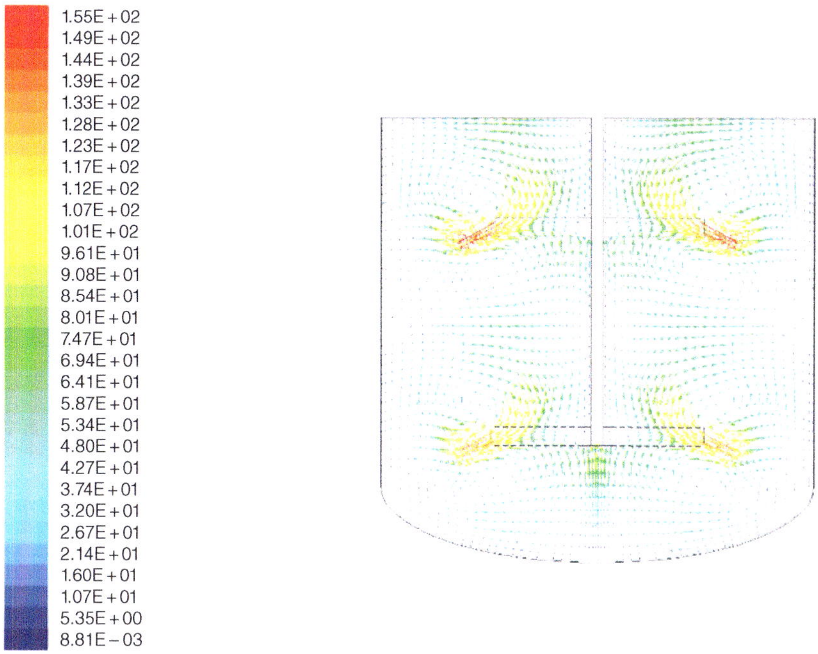

Figure 14.3 Vector velocity plot. *Source:* Permission of CEP.

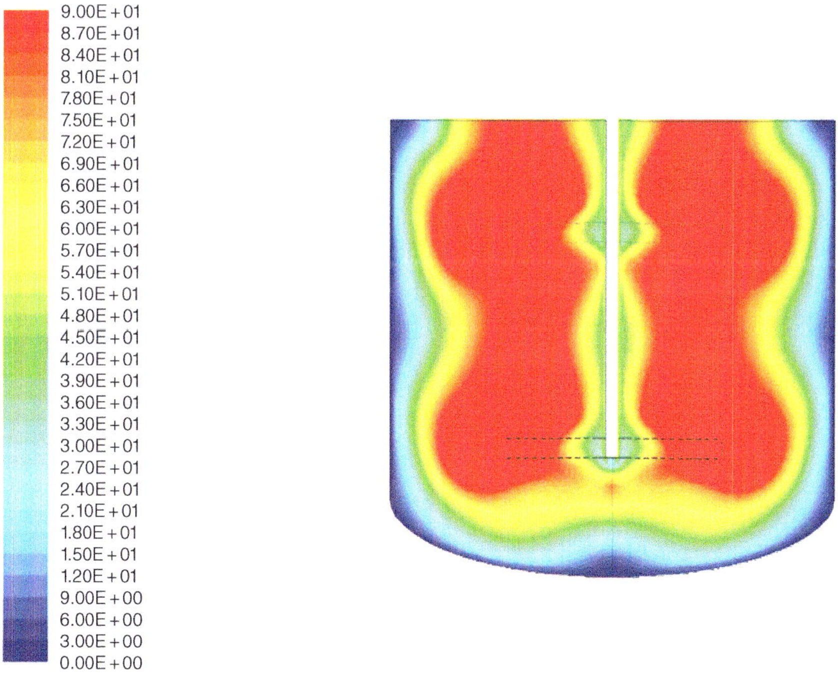

Figure 14.4 Raster velocity plot. *Source:* Permission of CEP.

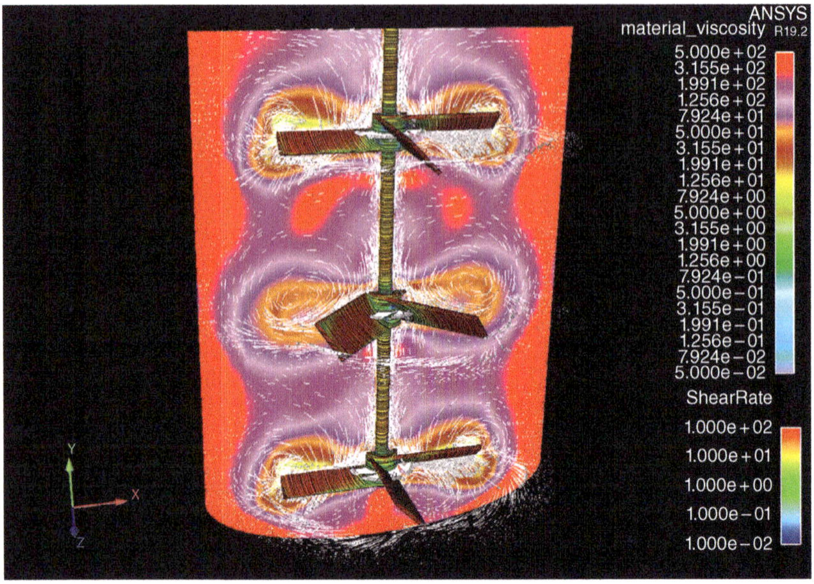

Figure 14.5 Cavern raster plot. *Source:* Tim Laska.

shear rates tend to be low in stagnant zones, he modeled caverns using apparent viscosity distribution as a proxy for velocity. The results turned out very well.

Figure 14.5 is a simulated 3-D raster plot of a shear-thinning fluid with [3] pitched blade turbines, with some particle tracks thrown in for good measure. It is possible to create sectional views of the output wherever needed.

Figure 14.6 shows virtual surfaces of the cavern boundaries. We would really like to have the caverns extend to the wall and have little dead space.

Reference [2] deals with validating similar simulations by comparing with fluorescing dyes and velocity measurements. The experimental results using DPIV were compared with the CFD plots, and the maximum error was less than 10%, so we would consider the CFD modeling techniques validated.

These plots also do not include the effect of gas bubbles on the flow pattern. But, it is expected that gas bubbles may help agitate some of the predicted stagnant areas.

Blending Progress

Using algorithms applied to the velocity plots, one can combine folding and turbulence to predict the progress of blending of miscible liquids. Figure 14.7a through d is an example of such a simulation, using a single impeller. In Figure 14.7a, the

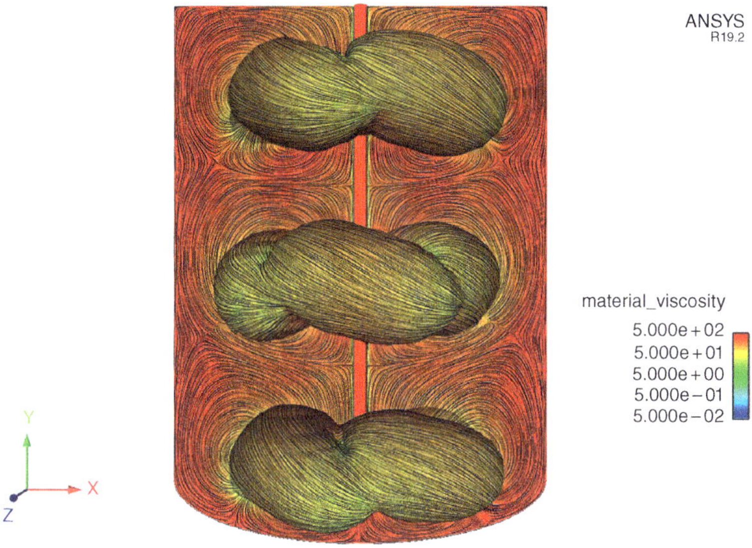

Figure 14.6 Cavern surface plot. *Source:* Tim Laska.

tank is completely stratified, with a high concentration in the red zone and a low or zero concentration in the blue zone. This is at time = zero (seconds). Figure 14.7b–d is at 4, 10 and 20 s, respectively, after starting the agitator. The tank would be almost uniform when the colors are all in the green range, but the simulation was not carried out quite long enough to show that. It would have been about 30 s for this case.

Actual blending tests have validated such simulations for simple blending of miscible liquids of similar densities and viscosities. However, as fermenter designers, we are more interested in blending of tanks when gas is being introduced. Alas, I was unable to find such simulations. However, I know from experience it makes a huge difference when gas is introduced.

As an example, I once did a series of "wet" tests simulating blending performance in a laboratory scale system intended to mimic typical full-scale fermenter designs. The laboratory scale system had an aspect ratio of about 3:1 and was equipped with a lower concave radial impeller and [3] upper axial flow impellers. The system was tested in both up- and down-pumping mode, with and without gas. Though I do not have the original data, the relative results tabulated in Table 14.1 are approximately correct, going from memory. The blend time was based on the time required to clarify an alkaline solution with phenolphthalein indicator, using the techniques described in Ref. [3].

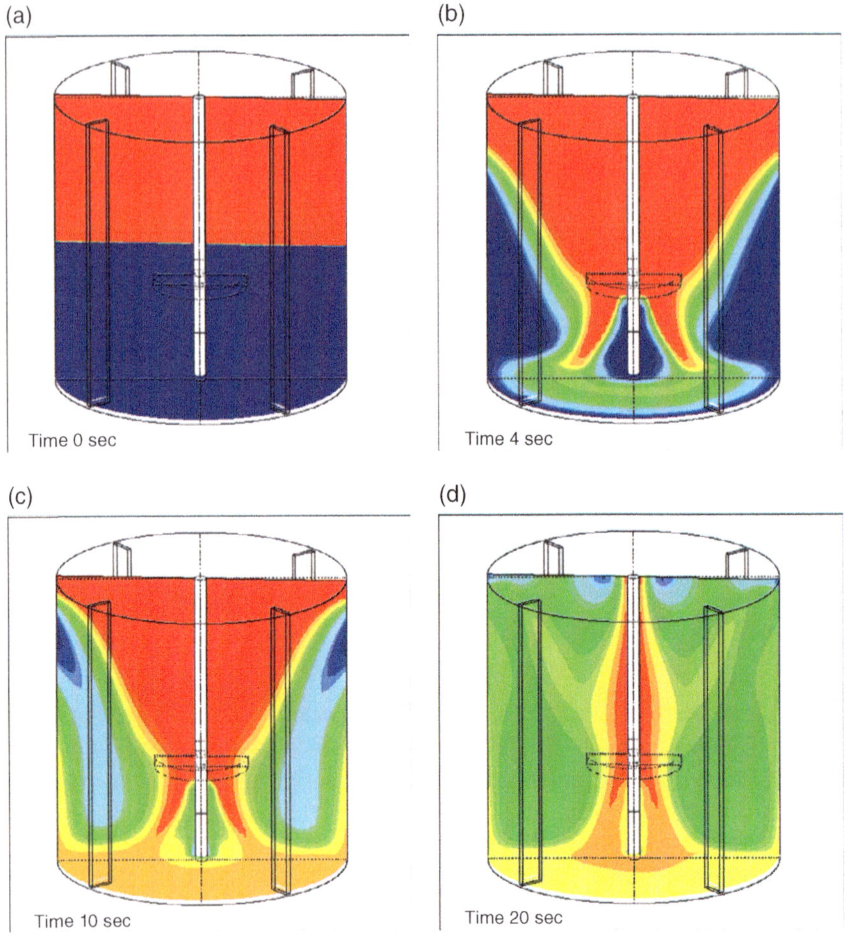

Figure 14.7 (a–d) blending progress at 0, 4, 10, and 20 s. *Source:* Courtesy of Bakker.org.

The main takeaway is that the presence of gas has a big impact. Specifically, it hinders blending when the impellers are pumping down and fighting the rising gas bubbles, and substantially aids blending when the impellers and the rising gas are working together to create more turnover, as they do when the axial impellers pump up. This is one of the reasons why up pumping is recommended for fermenters.

It is not clear whether CFD simulations of blending of gassed systems can be considered to be validated, as I have not personally seen any, nor am I aware of the methodology that would be needed to verify the simulations.

Table 14.1 Relative blend time.

Pumping direction	Gas on/off	Relative blend time
Down	off	1
Up	off	1.2
Down	on	2
Up	on	0.7

Flow Around Coils

One use of CFD is to calculate flow velocities around solid objects, such as coils in the tank. An example of this follows.

A client was designing a fermenter with 340 m^3 working volume, with a 750 kW motor, and planned an air distribution system capable of handling 1.5 VVM. They intended it to be a multi-product facility, capable of producing enzymes, amino acids, and many other products. Flexibility was key. One concern was that they might be heat transfer limited, and they wanted to avoid that. So, they planned for a lot of heat transfer area. Specifically, they designed for 12 vertical tube bundles, each one consisting of two rows of four tubes each. The tubes were of relatively large diameter to make sure they were strong enough to withstand the fluid forces imposed by the agitator as well as vibrations imposed by the large volume of air. A conceptual layout is presented in Figure 14.8, courtesy Bakker.org.

With so many tubes, a concern was that the velocity between the tubes might be low. To check on this, a CFD model was made of the system, and the results were displayed as raster plots across planes normal to the vessel axis, at several elevations. Figure 14.9 shows the views at 0.8, 0.6, 0.4, and 0.2 fraction of the liquid level off the bottom, going roughly from the upper impeller location on down. These are shown in order of left to right, top to bottom. In other words, the top left is the highest off-bottom plane studied and the bottom right is the lowest plane.

As one would expect, the velocities near the plane of the bottom impeller are fairly high across all rows of tubes. This is because the lower impeller is radial flow, discharging directly at the tubes. The upper impellers are all axial flow. They create higher velocities than the radial impeller along the vessel wall, which is useful for jacketed vessels. They also create higher velocities on the outermost row of tubes, about the same on the next two rows, and lower velocities on the innermost row. Overall, the results are similar. The velocities on the innermost row are on the order of 0.2 m/s, which is low but ample to promote heat transfer. Also, these fig-

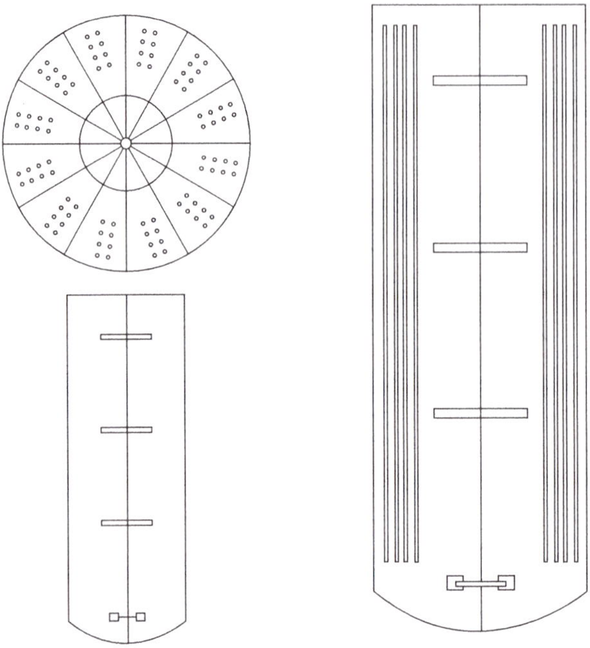

Figure 14.8 Coil layout. *Source:* Courtesy of Bakker.org.

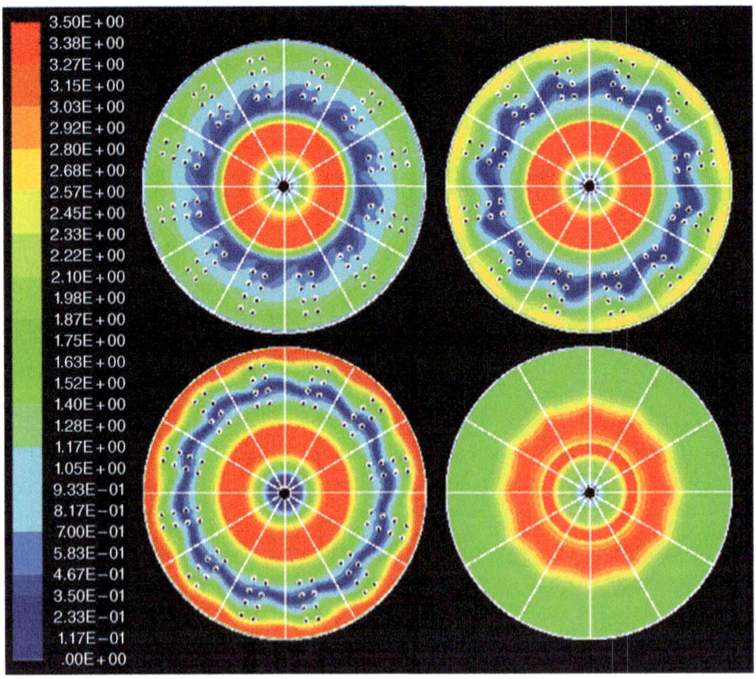

Figure 14.9 Velocities around coils. *Source:* Bakker.org.

ures are time-averaged and do not account for the turbulent velocity fluctuations, which will be quite high in a fermenter.

Moreover, this simulation did not take into account the effect of gas flow on the local velocities. Though such an advanced simulation may be possible, it would be almost impossible to validate the results. However, for this particular client, the end result was that they had ample heat transfer capacity for all of the fermentations they had initially planned to run.

The above simulation was for low viscosity liquids. For viscous fermentations, flow between the coils could prove challenging. Wet simulations of a similar geometry for a gum fermentation revealed little flow at the wall unless the two inner rows of tubes were removed. So, we would recommend wet testing for viscous, non-Newtonian fluids to be sure.

Bubble Size, $k_l a$, Holdup

Various hydrodynamic models can be developed to predict bubble size distribution, local volume fraction (holdup), and $k_l a$ distribution. Most such models are derived from turbulence theory and ungassed flow velocity distribution. Rarely are they calibrated to match actual experimental results. Of course, it would be difficult or impossible to fully measure these distributions in an actual fermenter vessel or even an air/water test in a clear pilot scale vessel. What is possible is to measure total gas holdup and average $k_l a$. If the CFD model agrees with the total result, then the distribution may in fact be reasonably accurate. Figure 14.10 from Bakker.org illustrates CFD raster plot results for the same fermenter described in the previous section. The results are displayed for gas volume fraction, three times the $k_l a$ value (in 1/s) and 30 times the bubble size in m (the scalar multipliers are to get all magnitudes of results on the same legend on the left of the figure).

Because the model was not based on actual experimental data, the results could be substantially off. It also did not account for the effect of gas on velocity distribution, so the apparent flow pattern may be less axial than shown. Nonetheless, we can learn a few important things from observing the plots.

One of the most important things we can learn is that volume fraction, $k_l a$, and bubble size are not evenly distributed. Volume fraction is high near the impellers and lower everywhere else except at the very surface where gas is coalescing. Likewise, $k_l a$ is much higher near the impellers than elsewhere, and bubble size is smaller near the impellers. Though this system, if I recall correctly, was modeled as a coalescing system, and therefore, the distribution might be less extreme in an ionic broth, it is likely that in any system, most of the mass transfer takes place in a small zone near each impeller. Probably, at least 80% of the mass transfer takes place in 20% of the volume. It could be closer to 90/10 in some systems. This highlights the importance of a mixing system that can circulate oxygen-rich

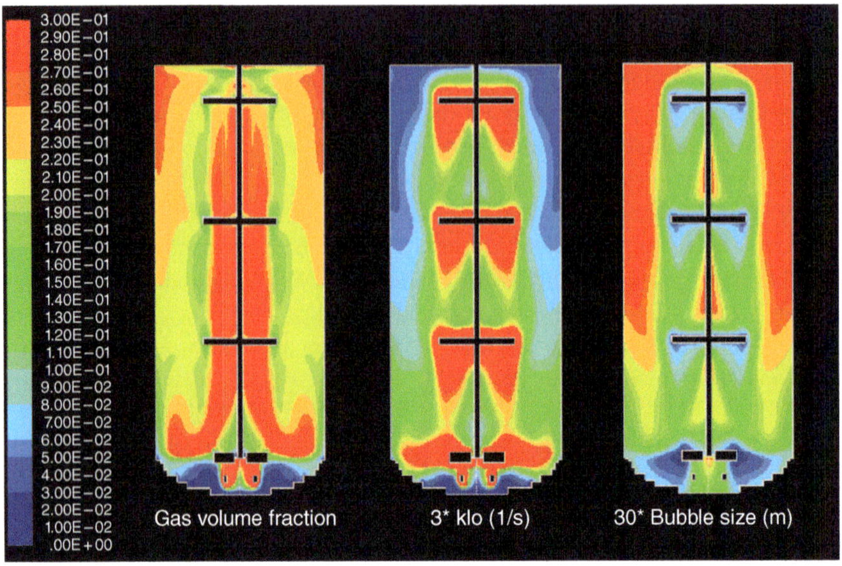

Figure 14.10 Volume fraction, etc. *Source:* Bakker.org.

(or methane or hydrogen rich; whatever gas is being dispersed) rapidly throughout the vessel, so that the dissolved gas concentration distribution is as uniform as possible. This segues into our next section.

DO Distribution

In 2014, I co-authored a paper (Ref. [4]) with several co-authors from Tridiagonal Solutions. The purpose of the paper was to evaluate feasibility of using CFD to predict DO distribution using various combinations of power and airflow as well as various OUR values. A sample of the output is shown in Figure 14.11. It shows predicted DO distribution at OUR levels of 10 and 25 mmol/l-h. The model used an up-pumping/radial flow impeller combination, in a 500 m^3 fermenter.

The model used fundamental equations and some simplifying assumptions to save computation time. Unfortunately, the overall mass transfer rates predicted were much lower than what would be expected based on the empirical k_la model that was to be the basis of the article, but was rejected by the co-authors in favor of the simplified approach.

Nonetheless, we can see that the DO distribution can be substantially unequal. Near the impellers, the DO can easily be 20–30 times as high as it is in the further regions. Of greater importance is whether DO in the lowest regions is still high

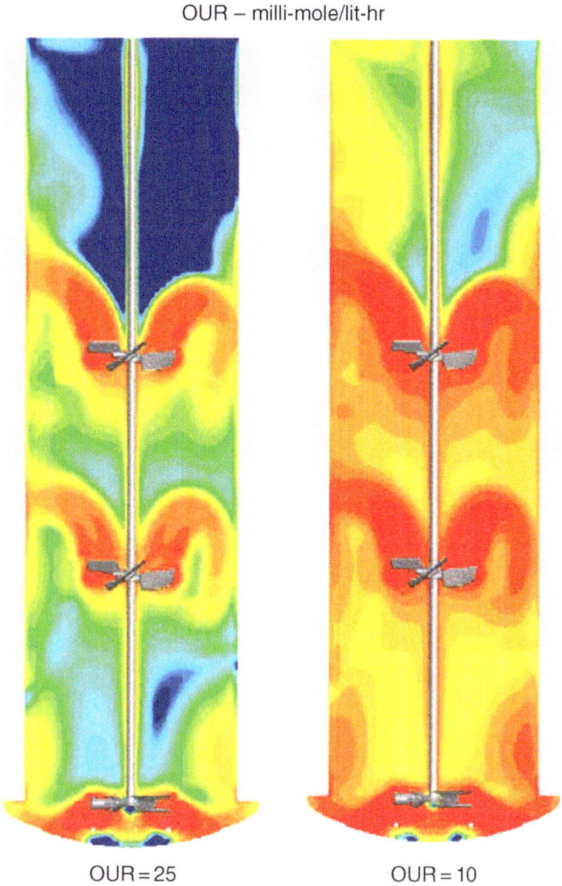

Figure 14.11 DO distribution.

enough to sustain the organism. In the referenced article, the DO was as low as 0.01 mg/l in some areas, which would usually be insufficient. However, these results are not consistent with predictions of overall mass transfer using popular k_la models such as the V'ant Riet ionic model referenced in several chapters, such as Chapter 8.

In the years following the publication of that article, modeling has gotten much better. Though I do not have more recent modeling results, I believe they are closer to the truth. Still, it is not possible to do a full experimental verification of DO distribution. All that we can really measure is DO near the vessel wall. However, this at least lets us see where the DO is likely to be lowest: near the wall and near the liquid surface. For process control, the DO in the most likely lowest DO locations must be controlled at the minimum value the organism needs. Using 2 or 3 DO probes may serve as a crude check on the CFD model.

Summary of Chapter

We have shown that CFD techniques can be used to predict a multitude of things, limited only by one's imagination and budget. CFD is very accurate for simple things such as velocity distribution. For more complex phenomena that are especially valuable, such as k_la and DO distribution, the models are more qualitative in nature, and the absolute results have not been validated. Future work needed includes improving the models, calibrating them to match experimental results and maybe some creative ways to experimentally validate the models.

List of Symbols

k_la	Overall mass transfer coefficient
OTR	Oxygen transfer rate
OUR	Oxygen uptake rate
VVM	Volume of gas per volume of liquid per minute

References

1 Paul, E., Atiemo-Obeng, V., and Kresta, S. (2004). *Handbook of Industrial Mixing*. Wiley Interscience.

2 Arratia, P. et al. (2006). Mixing of shear-thinning fluids with yield stress in stirred tanks. *AIChE Journal* 52 (7): 2310–2322.

3 Fenic, J. and Fondy, P. (1966). *Application of Similarity Analysis to the Blending of Miscible Liquids*. Presented at the AIChE meeting in Atlantic City, Paper 4 d, (18–21 September 1966).

4 Benz, G., et al. (2014). *Computing Dissolved Oxygen Profiles in Aerobic Fermenters*. Presented at NAMF, 2014.

15

Agitator Seal Design Considerations

Introduction

Because fermentation involves growing of a living organism, the process environment must be controlled. Among other requirements, this means that at the point where the agitator shaft enters the tank, it must be sealed. This chapter will illuminate what the functions of the seal are and discuss most of the many alternatives that have been used in fermentation.

Except where a magnetic drive mixer is employed, the normal configuration of an agitator consists of a motor, gear drive, shaft, and impeller (Figure 15.1).

Much of this chapter has been drawn from material previously published in an article in CEP, co-authored by me (Ref. [1]). Text material unrelated to typical fermenter seals has been removed, and other topics are covered in greater depth here than in the referenced article.

Terminology

To assure comprehension of specific terms that occasionally have different meanings in different industries, some important terms are defined here, rather than in a glossary at the end of the article.

Aseptic: Lacking any undesirable living organisms, such as in a fermenter or bioreactor, which must have some living organisms but only those needed for operation of the process. *Asepticity* means the state of being aseptic.

Containment: Prevention of dangerous substances, either biohazard or chemical hazards, from exiting the process unit, such as an agitated tank.

Contamination: Any undesirable material that has entered the process fluid, whether biological or chemical, and whether it comes from outside the process

Agitator Design for Gas-Liquid Fermenters and Bioreactors, First Edition. Gregory T. Benz.

Figure 15.1 Agitator with seal. *Source:* Photo courtesy Chemineer, a brand of NOV. Permission granted by NOV.

unit or from within (such as from wearing parts or trapped material from previous batches).

Isolation: The state of having a secure barrier between the process unit and the outside environment, preventing contamination from outside.

Sanitary: Easily cleaned, and not prone to trap material. This adjective refers to the equipment; it does not describe a state of the process, unlike *aseptic* or *sterile*. More detail on this is available in Chapter 18.

Sterile: Having no living organisms present. This is desirable in some applications; impossible in bioreactors. *Sterility* is the state corresponding to this adjective.

Main Functions of Fermenter Shaft Seals

Agitator seals used in fermenters or bioreactors may be called upon to perform one or several functions. The principle functions follow.

1) Containment of tank contents. This function is mainly concerned with protecting plant personnel or the surrounding community from biohazards, toxins, or otherwise reactive or toxic chemical agents.

2) Isolation of tank contents. This function is mainly concerned with protecting the contents from the outside environment, whether biological or chemical.
3) Avoidance of contamination. This concern is not quite the same as isolation, as it includes sources of contamination from the seal itself (particle shedding) and material left over from previous batches or earlier times in a continuous process.
4) Static and dynamic sealing of agitator. Holding the vessel pressure either in vacuum or in above atmospheric, whether operating or stationary.
5) Safety for personnel, such as avoidance of sudden pressure venting.

Agitator seal applications for fermenter service are typically characterized by and concerned with:

- Strict leakage requirements.
- Process purity – low or zero contamination.
- Hygiene, "Clean in Place,", "Sterilizing in Place" procedures, etc.
- Hazardous or volatile fluids.
- Corrosives, especially chlorides.
- Large shaft size.
- Low speed (typically 30–350 rpm).
- High radial run out (up to 1.5 mm TIR), high vibration velocity (up to 20 mm/s on large fermenters).

The position of the agitator mounting has a major influence on the seal selection and design. Agitator mounting is classified into top, side entry, or bottom entry (Figure 15.2).

Top is the most popular type as it has the following advantages:

1) Viscosity and abrasion properties of the process liquid have no direct impact on the seal operation and performance.
2) The seal is not wetted by the process liquid; only vapor or gas has contact with the seal so corrosion problems are reduced.
3) Mechanical seal face wear can be contained with optional construction features such that zero contamination of wear particles to the product can be achieved.

A bottom entry seal is fully immersed in the process fluid. In some cases, the seal can be lubricated by the process fluid. However, if the process fluid contains abrasives, these abrasive particles will accumulate at the bottom of the agitator at the seal area, in which case an external flush of clean liquid, compatible with the process fluid, is required. Bottom entering agitators are uncommon in bioprocessing but are occasionally used on small fermenters, when there is a need for very many nozzles on top, leaving insufficient room for the agitator. In such a case, bottom entry may be the only option, but it usually requires seal flushing, which can seriously dilute the product. Unlike top entry agitators, bottom entry units need only a short shaft. However, zero contamination due to seal face wear particles is not possible.

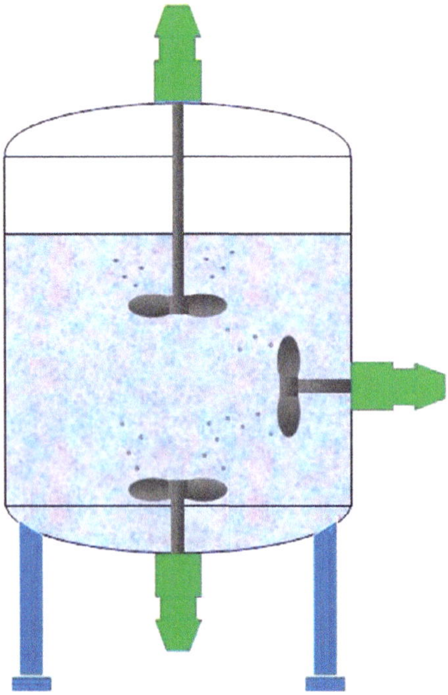

Figure 15.2 Shaft entry position.
Source: Permission CEP.

Side entry agitators are normally fully immersed in the process liquid and operate similarly to a horizontal pump mechanical seal. They are not normally used for fermenters and will not be discussed further herein.

Common Types of Shaft Seals

A number of different sealing methods have been used in fermenters. Some are more common than others. We will cover all that the author is aware of having been used in such service, with respect to each of the major functions.

1) Lip seal (Figures 15.3 and 15.4). Lip seals are low pressure seals which use either a stationary elastomeric member to apply radial pressure against a shaft (Figure 15.3) or a rotating elastomeric member using axial spring pressure against a stationary surface such as a flange (Figure 15.4). Such seals are inexpensive and require no lubrication. However, they are mainly useful to keep large objects out of the tank and diminish odor release. They cannot handle more than a few mm of pressure and are poor at containment and isolation. They create wear on the shaft or sleeve, which can eventually lead to expen-

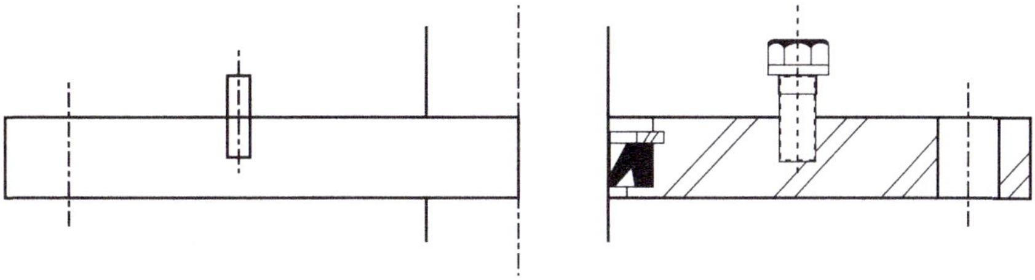

Figure 15.3 Radial lip seal.

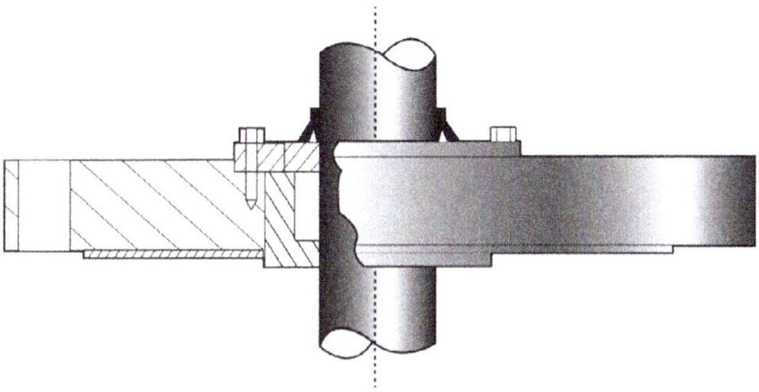

Figure 15.4 Axial lip seal.

sive repair. By adding a steam gland, they can assure no living organisms pass through the seal area. One major pharmaceutical manufacturer has used stacked radial lip seals with a steam gland on many of its large fermenters.

2) Compression packing (Figure 15.5), also called a stuffing box. This consists of a series of rings of packing material arranged between a stationary tube attached to the mounting flange and the rotating shaft. Pressure applied axially by a seal gland (often split) becomes radial pressure by flexion of the packing. This type of seal can either be lubricated or dry. Pressure rating depends on how many rings are used. Generally, pressure is limited to 6 bar-g, though there have been cases as high as 1000 bar-g. Packing can be sterilized by steam injection. There can be some shedding of packing into the process fluid. Packing acts mainly as a throttling device; it cannot provide complete isolation or containment. However, it can provide a sterile barrier if steamed.
3) Single mechanical seal (Figure 15.6a and b). This consists of two sealing faces, one stationary and one rotating with the shaft. The faces are lapped nearly flat, so essentially they provide a very slow leakage rate, making them a throttling device but a couple of orders of magnitude more effective than packing. Initial cost is higher but seal life is longer, so reduced downtime normally

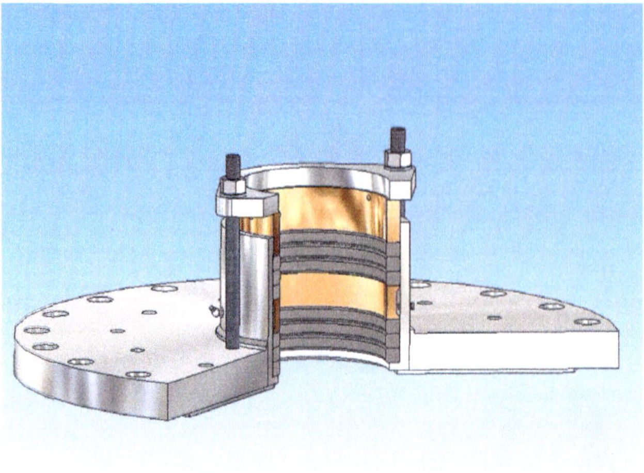

Figure 15.5 Compression packing. *Source:* CEP blanket permission.

pays for the investment. They provide sufficient containment and isolation for non-critical applications. They should not be used on sensitive cultures, ones containing pathogens, or on agitators containing highly toxic reagents.

A single mechanical seal can be either wet running or contact dry running. (Figure 15.7) Wet seals require lubrication either from the process fluid or a fluid from an external source. A contact dry-running seal operates in the absence of lubrication; the seal ring and mating ring are in direct rubbing contact. Due to the absence of a lubricating fluid, a dry-running seal is limited to low temperature and low speed operation. The seal operating life is lower than wet seals due to higher face wear resulting from full contact operation. If no carbon particles can be tolerated in the process fluid, the wear particles from the carbon face will have to be contained by a debris well (also called "carbon catcher"). A dry-running seal offers the advantage of simplicity and eliminates the need for an external fluid source. In some cases, the use of a single dry-running seal eliminates the need for a higher cost dual wet running seal.

4) Single fully split mechanical seal (Figure 15.8). Operating on the same sealing principles as above, the single fully split mechanical seal adds a longitudinal split line for ease of maintenance. With such a design, it is neither necessary to remove the agitator for servicing nor to provide shaft support, couplings, and spacers to effect seal change. Performance is similar to the single seal; however this design does create an additional leak path. The pressure rating will be lower and the leakage rate higher than an equivalent non-split seal.

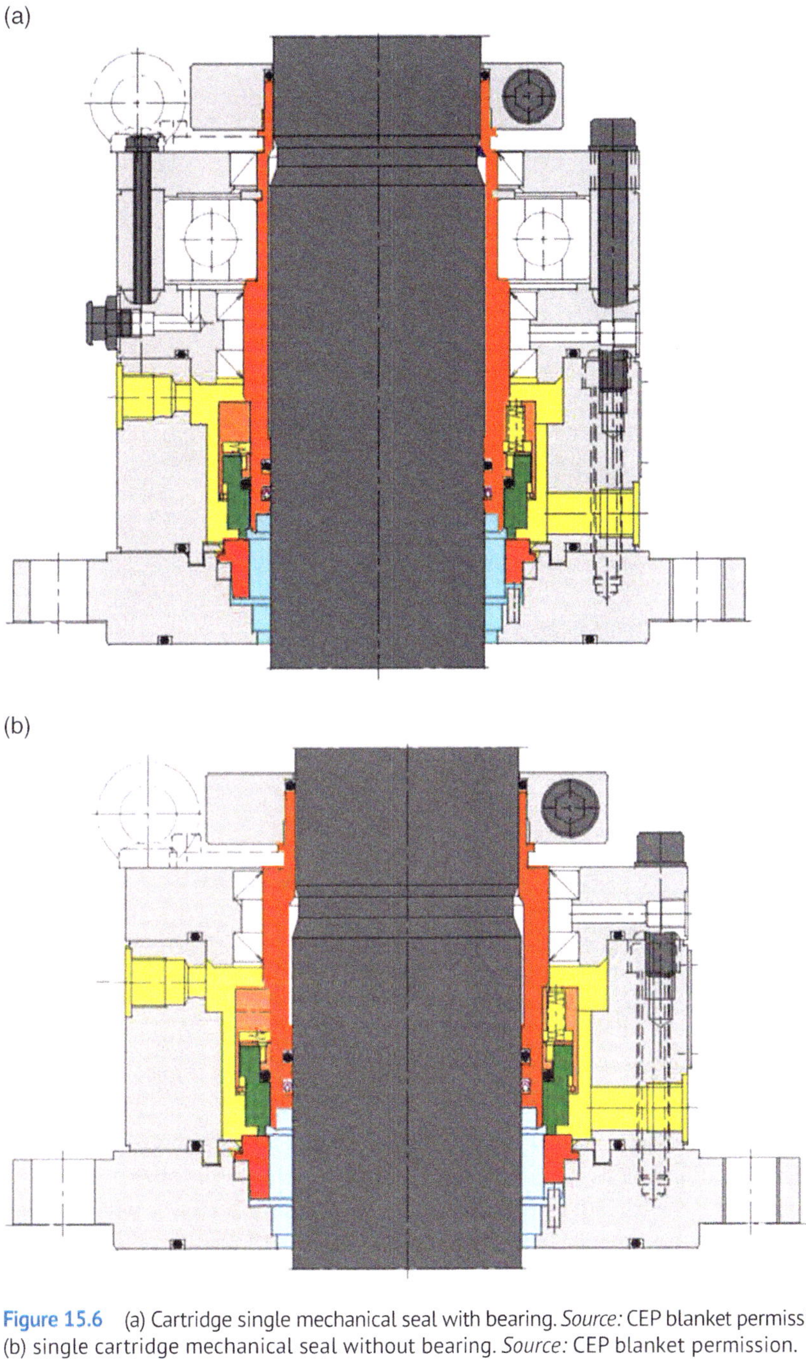

Figure 15.6 (a) Cartridge single mechanical seal with bearing. *Source:* CEP blanket permission. (b) single cartridge mechanical seal without bearing. *Source:* CEP blanket permission.

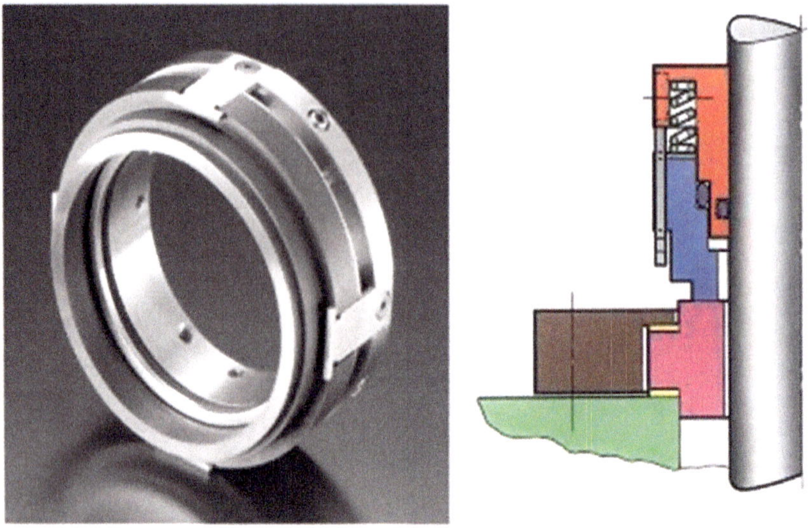

Figure 15.7 Dry-running contacting single seal. *Source:* CEP blanket permission.

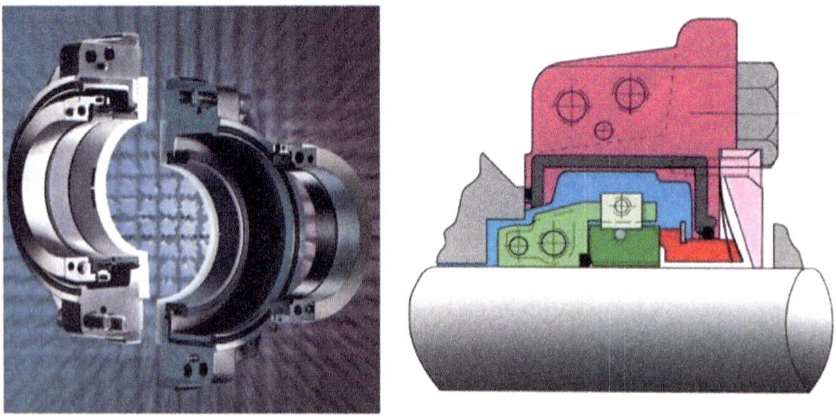

Figure 15.8 Fully split single mechanical seal. *Source:* CEP blanket permission.

5) Dual pressurized (Double) mechanical seal (Figure 15.9). Although several different configurations of dual pressurized seal are possible, the most common for agitators is the back-to-back arrangement, so it is the only one we will illustrate. Essentially, it is two single seals arranged in such a way that there is a chamber between the two, so that one seal (the outboard seal) seals the chamber contents from the outside environment, and the other (the inboard seal) seals the chamber contents from the tank. In such an arrangement, a barrier liquid is introduced between the dual pressurized seals for lubrication. This liquid also completely isolates the process fluid and vapors from the environment. The pres-

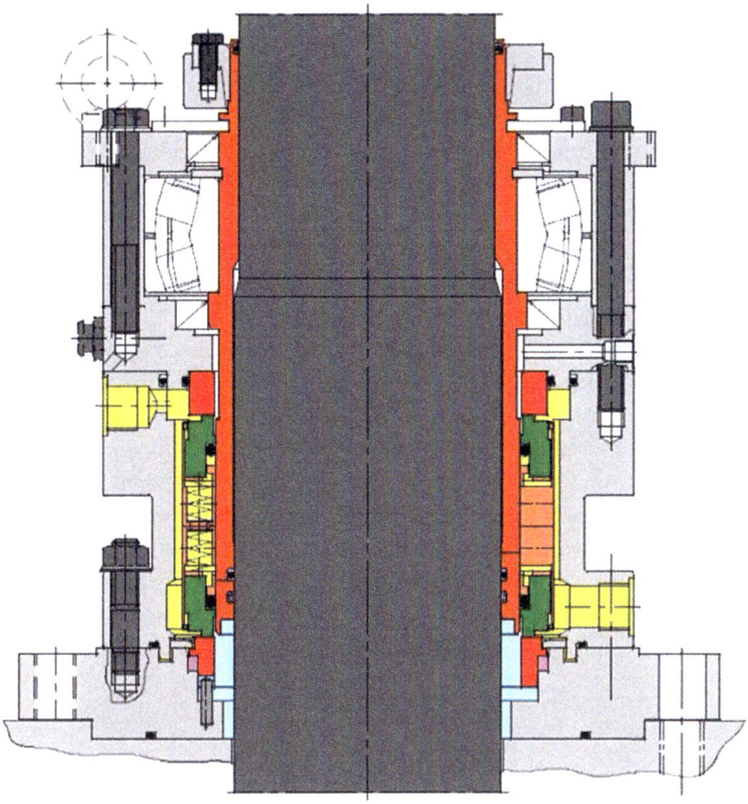

Figure 15.9 Double mechanical seal. *Source:* CEP blanket permission.

sure of the barrier liquid is always higher than the vessel pressure being sealed. The mechanical seal chamber is pressurized above the tank pressure, so any leakage which occurs is from the pressure chamber into either the tank or the outside environment. If all is working properly, the tank contents cannot leak out and outside contaminants cannot leak in. Therefore such a seal both isolates and contains the tank contents. A pressurizer or lubricator is used to pressurize the barrier fluid. The seal cartridge or the lubrication reservoir may also be equipped with a heat exchange jacket or coil for removing the heat generated by the mechanical seals. This is known as a "thermosyphon" arrangement. The barrier fluid must be compatible with the process fluid and the mechanical seal materials. The jacket or coil can also be used to maintain the sealing fluid at temperatures of about 90 °C to assure a sterile barrier, or steam condensate can be used as a seal fluid. Seal life is much longer with sealing fluids such as mineral oil, glycerin or glycols than it is with steam or steam condensate.

Some of the modern dual pressurized seals are designed with reverse pressure capability on the inboard seal and mating ring. Reverse pressure is caused by the failure of barrier fluid pressure or unexpected tank pressure surging higher than the barrier fluid pressure. Reverse pressure capability of a seal is only capable of holding the seal parts at their correct location to prevent catastrophic failure; the inboard seal will still have face opening and process fluid or vapor will flow from the tank into the pressurized seal chamber. A dual pressurized seal should not continuously operate with reverse pressure as the fluid between the seal faces will change from barrier fluid to process fluid or vapor. The seal will prematurely fail due to abrasives, clogging, or vaporization of the process fluid. For top entry dual pressurized seals, the inboard seal will run dry resulting in premature failure.

This is the first seal type discussed so far which can provide complete containment and isolation when working properly. Its use is mandated by law for containing pathogens in many cases, as well as for toxins. Only hermetically sealed devices, such as magnetic drive or immersed agitators, are safer for containment.

Double seals are almost always mounted in cartridges for ease of maintenance and usually have a bearing to assure alignment and increase shaft critical speed.

6) Gas lubricated seal (Non-contacting) (Figure 15.10).

A wet dual pressurized seal requires a barrier liquid between the seals for lubrication and the complete isolation of the process fluid from the environment. However, leaking barrier liquid into the process potentially can have a

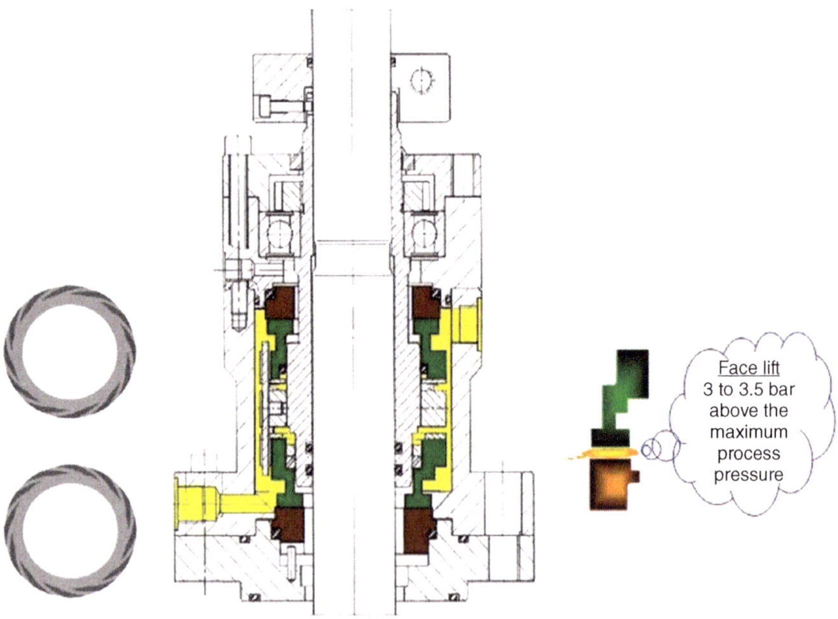

Figure 15.10 Gas lubricated seal. *Source:* CEP blanket permission.

negative effect on operations. The barrier liquid may contaminate the product, possibly resulting in lost revenue for the company.

A more recent innovation in agitator sealing is the gas lubricated seal. Available as either single or dual pressurized, these seals have the same functionality as other single or double wet seals. However, their design is such that they run dry exclusively, and they are non-contacting, resulting in zero wear. This is done by establishing a thin gas film to separate the sealing faces. The thin gas film is achieved by specially designed seal face geometry that produces a pressure gradient across the seal face (Figure 15.11). The face separation is developed by a combination of static opening forces generated by the barrier gas pressure and dynamic opening forces generated by the face geometry while the seal is rotating. At slow speed, the face separation development is dominated by static pressure lifting force; hence gas seals run non-contacting even at very slow speed. The gas lubricated seal can be either a unidirectional or bidirectional design.

A gas lubricated seal must be designed to give high operational safety. The face groove is designed to give the best stable gas film and hydrostatic lift (no contact at low speed) as well as being optimized for minimum barrier gas consumption.

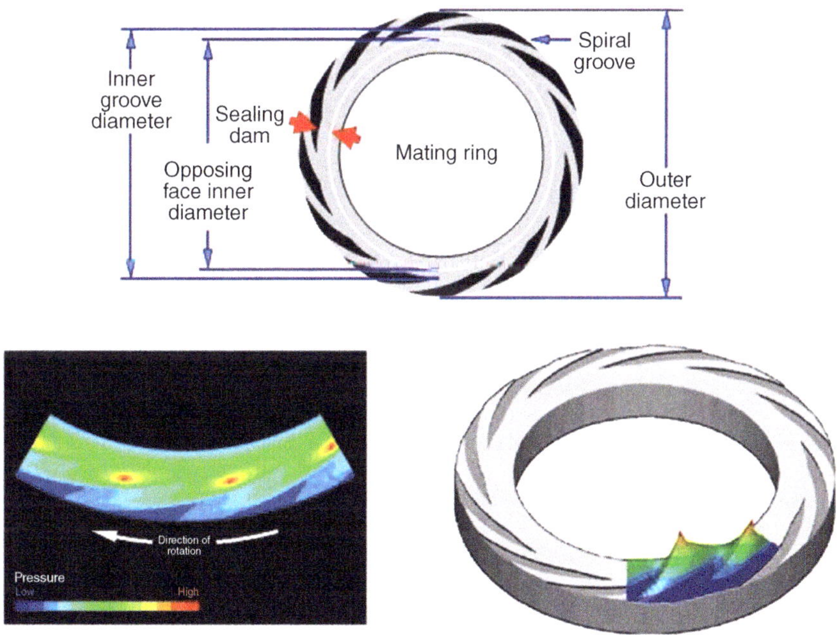

Figure 15.11 Pressure gradient. *Source:* CEP blanket permission.

This latter option dual pressurized gas lubricated seal only requires a simple gas control panel (Figure 15.12) to maintain a positive sealing pressure against the tank pressure. Dry nitrogen is commonly used as the barrier gas. It is normally supplied by a pressurized nitrogen cylinder. The outboard seal maintains control of leakage into the atmosphere.

Unlike dual pressurized wet seals which require a reservoir or a lubricator system, a dual pressurized gas lubricated seal needs very little space for the gas control panel and is easy to operate. The benefits of dual pressurized gas lubricated seals are a longer operating life, higher reliability and low maintenance costs in the long run despite the higher initial cost.

The present limitation of gas lubricated seal is the pressure rating at around 20 bar sealing pressure; wet dual pressurized seals are needed for higher sealing pressures. But most fermenters operate at less than 1 bar back pressure.

7) Debris catcher option (Figure 15.13, with dual pressurized seal). This is also referred to as a carbon catcher, debris well, or sanitary gland. While not a seal per se, this option can be added to single or double seals to prevent seal wear particles from entering the process, on top-entering units only. Accumulated particles can be cleaned during seal change or by draining during operation.
8) Cartridge seal (Figure 15.6a and b) Mechanical seals consist of many components. Assembling individual components onto the agitator shaft is not only time-consuming but will also not be able to ensure an error-free installation. A cartridge seal is a preassembled, completely self-contained unit consisting

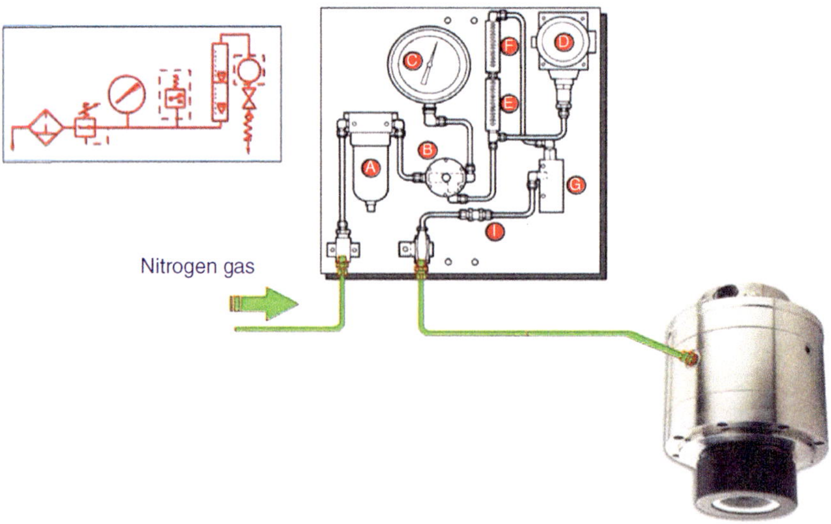

Figure 15.12 Gas control panel. *Source:* CEP blanket permission.

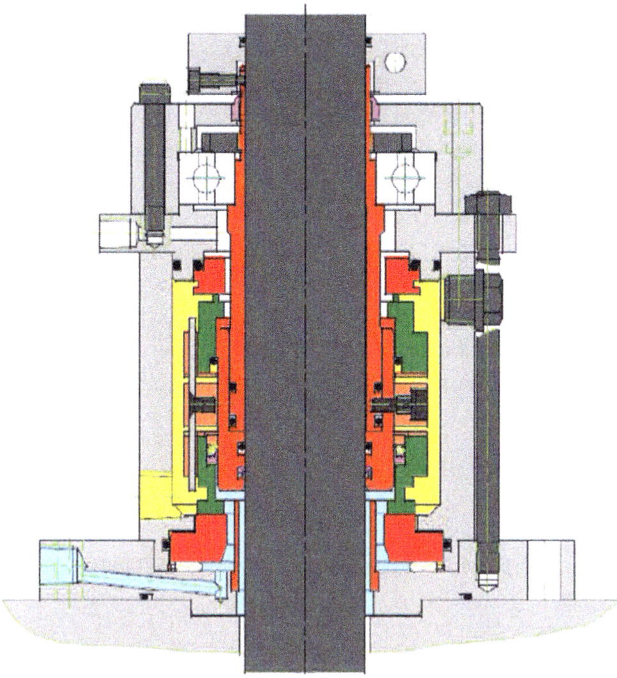

Figure 15.13 Debris catcher. *Source:* CEP blanket permission.

of mechanical seal parts with adaptive parts including sleeve, gland, and drive collar. The advantage of cartridge seals is that the components can be assembled in a clean environment at the factory or maintenance shop and can be leak tested to ensure all the components are correctly assembled. Installing a cartridge seal onto an agitator shaft is simple, fast, and error-free, though some designs require "floating" the cartridge into place for best alignment. End users can enjoy a significant savings in installation time, shorter maintenance time, and lower cost together with an improvement in reliability, when compared to non-cartridge designs.

9) Bearing options (Figure 15.6a).

In agitator applications, especially when equipped with steady bearings, the radial load on the shaft is generally small (though it can be high if the shaft bending moment is high), but the axial thrust load is often very high, especially at high tank pressures. Though traditional fermenters rarely operate with more than 1 bar-g back pressure, some extremophiles can operate at more than 5 bar-g. However, thrust bearings will usually not be incorporated in the mechanical seal cartridge to take up the axial loads. This is because almost all agitators are already designed with a thrust bearing in the drive to take up the full axial load requirements. A second thrust bearing in the seal could induce unnecessary stress between the two thrust bearings. Agitator OEMs design the detailed construction of the equipment to ensure the effective operation of the equipment according to process requirements. The loading expected on the mechanical seal is also calculated to ensure that the seal manufacturer has the relevant information.

A seal cartridge bearing is used to steady the shaft and reduce radial run out to create a favorable condition for good seal performance so the bearing is used to withstand radial loads and sometimes the thrust load from the seal sleeve. Mechanical seals with bearings must then be engineered on an individual basis to meet the service conditions. The factors to consider for bearing selection are as follows:

- Minimum/maximum radial loads.
- Minimum/maximum operating temperature.
- Minimum/maximum shaft speed.
- Mode of operation, changes of temperature, speed, and radial load.
- Service life expectation.
- Bearing materials.
- Suitable lubricant to lubricate the bearing.

The kinds of bearings which are most widely used are deep groove ball bearings and double row spherical roller bearings, and, occasionally, double row tapered roller bearings. Deep groove ball bearings are used for relatively low load magnitudes only. Double row spherical roller bearing can be used for heavy duty conditions. Double row spherical roller bearings can also tolerate angular movement of the bearing axis without damage to the cage or rollers. Double row tapered roller bearings, though rare, are used when there is high thrust load but little angular misalignment, such as typically the case on units equipped with steady bearings.

Since the speed is low on agitator applications, a suitable grade of grease must be used in the bearing based on the load, speed, and operating temperature. These factors are important to ensure that there is sufficient friction for the roller to rotate in the bearing cage and sufficient lubricating film thickness to minimize metal to metal contact. Bearing suppliers should be consulted if in doubt.

10) Magnetic drive (Figure 15.14). Again, though not a seal, we include this design for completeness, as the end result is similar to sealing a tank. This design, by virtue of having no shaft penetration, has absolute containment and isolation. However, due to the need to support the in-tank shaft with some kind of in-tank bearing system, it is subject to contamination caused by particle shedding from the bearing (which is usually some kind of food grade

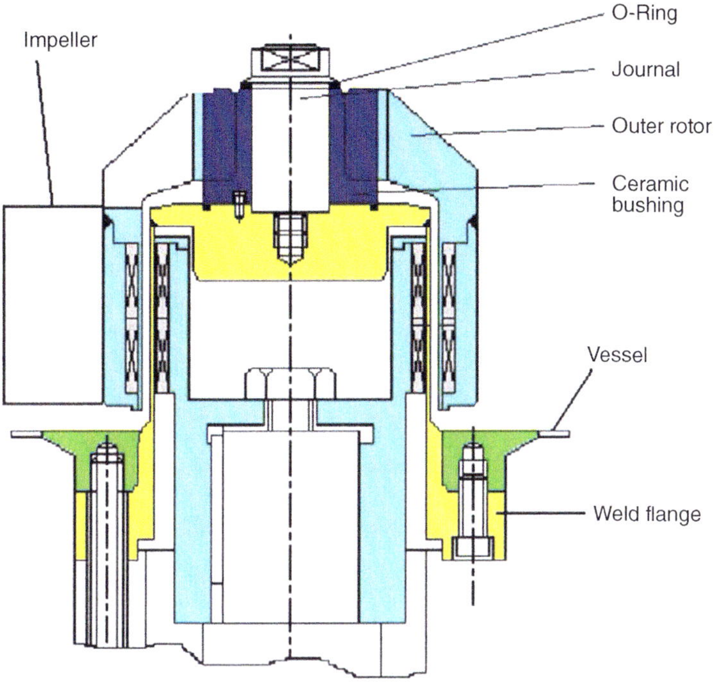

Figure 15.14 Magnetic drive. *Source:* CEP blanket permission.

Teflon or ceramic material) and also from trapped process material remaining between batches. It is quite difficult to make these designs sanitary. The high initial cost and long-term maintenance cost are important considerations for using magnetic drives.

Most magnetic drive systems are bottom entry. It is difficult to get more than one impeller with such systems. Therefore, they are almost never used in fermenters.

Material Considerations

The purity of the process and hence product is always an important consideration. FDA-recognized material grades (specifically elastomers) may be required.

The FDA requirements are covered in Title 21 of the USA Code of Federal Regulations.

The FDA will advise that a substance is:-

- A food additive
- It is generally recognized as safe

- Has prior sanction or approval under amendment
- Is not a food additive under the conditions of use

In general, silicon carbide can be considered as Generally Recognized As Safe (GRAS). Reaction bonded silicon carbide has free silicon, which is dissolved by weak alkalies. For process involving an alkaline cleaning cycle, self-sintered silicon carbide which is resistant to acid and alkalies is to be considered.

With respect to carbons, the FDA considers these materials to be Generally Recognized As Safe. They base this opinion on the ingredients and the fact that only small amounts of wear debris will pass into very large amounts of product and hence would not be a problem. Resin impregnated carbon–graphite grades are considered to be Generally Recognized As Safe. Many pharmaceutical plants do not allow carbon black powder present in the product; a debris well option should be incorporated in the mixer seal for this case.

The term GRAS cannot be used in connection with elastomers. The FDA is very specific and has approved the ingredients which can be used in elastomers as well as specified some extraction limits. They have considered that elastomers may contain substances that can be injurious to public health, and therefore, the ingredients have to be approved. Paragraph 177.2600 covers the requirements for elastomers and gives lists of permitted ingredients together with extraction requirements. FDA-approved nitrile, ethylene propylene, fluoroelastomer, and perfluoroelastomer are commercially available. When seal manufacturers purchase elastomers for food applications, they may require a statement that the elastomers supplied "comply with the requirements of Paragraph 177.2600 of the Code of Federal Regulations, Title 21."

The material for metal parts is the same as the agitator shaft material or may be of higher corrosion resistant material. Some common agitator shaft materials in fermenters are stainless steel, Alloy C-276 or other nickel-rich alloys.

Pharmaceutical operations usually involve a cleaning cycle during start-up or between batches. The compatibility of mechanical seal component materials with the cleaning medium and its temperature should be carefully reviewed.

For contacting dry-running seals and gas lubricated seals, the carbon face material is a special grade carbon conforming to FDA requirements (GRAS). The carbon material has a special impregnation which enhances dry-running performance and provides good lubrication while running dry in a nitrogen environment without emitting a squealing noise (if present, the squealing noise can be attenuated by insertion of a soft acoustic absorption ring between the spring ring and the seal carbon). The carbon material also has negligible generation of particles under steady conditions. The face materials of a gas lubricated seal must be able to run in contact in case of failure in barrier gas supply or a large misalignment of the shaft. Diamond-coated silicon carbide is used if two hard faces are selected. The diamond coating improves the dry-running properties in case of contact for a limited period of time.

Methods of Lubricating Seals

Different types of seals can use different types of lubricants or none at all. Choice of a lubrication method must balance such factors as safety, sterility, lubricity, and seal life, boiling point, pouring point, heat transfer properties, contamination, compatibility with the process and seal material, availability, and cost. The list below is by no means exhaustive, but the general characteristics apply to similar lubrication fluids.

1) Air or inert gas lube, AKA “dry.” This method has the lowest utility cost and has no process compatibility issues. Sterilization requires intermittent exposure to steam or steam condensate, so such seals will effectively run “wet” at times. Dual pressurized gas lubricated seals or contact dry-running seals must maintain the barrier gas pressure during the sterilization process. Dry-running seals in general do not last as long as lubricated seals, with the exception of non-contacting gas lubricated seals, which may have very long life.
2) Steam or steam condensate. This method assures continuous sterilization and rarely has any process compatibility issues. Seal life depends on whether the condensing fluid is sub-cooled enough below saturation to avoid “flashing” between seal faces, which is detrimental to seal life. It is best to use condensate chilled to below 80 °C to avoid boiling between seal faces, so a temperature control system is desirable. Utility cost is higher than with other sealing fluids.
3) Grease. Used only on packing. Enhances seal packing life and quality of seal (lower leakage rate than dry). However, it is difficult to sterilize and there may be process compatibility issues.
4) Good lubricant barrier liquid (mineral oils, glycerin, glycols, or similar). These provide the best seal life for contacting seals and a low seal leakage rate. Usually, a good lubricant barrier liquid can be chosen that has process compatibility. In the case of fermenters, leaking lubricant barrier liquid is often consumed by the broth’s microorganisms. Sterility can be provided by using a steam jacket or coil to heat the lubricant to sterilization temperature. Mineral oil has to be an FDA-accepted grade.

Seal Environmental Control and Seal Support System

After the seal type, materials, and lubricant are selected, the environmental controls will be the next important factor determining the proper operation of and service life of the seal. Many seal failures are due to incorrect or unreliable seal support systems. Many users pay attention to the seal selection itself but neglect the design of the seal support system to meet the process operating conditions which might vary during pre-commissioning, start-up, normal operation, or shutdown.

The amount of heat generated by the mechanical seal may not be high, but unless the heat is properly dissipated, the heat will accumulate to a point of raising the temperature of the liquid in the seal to its boiling point. If this occurs, the seal will not have proper lubrication and the seal may have premature failure. The seal environment control plays a very important role in the reliability of a seal.

API (American Petroleum Institute) piping plans are commonly used to describe or specify the method of environment control for mechanical seals. API piping plans can be found in ANSI/API Standard 682 Third edition entitled "Pumps-Shaft Sealing Systems for Centrifugal and rotary pumps." The standard is also named as "ISO 21049," a standard with international approval.

We will describe only the most common seal control systems used for fermenter agitators. Consult the seal manufacturer for specific expert guidance.

1) **API Plan 32** (Figure 15.15); shown on a pump
   Plan 32 uses using an injection of clean cool liquid (the flush liquid) from an external source to the seal. A labyrinth bushing is installed at the bottom of the seal chamber to control the flushing flow and pressure as a barrier to isolate the agitator process fluid from the seal chamber. Plan 32 is used either for process fluids containing abrasives which might cause high wear to the seal faces or for highly viscous process fluids which impede free movement of seal components. It is important to choose a proper source of liquid to eliminate the potential for vaporization of the injected fluid and to avoid contamination of the process liquid with the injected liquid. User must be aware that product degradation or dilution will occur when using this plan. This plan is most commonly used on bottom entering agitators, as top-entering units are not in contact with the process fluid. Since flush liquid flow, though small, can accumulate over

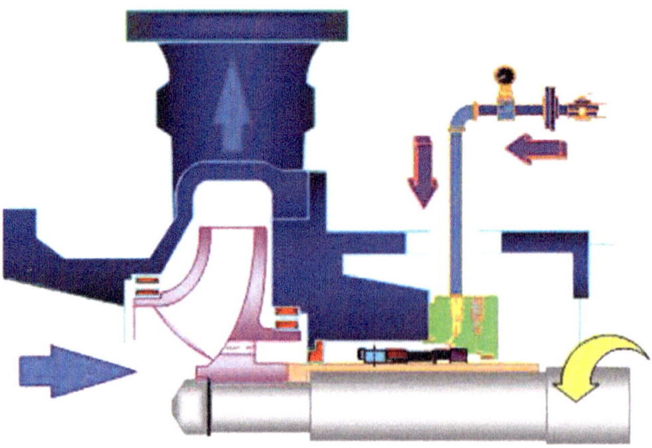

Figure 15.15 API plan 32. *Source:* CEP blanket permission.

time in a process batch, such a method can cause considerable product dilution. As an example, if flush liquid flows at 200 ml/min (not an unreasonable amount for a production fermenter shaft), over a 4-day fermentation cycle, it will add 1152 l of flush liquid to the batch.

2) **API plan 53A** (shown on a pump)
This plan is used with a pressurized dual seal and is the most common way of lubricating double seal cartridge units. During normal operation, since most agitator speeds are low and heat generation is also low, circulation is maintained by natural convection or "thermosiphon" design, though, rarely, a pumping ring is used. The circulation flow of barrier fluid is determined primarily by the temperature difference between the seal chamber and the reservoir. Liquid flow is produced by gravity through the effects of temperature on density. The barrier fluid viscosity and the total piping friction of the circulating loop act as a resistance to this flow. Users must be aware that the flowrates from thermosiphon operation are small and easily stopped by bubbles from vaporization or dissolved gases causing vapor-locking in the system. Large diameter piping connections and maximum drill through on the seal gland connection should be used to maximize flow and to prevent vapor-locking. The cooler or reservoir should be 0.5–1 M above the seal center. The liquid must enter the seal from the bottom and exit out the top of the seal.

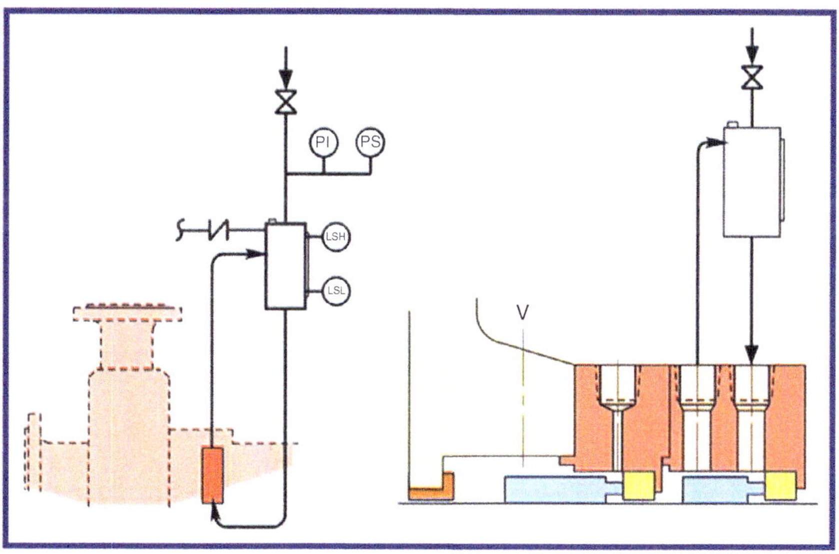

Figure 15.16 API Plan 53A. *Source:* CEP blanket permission.

If a pumping ring is used, the pumping ring design and shaft speed are primary variables in addition to viscosity and piping friction. Low shaft speeds produce little pumping action with a pumping ring. It is rare to see them used on fermenters.

The leakage from the inner seal enters the vessel and the leakage from the outer seal leaks out into the atmosphere.

Plan 53A has a reservoir to provide a barrier fluid for a pressurized dual seal. The barrier fluid is pressurized by a gas, usually nitrogen, at least 10% above the process pressure but a minimum of 1.4–3.5 bars above the maximum process pressure. The reservoir may have an internal coil or a jacket to remove heat (or add it to maintain sterility), or the seal chamber may have a coil or jacket. Sometimes atmospheric cooling of the reservoir is sufficient. If cooling water is used, the cooling water flows through the coils or jacket, and the barrier fluid flows past the external surface of the coils. The outlet connection of the reservoir to the seal should be located at least 5 cm from the bottom of the reservoir. Any wear particles that are denser than the barrier fluid will settle to the bottom of the reservoir. If the loop is contaminated for any reason, the contamination is isolated to a single unit. Due to direct exposure to the pressurizing gas, the barrier fluid in Plan 53A is subjected to gas dissolving in it. Due to this phenomenon, the maximum nitrogen pressure is normally limited to 20 bar-g as long as the barrier fluid temperature is less than 120 °C.

In lieu of such a seal plan, some users provide steam or steam condensate under pressure from a utility source directly to the seal and vent the fluid to a drain while controlling temperature. The fluid must not be allowed to recirculate back to a common reservoir serving several agitators. This has been known to cause cross contamination in a large facility

3) **API plan 74** (Figure 15.17a and b)

This is a new API piping plan for dual gas lubricated seals, dual contacting dry-running seals, and a combination of an outboard dry-running seal with an inboard gas lubricated seal. An external inert gas, usually nitrogen, is used to act as a barrier to prevent process fluid from leaking out into the atmosphere. Pressurized dual wet seals require circulation of a fluid between the seals but gas lubricated seals do not need circulation because the gas lubricated seal is non-contacting and the heat generation is very low. The piping is a dead end arrangement. The flow into the seal cavity is the leakage of the inert gas past both the outboard seal faces and the inboard seal faces. The leakage from the inboard seal into the process is an inert gas which can be separated easily from the process downstream with no contamination issue. This piping plan is a very simple and much lower cost system than wet dual pressurized systems. There is no drainage or clean-up issue, so the maintenance cost is also lower than wet dual pressurized systems that use a pressurized reservoir, accumulator, or heat exchanger, etc. However, the seals themselves are more expensive than those used for wet running operation.

(a)

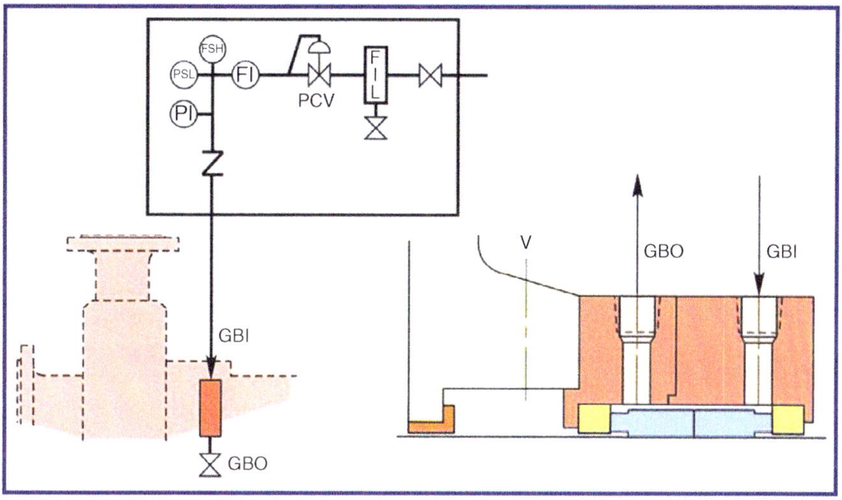

(b)

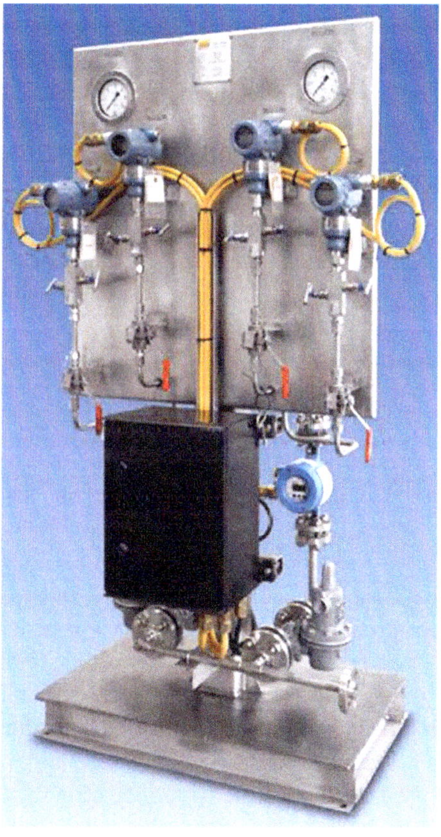

Figure 15.17 (a) API Plan 74. *Source:* CEP blanket permission. (b) API Plan 74 gas panel. *Source:* CEP blanket permission.

Plan 74 is usually organized on a control panel. It has a filter to remove moisture and fine particles. It has a regulator to regulate the pressurized gas to 2–3 bar above the tank pressure. The panel also contains a low pressure alarm to warn against problems with the gas supply to ensure the barrier gas pressure is always above the vessel pressure. A flow meter is used to monitor the seal leakage indicating normal or excessively high leakage. An optional flow alarm is used to give an alarm in the event of excessively high seal leakage.

This seal plan is becoming more common in pharmaceutical plants due to its low risk of contamination and long seal life. Seal materials used for this plan must still be capable of handling sterilization and cleaning media in many cases.

Seal Life Expectations

For the purposes of this chapter, we will divide seal life between replacement into three categories: low (<6 months), good (6 months to 2 years), and excellent (>2 years). These of course are just guidelines. Actual seal life depends not only on seal type and lubricant but also on seal materials, shaft size and speed, operating temperature and pressure, mode of operation, and the condition of the agitator.

Low: lip seals, dry or steam lubed packing, single steam lubed mechanical, double steam lubed mechanical (Note: steam lubed seals can be elevated to "good" if temperature-controlled steam condensate is used).

Good: grease lubed packing, single dry-running contacting mechanical, double dry-running contacting mechanical, sub-cooled steam condensate lubed mechanical.

Excellent: single mechanical with glycol or similar lube, double mechanical with glycol or similar lube, and non-contacting gas seals.

Special Process Considerations

Some process applications have special considerations which dictate seal construction.

For example, tanks containing solutions ready for packaging cannot have any biological contamination, so a sterile seal design (steam or hot lubricant lubed) is required. This is to avoid altering product strength, purity, or activity and is generally required even if the last processing step before packaging is a biostatic filter, which will remove all living organisms down to virus size. Inert particle contamination may or may not be a problem in tanks just prior to packaging, depending on whether filtration is used. If no filter is used, a debris catcher is essential. In such a case, magnetic drives cannot be used, as the in-tank agitator bearing sheds particles.

Table 15.1 Seal comparisons.

Seal type	Lubricant	Containment	Isolation	Sterilizability	Particle or other contamination	Seal life	Cost (1–10 scale)
Lip	Dry	Poor	Poor	Poor	Poor	Low	2
Packing	Dry	Fair	Fair	Fair	Fair	Low	2
Packing	Steam	Fair	Fair	Good	Good	Low	2
Packing	Grease	Fair	Fair	Fair	Fair	Good	2
Single mechanical	Dry-contacting	Good	Good	Fair	Fair	Good	4
Single mechanical	Dry-non-contacting	Good	Good	Fair	Fair	Excellent	4
Single mechanical	Steam	Good	Good	Good	Good	Low	3
Single mechanical	Glycol or similar	Good	Good	Good	Good	Excellent	3
Single split seal	Dry-contacting	Good	Good	Fair	Fair	Good	5
Single split seal	Dry non-contacting	Good	Good	Fair	Fair	Excellent	5
Single split Seal	Steam	Good	Good	Good	Good	Low	4
Single split Seal	Glycol or similar	Good	Good	Good	Good	Excellent	4
Dual Pressurized Seal	Dry-contacting	Good	Good	Good	Fair	Good	6
Dual Pressurized Seal	Steam	Excellent	Excellent	Excellent	Good	Low	9

(*Continued*)

Table 15.1 (Continued)

Seal type	Lubricant	Containment	Isolation	Sterilizability	Particle or other contamination	Seal life	Cost (1–10 scale)
Dual Pressurized Seal	Glycol or similar	Excellent	Excellent	Excellent	Good	Excellent	8
Dual Pressurized Gas lubricated Seal (non-contacting)	Dry; pressurized gas	Excellent	Excellent	Good	Excellent	Excellent	7
Magnetic drive	Excellent	Excellent	Excellent	Excellent	Poor	Excellent	10
Debris well option	N/A	N/A	N/A	N/A	Excellent	N/A	+1

There are some applications where filtration cannot be used prior to packaging, because the product consists of larger particles than the biostatic filters will pass. One case may be very large protein-based molecules, which are larger than some viruses. Another case is macro structures such as liposomes used as drug-delivery systems. These structures are often even larger than bacteria. Thus, neither biological contamination nor particle contamination is permitted. At present, the only seal design meeting this requirement is double mechanical with a hot lubricant and a debris catcher.

The above two examples are geared toward product quality. Another consideration is operator safety. If the tank contains strong pathogens or toxins, it may be mandated by OSHA or other regulations to use a double back-to-back mechanical seal (dry or wet) or a magnetic drive, in order to assure containment. In the former case, some agitator manufacturers also provide a seal shutoff to isolate tank contents during seal change. But, careful operator training is required to be sure the procedures are followed exactly. Incorrect seal change procedures with a pressurized or non-purged vessel may result in exposure to tank vapors during seal change.

Summary: Ratings of Seal Types

The following (Table 15.1) summarizes seal attributes in terms of containment, isolation, sterilizability, particle contamination, seal life, and cost. It may be used as a guide for most applications. Additional guidance is available from seal and agitator manufacturers.

Summary of Chapter

We have described the functions that a shaft seal must perform in fermenter service, an overview of the kinds of seals and support systems most commonly in use, and a rationale for their selection. When in doubt, it is recommended to consult the expert guidance of seal and agitator manufacturers.

Reference

1 Benz, G. and Teng, W. (2008). Selecting agitator seals for bioprocessing applications. *Chemical Engineering Progress* 104: 44–52.

16

Fermenter Agitator Mounting Methods

Introduction

As mentioned in Chapter 15, agitators used on fermenters can be either top or bottom entering. Essentially, they are never side entering, so we will discuss only top and bottom entering methods in this chapter.

Much of the discussion was inspired by Ref. [1], but we expand on the mounting issues here, with an orientation toward the unique issues for fermenters.

There are a number of different ways to mount an agitator, as we shall see. Some are better than others. Many of the mounting methods shown here are driven by the fact that, compared to other applications, fermenters are generally unique in two ways: (i) They have a much larger motor and gear drive than ordinary applications for the size of vessel they are mounted on, and (ii) the vessels are generally designed for low pressure, making them not very mechanically robust unless special vessel design considerations are applied. Many problems have been caused by a lack of vessel stiffness in fermenter service.

We will present here a fairly complete list of the mounting methods, and our opinion about what the pros and cons are.

Top Entering Methods

There are at least six methods used to mount large fermenter agitators on top of a vessel. These include:

1) Direct nozzle mount,
2) Beam gear drive mount with auxiliary packing or lip seal; beams tied into vessel sidewall,

Agitator Design for Gas-Liquid Fermenters and Bioreactors, First Edition. Gregory T. Benz.

3) Beam gear drive mount with auxiliary mechanical seal; beams tied into vessel sidewall,
4) Beam gear drive mount with auxiliary mechanical seal; beams tied into building structure,
5) Complete drive and seal mount to beams tied into vessel sidewall, with bellows connector, and
6) Complete drive and seal mount to beams tied into building structure, with bellows connector.

We will describe the pros and cons of each.

Direct Nozzle Mount

This approach, as illustrated in Figure 16.1, is the simplest method and is also the most common for smaller vessels and agitators. In my opinion, it is also the best way, except when it is simply not possible to do so. An example of where it may not be possible might be when the tank is made of a low-stiffness, nonmetallic material such as polypropylene. If the tank is made of metal (such as stainless steel, the most common class of material used for fermenter vessels), it is almost always possible to use direct nozzle mount. It just means more metal must be applied than needed for tank pressure alone.

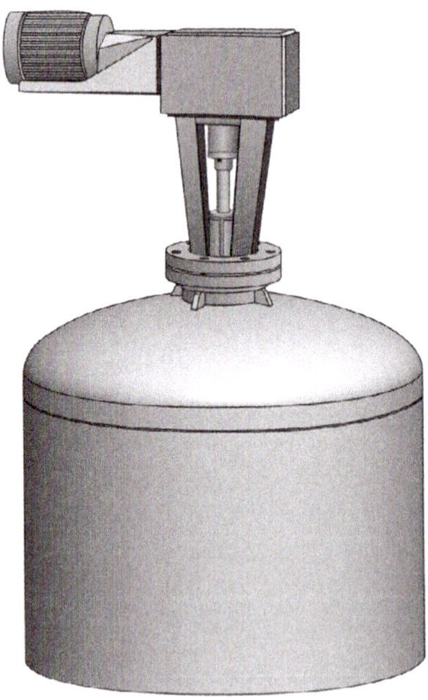

Figure 16.1 Direct nozzle mount unit.

This method of mounting has the advantage of keeping the gear drive and all seal components in line with each other, as well as having little head movement or misalignment of the steady bearing. Its disadvantage is that the vessel head must be designed to take the full weight of the agitator, its torque and bending moment, as well as any impeller thrust. But this is not really difficult or expensive to do. In this author's opinion, direct nozzle mounting is the preferred method.

In the ideal, all vessels with agitators directly mounted to the nozzle would have a Finite Element Analysis (FEA) done to check for local stresses, deflections, and effects on the system natural frequency. An example of such an FEA is presented in Figure 16.2.

However, the cost of running the FEA is often greater than the cost of simply following conservative guidelines for vessel head construction. Typically, the agitator is attached to the vessel either by a reinforced nozzle or a reinforced stud pad, as shown in Figure 16.3.

These can be used in conjunction with Table 16.1 to specify appropriate gussets and thicknesses, and Table 16.2 to specify vessel head and reinforcement pad thicknesses.

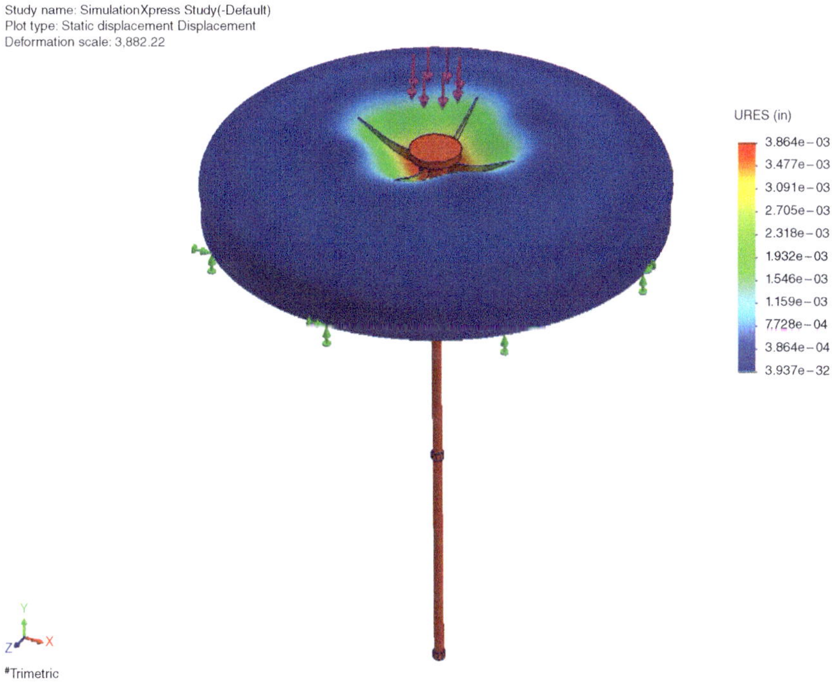

Figure 16.2 Agitator/vessel FEA. *Source:* Permission of Fusion Fluid Equipment.

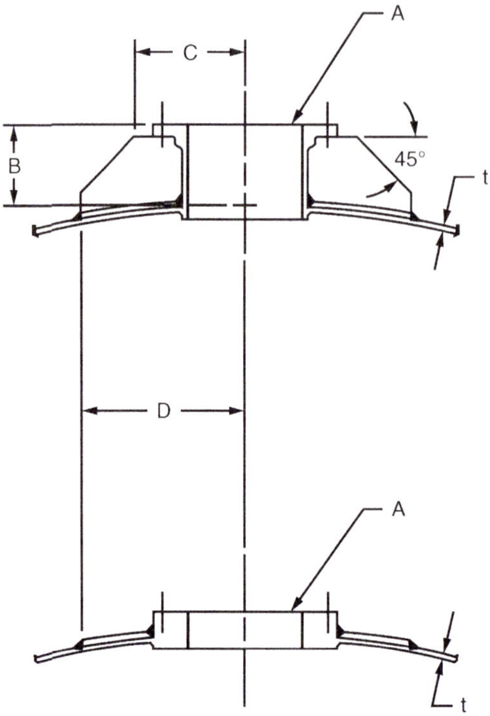

Figure 16.3 Mounting nozzle details.

Table 16.1 Agitator nozzle and pad mount reinforcement dimensions (typical).

Maximum torque, N-m	Nominal ANSI flange size, "A"	Nozzle projection, "B", mm	Radius to gusset taper, "C", mm	Gusset or pad radius, "D"
850	8–10″	150	200	300
3 560	12″	200	300	450
8 480	16″	200	350	500
14 500	20″	300	450	650
38 000	24″	300	500	710
87 000	30″	350	600	800

For head and reinforcing pad thickness, see Table 16.2.

The assumptions behind these tables are conservative for agitators equipped with steady bearings, as they are based on the bending moment that a typical maximum shaft diameter for the gear drive chosen can handle. For smaller vessels, the possible cost savings from using a less conservative design are small. For larger

Table 16.2 Vessel head and reinforcing pad thickness, mm.

Torque, N-m	ANSI nozzle size	Vessel diameter, m								
		1.5	1.8	2	2.5	2.8	3	3.7	4.6	6
850	8–10″	8	9.5	10	11	13	13	13	14	14
1700	12″	6.5	8	8	10	11	11	13	14	16
3400	12″	11	13	14	16	16	19	19	22	22
8500	16″	NR	13	14	14	16	17	19	22	25
15000	20″	NR	NR	11	13	14	16	18	22	25
25000	24″	NR	NR	NR	NR	14	16	19	22	25
34000	24″	NR	NR	NR	NR	14	16	19	22	25
50000	30″	NR	NR	NR	NR	NR	16	19	22	32
90000	30″	NR	NR	NR	NR	NR	NR	25	32	38

vessels and agitators than shown in the tables, there is enough money and risk involved that we suggest a thorough structural analysis be done.

Another issue is motor support. Agitator drives can have a right-angle configuration, with the motor horizontal (Figure 16.4) or a vertical motor (Figure 16.5). When a right-angle configuration is used with a large motor, it is sometimes necessary to add support under the motor, such as using a pipe leg or beam as shown in Figure 16.4. Right-angle designs require less overhead clearance and are easier to mount motors on. Vertical designs leave more room for vessel nozzles on top.

Other methods of reinforcement can be used also. For example, in the Corn Wet Milling industry, a common tank construction for atmospheric vessels is to use a light sheet metal cone top, with about eight substantial reinforcing gussets going all the way out to the vessel sidewall. This is not common for fermenters, as there is usually some back pressure, making a curved head more desirable.

I have also seen cases where the vessel head was reinforced using rib gussets on the inside. While this is structurally sound, it creates a place that is hard to clean.

Beam Gear Drive Mount with Auxiliary Packing or Lip Seal; Beams Tied into Vessel Sidewall

Because packing and lip seals are more tolerant of misalignment than mechanical seals, it is possible to mount them on a flange independently of the gear drive, which can then be mounted on structural steel tied into the sidewalls of the vessel, as illustrated by Figure 16.6. It is not OK to use structural steel tied into the building for this

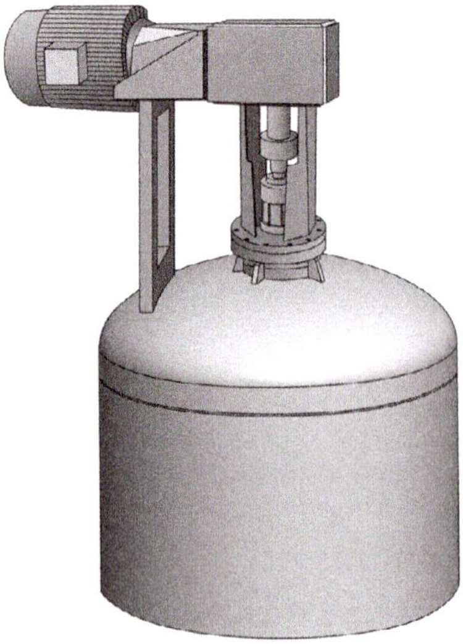

Figure 16.4 Right-angle mount with auxiliary motor support.

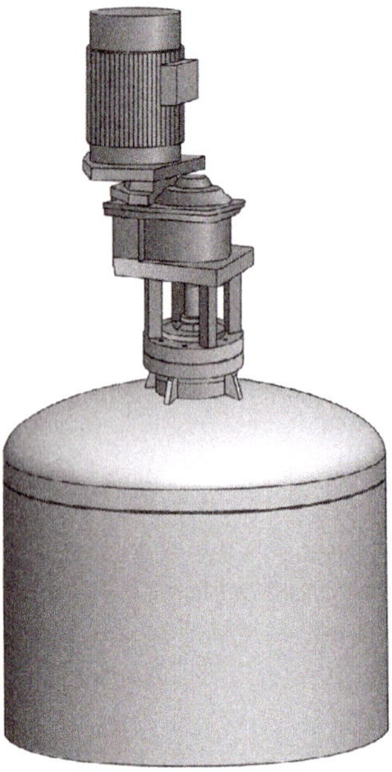

Figure 16.5 Vertical motor mount.

design, as differential movement of the vessel or the building can create misalignment beyond what even packing or a lip seal can tolerate. It is also not OK in general to use this method for mechanical seals, though there may be some exceptions in the case of split single seals designed with generous misalignment capability.

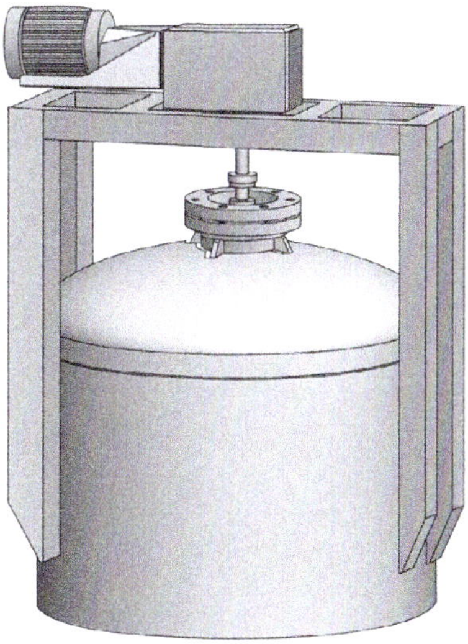

Figure 16.6 Auxiliary stuffing box or lip seal.

This method of mounting removes the agitator loads from the top head and passes them to the sidewall, so a thin top head can be used. There is little or no impact on the required vessel sidewall thickness as long as the beams are welded to a plate that spreads the load out over a reasonable area.

The main disadvantage of this method is that the beam structure takes up space that might be needed for vessel nozzles.

Beam Gear Drive Mount with Auxiliary Mechanical Seal; Beams Tied into Vessel Sidewall

Because the components of mechanical seals, especially with a double seal cartridge, must be precisely aligned for proper operation, it is not generally permissible to mount an auxiliary mechanical seal directly to the vessel nozzle with an independently mounted gear drive rigidly coupled to it. Instead, the auxiliary seal unit must be a complete assembly that includes full radial and axial shaft support bearings as well as the mounting flange and the seal cartridge. This is illustrated in Figure 16.7, with beams supporting the gear drive tied into the vessel sidewall.

Because the seal unit cannot be rigidly coupled to the gear drive without potentially causing a radial or axial binding force, it must be flexibly coupled to the gear drive. This coupling design must be carefully considered. It must transmit a lot of torque, allow for radial, axial, and angular misalignment, and must retain lubricant. When the tank is sterilized, there will be some differential growth between the top head of the vessel and the steel beam structure, so the coupling must allow

Figure 16.7 Auxiliary mechanical seal sidewall beams.

for that. One possible coupling design is a gear coupling, with one member having a sliding engagement. Several centimeters of sliding movement are possible.

Although this method removes the gear drive weight and the operating torque from the vessel head, the bending moment is still transmitted to the seal mounting nozzle, as is the weight of the shaft and impellers and any impeller thrust produced. Moreover, from a design standpoint, the vessel nozzle must be able to take the full agitator torque in case a bearing in the seal assembly ever seizes up. So, the cost savings to the vessel are minimal, the overall agitator cost is much higher, and the equipment reliability will be less due to adding several wearing parts. There is one advantage: since this design isolates the shaft load from the gear drive, it is not necessary to use a gear drive designed specifically for agitator service. A commercial gear drive with ample service factor will suffice, and that is much cheaper than a special purpose agitator drive. This is discussed in more detail in Chapter 17.

The beam structure may get in the way of nozzles on the top head.

Beam Gear Drive Mount with Auxiliary Mechanical Seal; Beams Tied into Building Structure

This is similar to the above, with the only difference being where the beams are supported. It has been used on very large installations (>1000 kW motor size). The choice to mount the gear drive on beams tied into the building means that the beams do not move with the vessel at all. When the vessel is sterilized, it can grow quite a bit. According to the Engineering Toolbox [2], the coefficient of thermal expansion for 316 SS is 1.6E-5 per degree K. That means a 25 m tall fermenter will grow about 0.04 m when its temperature is raised 100 °C above room temperature. Allowing for other misalignment, it may be necessary to have 10 or more centimeters of slide capability in the coupling. While this is theoretically possible in a sliding gear coupling, the author's experience has been that too much grease leaked out. Better results have been obtained using large double universal joint drive shafts, which would be our recommendation for this type of mount. These are considerably more expensive than gear couplings. Other than that, the pros and cons are similar to the above. This construction is illustrated in Figure 16.8.

Complete Drive and Seal Mount to Beams Tied into Vessel Sidewall, with Bellows Connector

This method, illustrated in Figure 16.9, uses an integrated drive and seal assembly supported by beams tied to the vessel sidewall. It essentially eliminates loads from the vessel head, save for the spring stiffness of the bellows. Thus, a thinner vessel head can be used. The main disadvantages are less room for nozzles on the top head due to the beam structure, and the introduction of a space that is hard to clean and sterilize, namely, the bellows area.

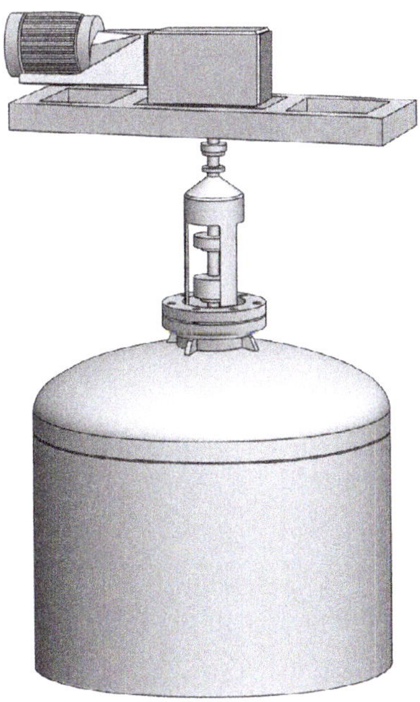

Figure 16.8 Auxiliary mechanical seal beams to building.

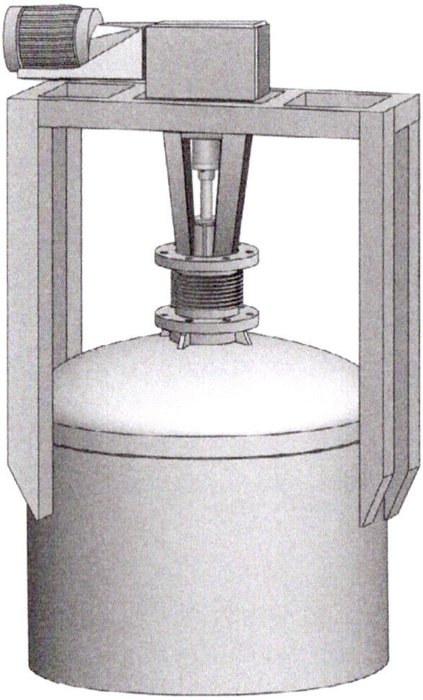

Figure 16.9 Bellows connector, beams to sidewall.

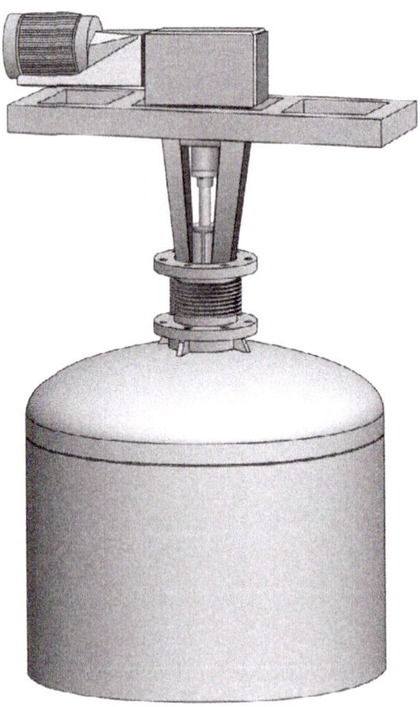

Figure 16.10 Bellows connector, beams to building.

Complete Drive and Seal Mount to Beams Tied into Building Structure, with Bellows Connector

This method is similar to the above, except that the beams are tied into the building, rather than the sidewall. Though this interferes less with available nozzle space on the top head, differential expansion potential and angular misalignment are both greater. A longer, more flexible bellows connector may be needed. All possible misalignment figures must be accounted for when selecting the bellows. Depending on bellows stiffness, there can still be a substantial load transmitted to the vessel nozzle. This construction is illustrated in Figure 16.10.

Bottom Entering Methods

Before delving into the common mounting methods, we will briefly describe why people might choose bottom entering over top entering, in spite of the disadvantages of doing so. The biggest disadvantage is the fact that the seal is immersed in the process fluid, which may not be a good seal lubricant, and could even be abrasive. If abrasive, it is recommended to flush the seal with clean liquid compatible with the process, as the flush fluid becomes part of the broth. Even with a tight throttle bushing, the flow rate of flush liquid can be 0.5–1.0 l/min. This may not sound like much, but if the fermentation lasts for days, which is often the case, thousands of liters of flush liquid may be added to the broth, reducing net fermenter capacity and increasing product recovery costs.

Another disadvantage is that if the vessel is tall enough to require a steady bearing, it will be at the top of the shaft. If the fermentation is fed batch, the liquid level at the start may not cover the steady bearing, and therefore, steady bearing flush liquid must be added to keep the bearing lubricated, further diluting the batch.

A third disadvantage is that most plants will require the tank to be empty to change the shaft seal on a bottom entry unit, which is not normally a requirement on top entry units unless the tank contents are toxic or pathogenic. Though most bottom entry units are equipped with a seal shutoff device that theoretically may allow seal change with a full tank, abrasive materials or even nonabrasive solids may prevent the device from functioning properly, and many plant managers choose not to trust such devices from a safety viewpoint.

So why do people choose bottom entry units? The most common reason seems to be to free up space on the top head for pipe nozzles. This tends to be more of an issue on small diameter fermenters (say, <2 m) than on large ones. However, extending the agitator nozzle on top entering units and using a smaller pipe size for that nozzle can often add space for other nozzles, which can be piped in below the footprint of the agitator assembly. This method does create an additional area that is hard to clean and sterilize.

Another reason for using a bottom entering unit is that the total shaft length is usually less, saving money on the shaft cost, and more importantly, it can sometimes eliminate the need for an in-tank coupling. In Chapter 18, we will see that in-tank couplings can be a source of contamination.

A somewhat unique reason I have seen for the use of bottom entry units is noise control. One of my clients built a facility in an area where plant noise was highly regulated. The fermenter was tall enough so that the vessel would extend above the facility walls, meaning if an agitator drive were placed on top of the vessel, the motor and gear drive noise would radiate into the surrounding community. By making the unit bottom entry, the noise would be mostly confined to within the building, and that which escaped would go upward and not have as direct a path into the neighborhood.

With the above reasons in mind, there are at least four relatively common ways of mounting a fermenter agitator below the tank:

1) Direct nozzle mount,
2) Floor gear drive mount with auxiliary packing or lip seal,
3) Floor gear drive mount with auxiliary mechanical seal, and
4) Floor integrated drive and seal mount with bellows connector.

In principle, there could also be bottom entering analogs of Figures 16.6–16.8. However, I have never seen this done, so I will not illustrate these options here. Detailed comments follow.

Direct Nozzle Mount

Just as is the case for top entering units, direct nozzle mounting (Figure 16.11) assures alignment of all seal components and is the simplest method to execute. It does transfer all agitator loads to the vessel nozzle, so reinforcement is needed. The same considerations apply as when designing top entering direct nozzle mount units, and the same figures and tables can be used. In this author's opinion, this is the preferred mounting method for bottom entry units. Caution: if the shaft seal is packing, tank contents will slowly leak down the shaft. This can result in damage to the gear drive by leaking past the gear drive top oil seal. This danger can be ameliorated by using a slinger and catch basin with drain within the pedestal, above the gear drive.

Floor Gear Drive Mount with Auxiliary Packing or Lip Seal

Similar to a beam mounted gear drive on a top entering unit, in terms of alignment of the shaft and seal with the vessel. It would be unusual to mount a gear drive on beams for a bottom entry unit, unless used as a common baseplate for the motor and

Figure 16.11 Direct nozzle mount bottom entering unit.

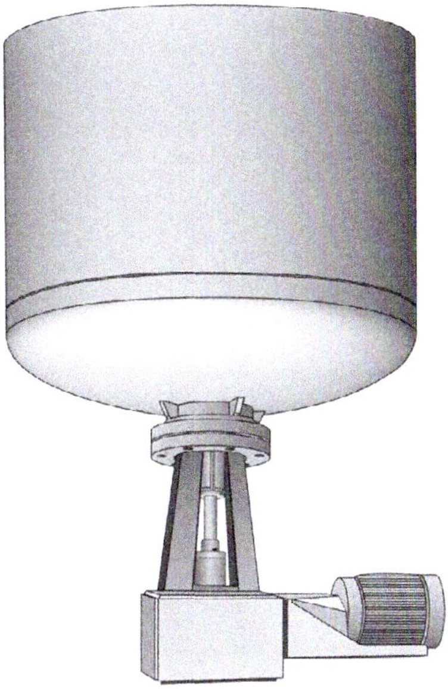

gear drive. In that case, the beams would be attached to the floor, so it would have the same effect as mounting the gear drive directly to the floor. Since the vessel can move somewhat relative to the floor, this method is least risky when the vessel is attached to the floor by a skirt, or, if supported by structural steel to the building, the support lugs should be close to the bottom of the vessel. If the vessel is "hung" from supports near the top of the straight side, more thermal expansion must be accounted for, and there is also a greater risk of angular misalignment and lateral movement. The above comments about seal leakage onto the gearbox apply here also.

This mounting method is illustrated in Figure 16.12.

Floor Gear Drive Mount with Auxiliary Mechanical Seal

As with top entering units, the precise alignment required in a mechanical seal precludes simply mounting the seal on the vessel and rigidly coupling it to the gear drive shaft. Instead, the seal must have its own shaft support system built-in. It can then be coupled to the gear drive with a flexible coupling, as shown in Figure 16.13. It is strongly suggested to support the vessel with a skirt or with lugs near the bottom of the straight side, so as to minimize potential misalignment. In such case, a single gear coupling may suffice. The amount of axial sliding movement needed will be based on thermal expansion plus any anticipated construction tolerances. The gear drive placement should be adjusted for minimum misalignment after the seal assembly is in place.

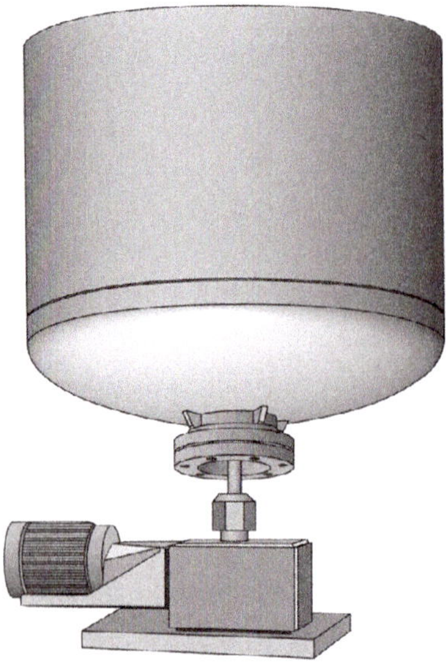

Figure 16.12 Floor mount with auxiliary packing.

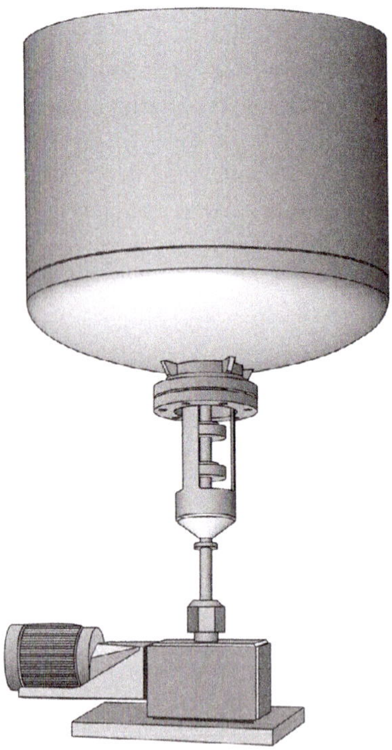

Figure 16.13 Floor mount with auxiliary mechanical seal.

If the vessel is hung from near the top of the straight side, it may be necessary to use a double gear coupling arrangement, or even a double universal joint drive shaft, at considerable extra expense. That arrangement also requires more clearance under the vessel, making the whole building taller.

The seal is changed within the pedestal assembly that contains the seal, support bearing, and any necessary shaft spacer couplings. There are some designs in use that require draining the vessel and removing the gear drive to change the seal. We do not favor this approach, and have not illustrated it here.

This bottom entry mounting method does have the advantage of not requiring an "agitator drive," as none of the weight or bending moment produced by the action of mixing is transmitted to the drive bearings. That partially offsets the cost of the additional mechanical components such as bearings in the seal pedestal and gear or universal joint couplings. However, with many more wearing parts, this design will have higher overall maintenance costs than a direct nozzle mount unit, or the next option.

Floor Integrated Drive and Seal Mount with Bellows Connector

This is similar to the top entering bellows connector design. It is illustrated in Figure 16.14. It mostly isolates the agitator loads from the vessel head, potentially allowing the use of thinner material for the vessel head. However, it adds a space

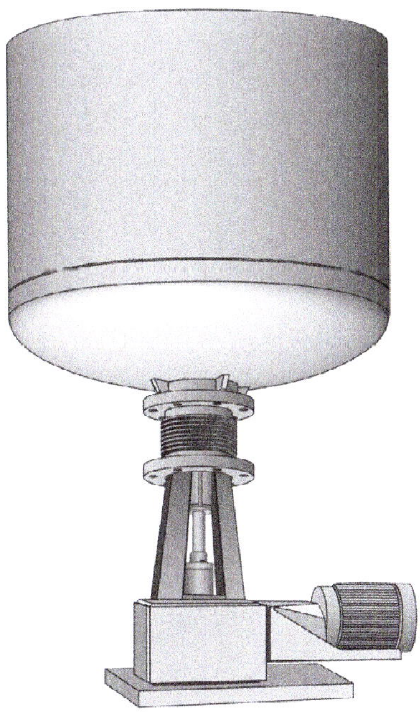

Figure 16.14 Floor mount with bellows connector.

that is difficult to clean and sterilize. The selection of the bellows must account for expansion and misalignment, so it is still preferred to mount the vessel with a skirt or lugs close to its bottom.

If the seal is packing, the same comments about tank contents leaking down the shaft apply here as well.

Summary of Chapter

Mounting large agitators on low-pressure vessels can be challenging. While this author believes that in most cases, direct nozzle mounting is the best way to do it, there are a number of other ways to do it that are structurally sound and have a substantial experience base. We have illustrated the most common ways to do it here and offered commentary on their pros and cons.

References

1 Hill, R. and Kime, D. (1976). How to specify drive trains for turbine agitators. *Chemical Engineering Magazine*: 46–51.

2 Engineering Toolbox Website. *Coefficients of Linear Thermal Expansion*. https://www.engineeringtoolbox.com/linear-expansion-coefficients-d_95.html. (accessed 11 October 2020).

17

Mechanical Design of Fermenter Agitators

Introduction

For any process to work profitably, the equipment used in it must be reliable. This is certainly true of fermenter agitators. In general, agitators are reliable pieces of equipment. In normal service, such as blending of miscible liquids with a mild-to-medium degree of agitation, the products made by the leading agitator suppliers have a useful service life of more than 20 years and have modest maintenance requirements, consisting largely of changing the oil, greasing the bearings, and changing the seals when needed. Alas, fermenter agitators have not been quite so reliable. Historically, based on the author's knowledge and experience, they have had more mechanical problems than any other class of agitation equipment. Some of this may be due to the fact that the combination of agitator power and gas expansion power imparted to the vessel results in much higher power/volume than most other agitator applications, resulting in highly fluctuating loads. Some of it may be because the low-pressure vessels fermenter agitators are mounted on are insufficiently stiff to adequately support the agitator. Some of it may be due to difficulty quantifying the actual loads that must be designed for. Some of it may be due to the size of the equipment ("Large" agitators tend to have a larger vibration velocity and a greater variation in loads than "small" agitators.) And, frankly, some of it may be due to competitive pressure to offer an economical design that at least will make it past the warranty period. That is why the buyer should be cognizant of agitator mechanical design and be actively involved in the decision process regarding mechanical integrity and safety factors.

Regardless of the reasons, the mean time between failures of many large fermenters has typically been in the range of 5–10 years rather than 20+. This need not be the case. We can do better, but it costs more. I believe it is money well spent if it improves reliability.

Agitator Design for Gas-Liquid Fermenters and Bioreactors, First Edition. Gregory T. Benz.

The intent of this chapter is not to teach the user how to do detailed mechanical design of agitators. That is still best left up to the equipment vendors, or to consultants with very specialized knowledge in this field. (That does not include me. My expertise is primarily in the chemical engineering aspects of agitator design, though I have some basic mechanical engineering knowledge about the subject. But I leave the tough stuff and detailed design to the mechanical engineering experts.) Rather, the intent of this chapter is to familiarize the reader with the basic aspects of agitator design so that he or she can understand the issues involved, can understand what the equipment vendor needs to know from the user, and have the ability to participate in some decisions about cost versus reliability. This will become especially evident when we discuss gear drives.

Some of the inspiration for this chapter came from Ref. [1]. Other information came from a variety of sources, including internal communications from Mr. Bill Fryers, who is one of those mechanical engineering experts to whom I am referring who has a great deal of agitator design experience. General mechanical engineering handbooks also supplied information, but will not be referenced by line item in the text because such information is commonplace.

Impeller Design Philosophy

Normally, agitator manufacturers will design each impeller for its share of the total motor power. For many applications, this is a reasonable approach, as the motor will shut down if the total agitator power exceeds full motor load for more than a few seconds. However, if there are multiple impellers and the agitator is operated when the tank is only partially filled, such that all impellers are not fully submerged in the liquid, this can be a dangerous practice if something changes the fluid properties so as to increase the impeller power draw above normal.

One example would be if the viscosity increased. This can happen in a polymer reactor but is unlikely in a fermenter. (Viscosity increases with time in a gum fermentation, but usually the peak viscosity happens with a full vessel, with all impellers submerged).

The more likely scenario that could cause such a thing in a fermenter would be if the partially filled vessel is accompanied by a reduced or even zero gas flow. That will cause the impeller power draw to increase. In the case of the older Rushton impellers, it is possible for the ungassed power to be 2–3 times as high as the gassed power. For example, suppose a fermenter were equipped with three Rushton impellers, each designed mechanically for 1/3 of the motor power when fully gassed, and the operators fill the vessel about 40–50% full such that only the lower impeller is covered. Then they start the unit at full speed with no gas flow. The power draw of the lower impeller in that case could be 60–90% of motor power. This would not overload the motor, but if the impeller were not designed for the ungassed power, it could very well fail. If it failed asymmetrically, that could also lead to a shaft failure.

Modern deeply concave impellers do not have as great a difference between gassed and ungassed power as Rushtons, so this is less of a problem and less of an expense to design for. Axial impellers have intermediate gassing factors.

The recommended design practice, in this author's opinion, is to design each impeller for ungassed power, starting with the bottom impeller, until the total ungassed power equals or exceeds motor power. If there are any remaining impellers after full motor power is reached, they can be designed for their share of the total *gassed* power. The user of fermenter agitators is well advised to have a dialog about this with each agitator bidder to assure that this is the design basis used. Agitator manufacturers do not warranty against improper operation of the equipment, and running a partially full tank ungassed would be considered improper operation unless explicitly stated in the design basis and warranty.

Impeller mechanical design is more complex than shaft design. Moreover, there are many kinds of impellers on the market, and many of those are proprietary. Even among well-known impellers such as the Rushton turbine, there are many variations of number of blades, blade height, and method of fastening to the disc and hub, making mechanical design complicated. Therefore, we will not get into detail on mechanical design of impellers. We will have to leave that to the equipment vendors.

Discussion on Hydraulic Force

As we will see later, the two major loads determining shaft design for stress are torque and bending moment. The origin of torque is obvious: it is the result of power being transmitted at a certain shaft speed. But where does the bending moment come from? It is the result of random unbalanced hydraulic forces acting on the impeller(s), and the moment arms those forces act upon. But where do the hydraulic forces come from? Over the years, I have often been asked why the manufacturers don't just make the impellers highly precise and fully balanced so as to avoid such forces. And, indeed, there are forces associated with mechanical imbalance and lack of perfect symmetry in impeller construction. But these are minor forces compared to the *hydraulic* imbalanced forces produced. And those forces do not originate from any imperfection in impeller construction, but from the nature of turbulence.

To understand this better, imagine riding along in a car at 100 Km/h, with your hand sticking out of the window. You will note there is a drag force on your hand. However, it is far from constant. In fact, your hand will be buffeted about as the forces fluctuate wildly. It is the same for agitator blades in a tank.

Each blade, just as your hand does, experiences varying loads. And, those variations are not synchronized or symmetric among each of the blades. Therefore, there is also a fluctuating net side load on the impeller. This load is highest under highly turbulent conditions and theoretically falls to zero in laminar flow. Having more blades per impeller reduces the peak magnitude of this load. However, there are practical limits to the number of blades that can be used, including expense and the fact that, for producing flow, fewer blades are better for a given power input.

The magnitude of this load can be empirically correlated by means of a concept called dimensionless hydraulic force:

$$F_h = F_D * T_Q / D \tag{17.1}$$

Where:

$$T_Q = P / 2\pi N \tag{17.2}$$

The value of F_D depends on the number of blades, the agitation intensity, the mounting arrangement (it is higher for off-center mount, for example), the flow direction and whether it is single phase or gas-liquid. High solidity impellers tend to have higher hydraulic forces when gassed than low solidity impellers. For our purposes, we will assume that all fermenter agitators are mounted in the center of a cylindrical vessel, fully baffled. I am aware of an installation that had horizontal vessels with two agitators per vessel. Based on the many problems it had, I do not recommend it.

Table 17.1 summarizes dimensionless hydraulic factors for typical fermenter impellers and should be adequate for high solidity impellers also. The values have not been previously published to my knowledge and come from several sources and my own experience. They are intended for design purposes and may be slightly conservative. However, irregular geometries could have much higher forces.

For example, let's suppose that we have a 1m diameter radial flow impeller with six blades operating at 125 rpm (2.083/s), and it draws 20 kW. The resultant torque is (20 000 W/(2.083/s * 2 * π) = 1530 N-m. (1W = 1 N-m/s). From Table 17.1, our dimensionless hydraulic force design value is 0.6, so our hydraulic force is 0.6 * 1530 N-m/1m = 918 N. For those who think in English units, this equates to 206 pounds force. Some would use 93.8 Kg-f, but that is not an SI unit.

High hydraulic forces also increase the stress on individual blades. Some vendors do not take this into account. While we do not delve into detailed impeller calculations here, it may be useful to ask vendors to show their blade stress calculations and show how they take hydraulic force into account.

Table 17.1 Dimensionless hydraulic forces.

Impeller number of blades	Direction of pumping	Problem class	Dimensionless hydraulic force
1	Does it matter?	Any	2.0
2	Any	Mild blending	0.6
2	Any	Vigorous blending	0.75
2	Radial	Blending	0.6
2	Radial	Gas/liquid	0.75
2	Down	Gas/liquid	0.9
2	Up	Gas/liquid	0.75
3	Any	Mild blending	0.45
3	Any	Vigorous blending	0.5
3	Radial	Blending	0.45
3	Radial	Gas/liquid	0.7
3	Down	Gas/liquid	0.9
3	Up	Gas/liquid	0.65
4	Any	Mild blending	0.3
4	Any	Vigorous blending	0.4
4	Radial	Blending	0.3
4	Radial	Gas/liquid	0.7
4	Down	Gas/liquid	0.9
4	Up	Gas/liquid	0.6
5	Radial	Gas/liquid	0.75
6	Radial	Gas/liquid	0.6

Note: the above applies to center-mounted agitators in cylindrical vessels only.

Shaft Design Philosophy

Similar to the above philosophy, the shaft should be designed for 100% of motor power and peak bending moment, at least at the upper part of the shaft. However, most large-scale fermenters have long enough shafts that they must be made in several sections. With multiple impellers, it is often the case that some impellers are mounted above some shaft sections. In such a case, it is permissible to design the lower shaft sections for the maximum power they could possibly see, which is generally the sum of the ungassed power of all impellers below the coupling at the top of the respective shaft section, up to a maximum of motor nameplate power. For example, if a shaft is built in three sections and has four impellers, but the bottom section has only two impellers mounted on it, that section need only be designed for the sum of the ungassed power of those two impellers.

For bottom entry units, the power on each section is summed from the top impeller down. However, it is unusual for bottom entering units to have long shafts and more than two impellers, as that may require an upper steady bearing, which needs to be kept submerged in liquid, reducing operational flexibility.

Shaft Design Based on Stress

Most mechanical system designs for stress start out with a load diagram. For an agitator shaft, the two principal kinds of loads are torque and bending moment. The torque arises from the power draw of the impellers, delivered at a certain shaft speed. The maximum torque for design is generally based on 100% motor nameplate power, as in Eq. (17.2).

The bending moment arises from the hydraulic forces described above, applied to a moment arm. For a cantilevered shaft (no steady bearing), the bending moment is the sum of the hydraulic force of each impeller times its distance from the lower gear drive bearing, or the mechanical seal bearing if there is one. (This is slightly conservative, as the hydraulic forces of each impeller are not necessarily aligned at any given moment. However, they can be on occasion, so we design as though they were.) For a unit with a steady bearing (which is most commonly the case for fermenters), the analysis is more complicated, as there are at least two bearing points: one at each end of the shaft. Though computer programs are the best way to design for such a case, a simplified way of doing it manually is feasible and conservative. Refer to Figures 17.1 and 17.2 for the dimensional variables used in the below equations.

For both design concepts, we would normally design for torque calculated based on motor power, using Eq. (17.2). However, for multiple shaft sections, we may not have all impellers contributing power, so we could look at that section summing the torque of impellers that actually contribute torque to that section, and use ungassed power for that calculation unless it is more than 100% motor power. For the bending moment, we use Eq. (17.1) to calculate the hydraulic force for each impeller, then calculate bending moment as follows.

For a cantilevered unit, bending moment is the sum of each impeller's contribution, or:

$$M = \sum F_h * L = F_{h1} * L_1 + F_{h2} * L_2 \ldots \tag{17.3}$$

For steady bearing units, each impeller contributes a bending moment in proportion to its distance from each bearing, that is:

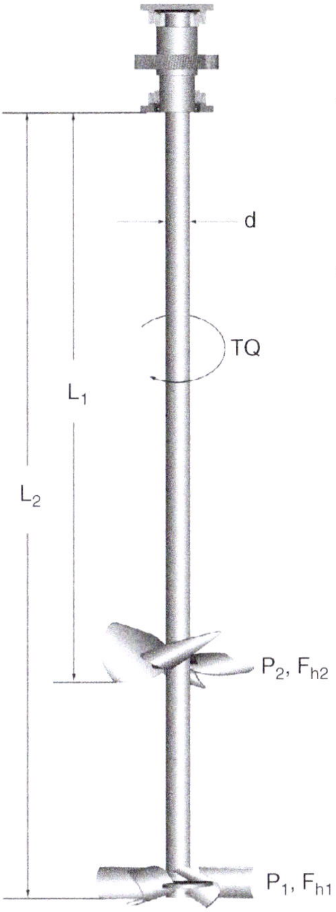

Figure 17.1 Cantilevered load diagram.

$$M_i = F_{hi} * B_i * C_i / L \tag{17.4}$$

Though the bending moments are not totally additive and occur at different places along the shaft, we can treat them as though they were by summing up the above equation for all impellers:

$$M = \sum F_{hi} * B_i * C_i / L = F_{h1} * B_1 * C_1 / L + F_{h2} * B_2 * C_2 / L\ldots \tag{17.5}$$

Once we have the torque and bending moment, we can calculate shaft tensile and shear stresses. First, we calculate the combined moment resulting from torque and bending moment:

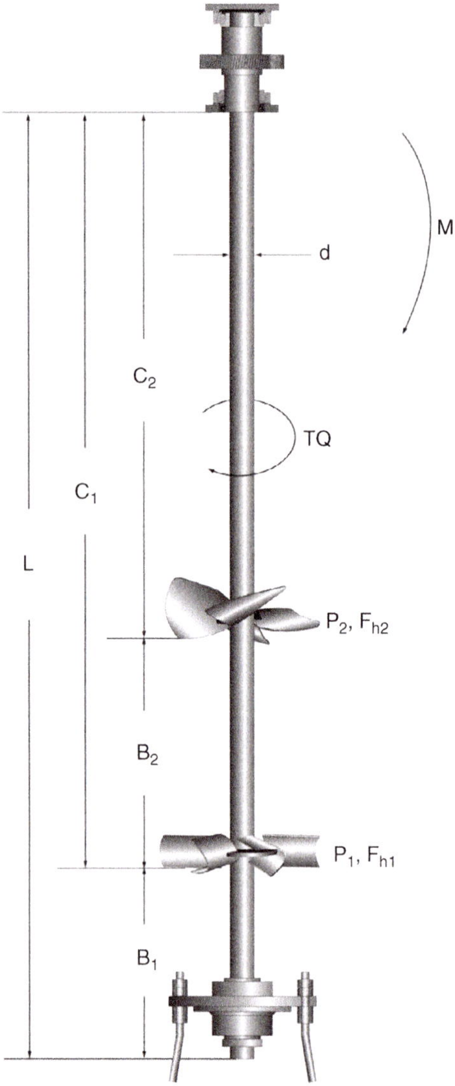

Figure 17.2 Steady bearing load diagram.

$$M_C = \left(M^2 + T_Q{}^2\right)^{0.5} \tag{17.6}$$

Then we can calculate shear stress as follows:

$$\sigma_s = M_C * 16 / (\pi d^3) \tag{17.7}$$

Tensile stress is similar, but adds the bending moment to the combined moment:

$$\sigma_t = \left(M + M_C\right) * 16 / (\pi d^3) \tag{17.8}$$

In case of hollow shafts, which will be discussed later, instead of d^3, use $(d_0^3 - d_i^3)$. Though other cross sections are possible, such as square shafts, I have never personally seen them used in a fermenter so they will not be discussed here.

To choose or evaluate shaft size using these equations, it is necessary to know what the allowable shear and tensile stresses are. These vary with material of construction and temperature. To allow for stress concentration and safety factors, it is customary to use yield stress/4 as a maximum tensile stress and about a five divider for shear stress. Table 17.2 lists typical values for some alloys used in fermenters. Since I have yet to see a fermenter with a carbon steel shaft, I have not included it. Fermenters typically operate at temperatures below 40 °C. However, I have used 125 °C as a design basis as some people operate fermenter agitators while cleaning with hot caustic. Though about 99% of fermenters use 304 or 316SS, or the low carbon variants, I have included some higher alloys for future reference, and because if the tank is ever cleaned with hot diaphragm grade caustic, there may be chloride ion concentrations in the tank of several thousand parts per million, which could lead to stress corrosion cracking of stainless steel, especially for the fasteners. Consult a corrosion specialist for detailed evaluation and material selection.

Table 17.2 Allowable stresses.

	Allowable shear stress		Allowable tensile stress	
Material	**psi**	**Pa/1000 or kPa**	**psi**	**Pa/1000 or kPa**
304SS	5 500	37 923	9 150	63 089
316SS	5 935	40 922	9 890	68 192
304LSS	5 081	35 033	8 470	58 401
316LSS	5 240	36 130	8 730	60 193
316LSS dual certified	5 935	40 922	9 890	68 192
Alloy 20	6 340	43 714	10 500	72 398
Alloy C-276	7 960	54 884	13 300	91 704
Alloy C-22	7 900	54 471	13 200	91 014
AL-6XN	7 950	54 815	13 250	91 359
Titanium Grade 2	3 150	21 719	5 250	36 199

The above based on a process temperature of 125 °C.

A very commonly used material these days is dual certified 316/316L stainless. It has the low carbon feature of an L-grade material but maintains the strength of the standard 316SS. It is included in the table.

Simple Example Problem

Suppose we have an agitator assembly consisting of a 261 kW motor, gear reduced to a shaft speed of 100 rpm. It has three impellers. The lowest impeller is a 1.6 m diameter deep concave disc turbine. The upper two impellers are 1.7 m diameter up-pumping high solidity axial flow type. Though they are sized to draw a total of about 85% of motor nameplate in the gassed condition, we will design for 100% of the motor power, prorated for each impeller. On that basis, we will design for 140 kW for the bottom impeller, and 60.5 kW for each of the upper impellers. The shaft extension from the lower gearbox bearing is 13.6 m. The impeller locations are 0, 3.7, and 7.4 m from the bottom end of the shaft, which is cantilevered. The proposed shaft diameter is 0.178 m, solid. The shaft material is 316SS. Our problem is to determine if this is strong enough.

First, we look at the total torque, based on motor power, using Eq. (17.2):

$$T_Q = P/2\pi N = 261\,\text{kW} * 1000\,\text{W/kW}/\left(2\pi * 100\,\text{rpm}/60\,\text{s/min}\right)$$
$$= 24924\,\text{N-m}$$

Next, to determine hydraulic force, we first must calculate the torque of each impeller, using the same equation but with the prorated power:

$$\text{Impeller 1 } T_Q = 13369\,\text{N-m}$$

$$\text{Impellers 2 and 3 } T_Q = 5777\,\text{N-m}$$

Now we can calculate the hydraulic force of each impeller, using Eq. (17.1) and dimensionless hydraulic force values from Table 17.1. For both types of impellers used here, Table 17.2 suggests a dimensionless hydraulic force of 0.6. Therefore, for the lower impeller:

$$F_h = F_D * T_Q / D = 0.6 * 13369\,\text{N-m} / 1.6\,\text{m} = 5013\,\text{N}$$

For each upper impeller:

$$F_h = F_D * T_Q / D = 0.6 * 5777\,\text{N-m} / 1.7\,\text{m} = 2039\,\text{N}$$

Now we can calculate the bending moment at the bearing at the top of the shaft. We must first calculate the impeller moment arm from that bearing, which equals the shaft length minus the impeller spacing from the bottom end of the shaft. Thus, for impeller one we have:

$$L_1 = 13.6\,\text{m} - 0 = 13.6\,\text{m}$$

$$\text{Similarly, } L_2 = 13.6 - 3.7 = 9.9\,\text{m and } L_3 = 13.6 - 7.4 = 6.2\,\text{m}$$

Now we can calculate the total bending moment using Eq. (17.3):

$$M = \sum F_{hi} * L_i = 5013\,\text{N} * 13.6\,\text{m} + 2039\,\text{N} * 9.9\,\text{m} + 2039\,\text{N} * 6.2\,\text{m}$$
$$= 101005\,\text{N-m}$$

Now we have the design torque and bending moment. To determine shaft shear stress and tensile stress, we calculate the combined bending moment using Eq. (17.6):

$$M_C = \left(M^2 + T_Q{}^2\right)^{0.5} = \left(101005^2 + 24924^2\right)^{0.5} = 104034\,\text{N-m}$$

We can calculate the shear stress from Eq. (17.7):

$$\sigma_s = M_C * 16 / (\pi d^3) = 104034\,\text{N} - \text{m} * 16 / \left(\pi * \left(0.178\,\text{m}\right)^3\right)$$
$$= 94\text{E}6\,\text{N}/\text{m}^2 = 94000\,\text{kPa}.$$

Tensile stress can be calculated by using Eq. (17.8):

$$\sigma_t = \left(M + M_C\right) * 16 / (\pi d^3) = \left(101005 + 104034\right) * 16 / \left(\pi * 0.178^3\right)$$
$$= 185\text{E}6\,\text{N/m}^2 = 185000\,\text{kPa}.$$

From Table 17.2, the maximum allowable shear stress for 316SS is 40922 kPa, and the maximum allowable tensile stress is 68192. So, our 0.178 m proposed shaft diameter is insufficient. The reader is encouraged to calculate the minimum shaft diameter required, which is almost 0.25 m. This would not be a practical design for stress, and it probably would not work for critical speed either. A steady bearing would most likely be required.

Sample Problem with Steady Bearing

Using the same problem as above, we will look at adding a steady bearing. Let us assume that we add it 0.5 m below the lower impeller. This changes L from 13.6 to 14.1 m. Noting that for each impeller, B + C = L, we have C_1 = 0.5 m, B_1 = (14.1−0.5) = 13.6 m; C_2 = (3.7+0.5) = 4.2 m, B_2 = (14.1−4.2) = 9.9 m; C_3 = (7.4+0.5) = 7.9 m and B_3 = (14.1−7.9) = 6.2 m.

Our torque remains the same, at 24924 N-m. The bending moment is obtained by summing the hydraulic forces times BC/L:

$$M = 5013N * (13.6 * 0.5)/14.1 + 2039N * (9.9 * 4.2)/ 14.1 + 2039N * (6.2 * 7.9)/14.1 = 15513\,N\text{-}m$$

Our combined moment, from Eq. (17.6), is:

$$M_C = \left(24924^2 + 15513^2\right)^{0.5} = 29357\,N\text{-}m$$

Our shear stress, using Eq. (17.7), is:

$$\sigma_s = M_C * 16/(\pi d^3) = 29357 * 16/\pi(0.178)^3 = 2.65E7\,Pa = 26511\,kPa$$

Our tensile stress, using Eq. (17.8), is:

$$\sigma_t = \left(M + M_C\right) * 16/(\pi d^3) = \left(15513 + 29357\right) * 16/\pi(0.178)^3 = 40519\,kPa.$$

As before, from Table 17.2, the maximum allowable shear stress for 316SS is 40922 kPa, and the maximum allowable tensile stress is 68192. So, our 0.178 m proposed shaft diameter is sufficient from a stress standpoint.

The next section discusses designing with steady bearings as part of a discussion on critical speed. We will illustrate a steady bearing design with this same example problem.

Shaft Design Based On Critical Speed

All physical systems have finite stiffness, inertia, and damping, making them possible to model as a combination of springs, masses, and inertial dampers. For simplicity, most models of agitator shaft systems assume zero damping, so the resultant deflections are "worst case." The primary mode of agitator shaft deflection is laterally, i.e., normal to the shaft axis. The resistance to deflection is the stiffness of the shaft and the resistance of the bearings supporting the shaft. The main inertial masses are the impellers, which are treated as concentrated loads, the in-tank couplings, and the distributed mass of the shaft itself. The shaft and impeller combined system has a series of natural frequencies at which they will tend to resonate if an oscillatory load is applied. The lowest such frequency is called the first lateral natural frequency, or, more commonly (and arguably, incorrectly), the shaft critical speed. However, the use of the term "critical speed" is so widespread and accepted that we will use it here for brevity.

There are many natural frequencies above the critical speed. However, they would not normally be a concern unless the operating speed of the agitator is above critical speed, which is often the case for small propeller agitators but almost never the case for production fermenters. So, we will concern ourselves only with the first lateral natural frequency, or critical speed.

There are also torsional natural frequencies. These tend to be much higher than the lateral natural frequencies, and I have yet to see a single case where they affected shaft design. However, they can affect phenomena within the gear drive. For example, if the torsional natural frequency matches a gear tooth meshing frequency, it can cause chatter and destroy the gears. I have actually know of cases where this happened. We will come back to this when discussing gear drive selection.

The key to avoiding critical speed problems is to avoid cyclical loads that have frequencies that get too close to the critical speed of the shaft. If a load has an oscillatory frequency (also known as an exciting frequency) that matches the natural frequency of the shaft, and if there is no damping, the shaft will resonate and increase in the amplitude of deflection until shaft failure. (The theoretical amplitude increase would be infinity, but the shaft will fail first). There is always some damping; some of it comes from the internal friction within the shaft material, but most of it comes from the fluid the agitator is immersed in. So, the deflection will not reach infinity in the real world, but could reach the point of shaft failure. If the load frequency is sufficiently far from the critical speed, the amplification factor of the loads will be small enough to allow safe and stable operation. Also, if the system has more intrinsic damping it can be operated closer to critical speed. We will provide guidelines on these two phenomena.

But first, we need to determine where the exciting frequencies come from. The most obvious one is the rotation of the shaft itself. If the shaft is not perfectly straight (and it never is), has a mechanical or hydraulic imbalance, is slightly eccentric due to coupling eccentricity or some other factor, it will create an oscillating load at an applied exciting frequency equal to the shaft speed. It is recommended that the shaft be straightened to a total indicator runout (TIR) of 0.003"/foot of length, or 0.25 mm/m, or a slope of 2.5E-4. The shaft speed must not be too close to the critical speed. But this is not the only source of oscillatory loads. Impellers create oscillatory pulses at a frequency of the number of blades times the shaft speed, and smaller pulses at any even dividend of the blade number times the shaft speed. For example, a 4-bladed impeller creates an exciting frequency of 4 times the shaft speed and a smaller magnitude oscillation at two times the shaft speed. 3-bladed impellers create an exciting frequency of three times the shaft speed. 6-bladed impellers create frequencies of 2, 3 and 6 times the shaft speed, with the largest magnitude at six times the shaft speed. It is important that none of these frequencies match the shaft critical speed. Based on large-scale lab measurements, there is no measurable interaction between the number of blades and the number of tank baffles.

How close can we come to critical speed? It depends on the amount of damping in the system. Impeller type and design affect this, as does whether the unit is cantilevered or has a steady bearing. We call the difference "stabilized" versus "unstabilized." Units with steady bearings are considered to be stabilized. Cantilevered units are considered to be intrinsically stabilized with some impellers, and for other impellers, stabilizing fins (no illustration shown) can be added to stabilize the system. These act somewhat as rudders, resisting lateral movement. This concept principally applies to pitched blade turbines, which are not recommended for fermenters except as upper impellers when gas is to be re-incorporated from the liquid surface, such as in hydrogenators or other systems where the gas is not vented. Stabilizing fins are not needed for any unit with a steady bearing. Table 17.3 presents guidelines for critical speed operating ranges.

Now we can discuss how to calculate critical speeds. Broadly speaking, there are two cases: cantilevered design and steady bearing design. Of course, there are other variants, such as stepped shafts, multiple steady bearings, etc. These more complicated cases are best handled by computer programs written for the purpose. But we will first discuss simplified hand calculations for basic cases.

Cantilevered Designs

For the case of a cantilevered design with constant shaft diameter, the critical speed may be calculated using Eq. (17.9) for stainless steel:

$$N_C = 27082 * d^2 / \left(L * W_e^{0.5} * \left(L + S\right)^{0.5}\right), \text{ where units are kg, m and s.} \quad (17.9)$$

This equation is straightforward except for calculating the equivalent mass, W_e. This has two components: impeller weight and shaft weight. For a constant diameter shaft, the shaft equivalent weight is the actual shaft weight/4, or:

$$W_e\left(\text{shaft}\right) = \text{shaft weight} / 4 = 1579 d^2 L \text{ for stainless steel.} \quad (17.10)$$

Table 17.3 Critical speed guidelines.

Shaft style	Impeller style	Number of blades	Stabilized?	N/N_C range	Blade passing frequency range
Cantilever	Pitched or radial	4	Y	$0.80 \le N/N_C; N/N_C \ge 1.25$	Avoid N/N_C between 1.67 and 2.4 as well as between 3.33 and 4.8
Cantilever	Pitched or radial	4	N	$0.65 \le N/N_C; N/N_C \ge 1.54$	Avoid N/N_C between 1.67 and 2.4 as well as between 3.33 and 4.8
Steady bearing	Pitched or radial	4	N	$0.80 \le N/N_C; N/N_C \ge 1.25$	Avoid N/N_C between 1.67 and 2.4 as well as between 3.33 and 4.8
Cantilever	Hydrofoil	3	—	$0.80 \le N/N_C; N/N_C \ge 1.25$	Avoid N/N_C between 2.5 and 3.6
Cantilever	Hydrofoil	4	—	$0.80 \le N/N_C; N/N_C \ge 1.25$	Avoid N/N_C between 1.67 and 2.4 as well as between 3.33 and 4.8
Steady bearing	Hydrofoil	3	—	$0.80 \le N/N_C; N/N_C \ge 1.25$	Avoid N/N_C between 2.5 and 3.6
Steady bearing	Hydrofoil	4	—	$0.80 \le N/N_C; N/N_C \ge 1.25$	Avoid N/N_C between 1.67 and 2.4 as well as between 3.33 and 4.8
Cantilever	Radial	6	—	$0.80 \le N/N_C; N/N_C \ge 1.25$	Avoid N/N_C between 2.5 and 3.6 as well as between 3.33 and 4.8 and 5 and 7.2
Steady bearing	Radial	6	—	$0.80 \le N/N_C; N/N_C \ge 1.25$	Avoid N/N_C between 2.5 and 3.6 as well as between 3.33 and 4.8 and 5 and 7.2

For the impeller, it is the sum of individual impeller weights ratioed to the cube of their proportion of their extensional position relative to the total shaft length, or:

$$W_e(\text{impeller}) = W_1 + W_2(L_2/L)^3 + W_3(L_3/L)^3 \\ +\ldots \text{or } W_e(\text{impeller}) = \Sigma W_i(L_i/L)^3 \tag{17.11}$$

Total equivalent weight is the sum of the shaft and impeller equivalent weights:

$$W_e = W_e(\text{shaft}) + W_e(\text{impeller}) \tag{17.12}$$

The obvious question now is, Where do we get the impeller weights? Given the vast array of proprietary impellers on the market, and the inherently greater complexity of designing them compared to shafts, the best place to get them is from your selected agitator vendor. I spent some time trying to come up with a simplified procedure for calculating impeller weights for this chapter and was unable to do so. What I decided to do instead is to use software I own or have access to in order to create a couple of tables for the most commonly recommended impeller types for fermenters: the deep concave radial style, and the high solidity axial. These weights may vary from the actual weights used by any particular manufacturer. They at least give a general idea of the impeller weights and let us do some sample problems here. The tables give impeller weights in kg as a function of impeller diameter in meters, and torque in kN-m. Note that the shaft size will also affect impeller weight, as it affects hub design. "Typical" shaft diameters for the torque were used.

For example, a 1.5 m deep concave impeller with a design torque of 12 kN-m would have an approximate mass of 273 kg, and a 2 m high solidity axial impeller with a design torque of 30 kN-m would have an approximate mass of 900 kg.

Example Problem

To illustrate a simple cantilevered shaft critical speed calculation, we will use the same example that we used for a cantilevered stress calculation. In that problem, we did not have the shaft support bearing span, so for this problem, we will assume a span of 0.5 m. We also did not have impeller weights, so we have to estimate those. We stated that the shaft speed was 100 rpm, or 1.67/s.

Our lower impeller is a 1.6 m diameter deep concave impeller, with a design torque of 13369 N-m, or 13.36 kN-m. From Table 17.4, we see that a 1.5 m impeller with a torque of 12 kN-m has a mass of 273 kg, and a 1.75 m impeller with a design

Table 17.4 Deep concave turbine weights.

	Weights in kg mass					
	Impeller diameter, m					
Torque, kN-m	1	1.25	1.5	1.75	2	2.5
0.75	69	132	186	—	—	—
1	71	135	190	347	—	—
1.5	78	136	190	347	495	—
2.5	98	151	205	348	495	955
4	107	153	208	362	505	971
6	151	194	218	365	512	974
9	—	206	269	376	524	981
12	—	236	273	385	535	983
16	—	283	355	402	538	1003
22	—	336	366	491	644	1006
30	—	—	484	532	655	1034
42	—	—	—	649	810	1060
60	—	—	—	787	961	1294
90	—	—	—	—	1143	1593
130	—	—	—	—	—	1878

torque of 16 kN-m has a mass of 402 kg. Our impeller will be somewhere between these two figures. For the purpose of this problem, we will use 337 kg.

Similarly, the upper impellers are 1.7 m diameter and have a design torque of 5.77 kN-m. From Table 17.5, we see that a 1.75 m impeller with a design torque of 6 kN m has a mass of 344 kg. Likely, our impeller will be somewhat lighter. We will use 330 kg for this problem.

We know that the impellers are 0, 3.7, and 7.4 m from the lower end of the shaft, and that the shaft length, L, is 13.6 m. To use Eq. (17.11), we need to know the extension from the lower support bearing. This is just (L-C), where C is the distance from the end of the shaft to the impeller. So, we have $L_1 = (13.6-0) = 13.6$, $L_2 = 13.6-3.7 = 9.9$, and $L_3 = 13.6-7.4 = 6.2$ m. Now we can solve Eq. (17.11):

$$W_c\left(\text{impeller}\right) = 337 + 330\left(9.9/13.6\right)^3 + 330\left(6.2/13.6\right)^3 = 495.6\,\text{kg}.$$

Table 17.5 High solidity axial turbine weights.

	Weights in kg mass					
	Impeller diameter, m					
Torque, kN-m	1	1.25	1.5	1.75	2	2.5
0.75	48	95	107	175	223	—
1	70	96	125	175	223	375
1.5	75	116	145	200	258	375
2.5	80	116	202	254	323	483
4	110	173	240	273	395	485
6	127	180	273	344	460	697
9	160	213	299	409	477	718
12	196	280	345	445	605	802
16	204	300	440	540	730	911
22	—	351	455	640	850	1153
30	—	398	532	677	900	1368
42	—	445	590	774	1068	1548
60	—	—	659	870	1202	1745
90	—	—	—	1065	1380	2145
130	—	—	—	1260	1514	2345

The equivalent weight of the shaft, which has a diameter of 0.178 m, is calculated using Eq. (17.10):

$$W_e(\text{shaft}) = 1579 * (0.078\,\text{m})^2 * 13.6\,\text{m} = 130.7\,\text{kg}.$$

The total equivalent weight is the sum of these two:

$$W_e = W_e(\text{shaft}) + W_e(\text{impeller}) = 130.7 + 495.6 = 626.3\,\text{kg}.$$

Now we can calculate the critical speed using Eq. (17.9):

$$\begin{aligned} N_C &= 1528 * d^2 / \left(L * W_e^{0.5} * (L+S)^{0.5}\right) \\ &= 27082 * (0.178)^2 / \left(13.6 * 623.6^{0.5} * (13.6+0.5)^{0.5}\right) \\ &= 0.673 / \text{s or } 40.37\,\text{rpm}. \end{aligned}$$

Our shaft speed is 100 rpm, therefore $N/N_C = 100/40.37 = 2.48$. That means the unit is operating above critical speed. It is not good practice to operate a large agitator above the first critical speed. Therefore, this design, which as inadequate from a shaft stress standpoint, is also inadequate from a critical speed standpoint. Addition of a bottom steady bearing will help for both stress and critical speed. We will see in the next section whether a solid 0.178 m shaft will be adequate with a steady bearing.

This was a fairly simple design. More complicated designs, such as those involving steeped shafts, hollow shafts, etc., are best handled with a computer program or FEA. Good general references on mechanical design can be found in Refs. [2–4]. A number of manual calculation methods are included in Ref. [5].

Units with Steady Bearings

Shaft design with steady bearings is rather complex. Though Ref. [5] has a number of cases presented, they are not that easy to use. It is best to use a computer program that has a simplified FEA routine built-in. We will show such a simulation here. But first, let us discuss a quick and reasonably accurate way to do it for the case that the shaft does not have too many impellers. There is an equation for the critical speed of the shaft alone that is based on the shaft being simply supported (imagine pivot points) at each end. Adding impeller weights would tend to lower the critical speed. However, the real way shafts are supported is with two bearings at the top of the shaft plus the steady bearing below. This is a much more rigid system than a simple support at both ends. In my experience, in most cases, that additional rigidity more than compensates for the impeller weight. So, for a quick estimate of the critical speed, we can use that equation for a constant diameter stainless steel shaft:

$$N_C = 2180d / L^2; \text{MKS units} \tag{17.13}$$

For the hypothetical sample problem above, we would calculate:

$N_C = 2180 * 0.178/14.1^2 = 1.95/s = 117\,rpm$. That would put N/N_C at 0.85, which would not be acceptable. However, we will check this again with our simplified FEA/Dynamic analysis model.

To model this, we must conceptually divide the shaft into sections. We can use as many sections as we want, but we must have at least a new section for each discontinuity of any type: lumped masses such as impellers and in-tank couplings; lateral forces; power draw; changes in shaft section, bearings, etc. We generally number the spans from the gear drive end to the bottom end. Each section or span begins with a station number and ends with the next higher station number. For boundary conditions, we add a fictitious short span at each end of the shaft. (This makes the algorithm stable and more likely to converge.)

Table 17.6 Dynamic analysis inputs.

Station number	Span length, m/in.	Shaft outside diameter, m/in.	Shaft inside diameter, m/in.	Spring constant to ground, (kN/m)/(lb/in.)	Lumped mass, kg/lb	Power, kW/Hp	Hydraulic force, N/lb
0	—	0.178/7	0/0	—	—	—	—
1	0.0254/1	0.178/7	0/0	1.75E5/1E6	—	—	—
2	0.5/19.7	0.178/7	0/0	1.75E5/1E6	—	—	—
3	6.2/244.1	0.178/7	0/0	—	330/726	60/81	2039/458
4	3.7/145.7	0.178/7	0/0	—	330/726	60/81	2038/458
5	3.7/145.7	0.178/7	0/0	—	227/741	140/188	5013/1127
6	0.5/19.7	0.178/7	0/0	1.75E4/1E5	—	—	—
7	0.025/1	0.178/7	0/0	—	—	—	—

The dynamic analysis program used for this example was developed exclusively for Benz Technology International, Inc. It uses English units only at this point. So far in this chapter, we have used SI units, but will illustrate the problem in English units. As an aid to understanding the program inputs, we have created Table 17.6.

The basic model is conceptualized in Figure 17.3. To keep it simple, we did not show any gears which might be between the upper two bearings, or any in-tank couplings. A gear would add some weight and a side load, but the effect on shaft design would be minimal due to the proximity to the shaft support bearings. A shaft of 14.1 meters total length would need an in-tank coupling for manufacturing purposes, and there would need to be a coupling between the gear drive and the extension shaft. Again, we did not include these for simplicity. The coupling weight would be small compared to the impeller weights; likely about 50 kg for this example. So, adding in the coupling weights to the model would possibly decrease the critical speed by a couple of percent.

The shaft is supported by bearings, which are modeled as linear springs to ground. As long as the bearing stiffness is high, the actual value has little impact on the calculations. It cannot be infinite, however, as the program would not converge on a result. The values shown are reasonable, in terms of order of magnitude. The steady bearing is shown as 1/10 as stiff as the gear drive bearings, as it is not mounted in as rigid a structure. This type of modeling has been validated by many field measurements, not only for agitators, but for many kinds of rotating equipment.

The program calculates stresses and deflections, as well as critical speeds. If a station had been added for a mechanical seal, it could predict deflection through the seal, which could be important in some cases. The program output is shown as Figure 17.4.

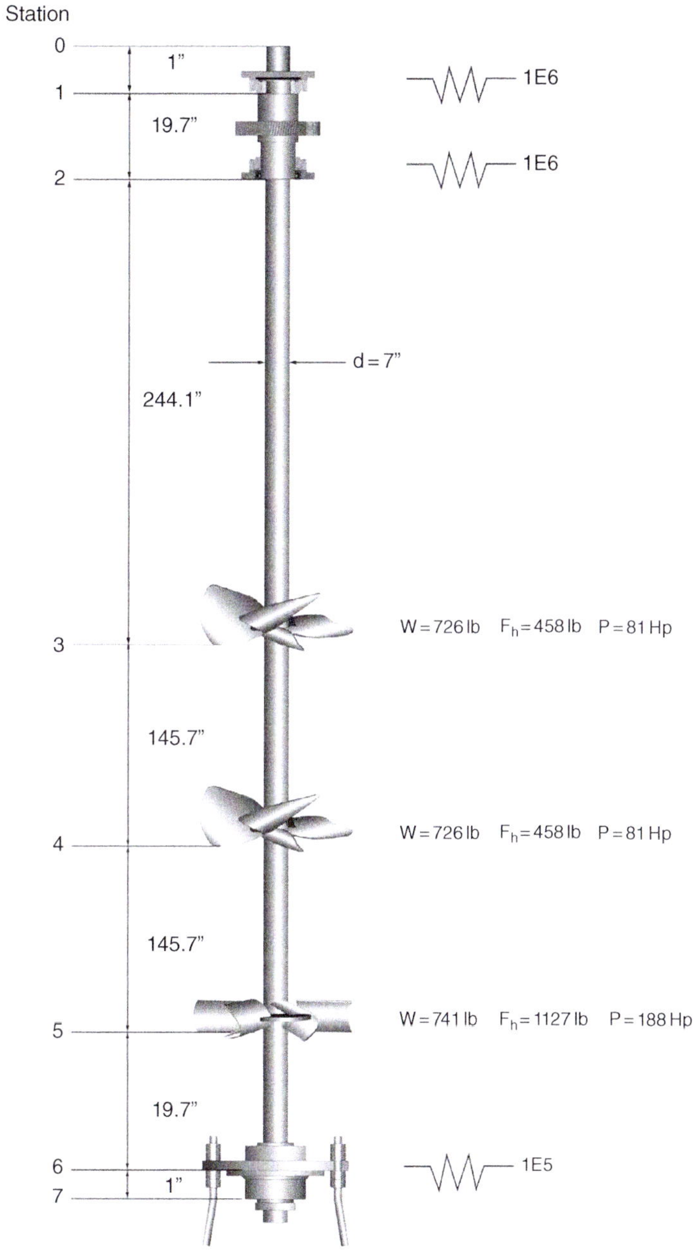

Figure 17.3 Critical speed model.

Figure 17.5 Steady bearing dynamic analysis example output

Carbon Steel
Stainless Steel
Titanium, Grade 2
Alloy C276

Top Entering
Side Entering

Calculate

		1st Critical Speed	2nd Critical Speed (RPM)
Customer			
Shaft Speed (RPM)	100	127.0 (78.7%)	427.7 (23.4%)
Temperature (F)	223		

Shaft Length (in)	Shaft O.D. (in)	Shaft I.D. (in)	Support Stiffness (lb/in)	Lumped Mass (lb)	Impeller Force (lb)	Impeller Power (HP)	Shaft Deflection (in)	Bending Moment (in-lbs)	Shear (lbs)	Tensile Stress (psi)	Shear Stress (psi)
1	7	0	1.00E+06	0.000	0.000	0.000	-4.232E-03	0.000E+00	-4.232E+03	3.584E+03	3.584E+03
19.7	7	0	1.00E+06	0.000	0.000	0.000	4.815E-03	-8.336E+04	5.832E+02	2.554E+03	3.792E+03
244.1	7	0	0.00E+00	726.000	458.000	81.000	4.690E-01	5.900E+04	1.252E+02	3.835E+03	2.959E+03
145.7	7	0	0.00E+00	726.000	458.000	81.000	4.839E-01	7.724E+04	-3.328E+02	3.512E+03	2.365E+03
145.7	7	0	0	741.000	1127.000	188.000	8.806E-02	2.876E+04	-1.460E+03	9.540E+02	5.270E+02
19.7	7	0	1.00E+05	0.000	0.000	33.000	1.460E-02	-7.749E-09	1.586E-09	0.000E+00	1.151E-10
1	7	0	0	0.000	0.000	0.000	1.084E-02	-1.586E-08	1.586E-09	0.000E+00	2.355E-10

The maximum tensile stress of 4.565E+03 psi occurs at 2.647E+02 inches.
The maximum shear stress of 3.792E+03 psi occurs at 2.070E+01 inches.

Figure 17.4 Steady bearing dynamic analysis.

Several key results are worth mentioning. The first critical speed is calculated as 127 rpm, which would have our agitator operating at 78.7% of critical speed. Though this is near the limit of 80%, it is acceptable. Adding in-tank couplings could possibly push it over 80%, so for an actual agitator design, this would need to be checked. Our manual simplified calculation, based on a simply supported shaft, gave a critical speed of 117 rpm, which is 8% below what the program calculated. As noted earlier, the simple calculation is usually conservative, except when there are many impellers or unusually heavy ones. Viscous fermenter designs may fall into that category, as they have about twice as many impellers as for low viscosity fermentations, and they are applied at a larger D/T ratio.

The maximum shear stress is calculated as 3792 psi, and the maximum tensile stress is 4565 psi. These are within the allowable numbers. Converting to kPa, the corresponding values are 26 146 and 31 476. Earlier, using hand calculations, we had calculated values of 26 511 and 40 519. The manual shear stress calculation agrees with the computer program within 1.5%. It is higher than the computer value for tensile stress by almost 29%, because the dual bearing arrangement at the top of the shaft does not allow as much shaft deflection as a simply supported shaft. So, on balance, the hand calculation is conservative. Maybe that is why I never had a shaft failure on any of the steady bearing units I designed early in my career!

On this printout, the maximum shaft deflection shown is about 0.48", or about 12 mm. This is based on all of the impeller hydraulic forces being at peak magnitude and all in the same direction. That does not happen very often. On the other hand, it does not account for amplification due to the N/N_C ratio, which would increase it, nor the effects of fluid damping, which would reduce the effect of the critical speed amplification ratio. On balance, the deflection numbers are probably close to a worst case scenario. However, they do not include deflection due to shaft straightness issues, errors in coupling squareness, or any other shaft manufacturing issue. Though we did not put in a station for a mechanical seal, we could easily do so, and thereby check for deflection in the seal area.

Shear force is also shown. It is listed as Shear (lbs). Its main use is to determine bearing reaction loads. For this simulation, the shear load for the span terminating in the steady bearing is 1460 lb, or 6494N. Typical design practice is to design the steady bearing support structure for three times the actual load, so we would design for roughly 20 000 Newtons side force.

Solid Shaft vs. Hollow Shaft

The above discussion and examples were based on using solid shafts, which is the most common case. But there are cases where a hollow shaft is sometimes desirable or necessary. We will list some of the reasons. But first, allow me to address one of my pet peeves about shaft terminology. Hollow shafts are often referred to in the industry as "pipe shafts". While hollow agitator shafting may be thought of as a pipe, commercial pipe is not a suitable material for manufacturing hollow shafts. This is because it is not round enough, not straight enough, and the common use of welded seams with filler metal that may not match the metallurgy of the pipe can increase corrosion problems. It is also rather rough, so if the shaft must be polished, a considerable material removal allowance must be part of the design calculation. Proper hollow shaft material should be ordered as "seamless" pipe, with careful controls on roundness, straightness and actual wall thickness, as well as appropriate surface smoothness. In some cases, it is better to use mechanical tubing than tightly controlled pipe.

With that out of the way, we can move on to when it is appropriate to consider using a hollow shaft. Here are a few reasons:

- When the solid shaft diameter needed to stay within critical speed limits would force the use of a larger gear drive than needed for the rest of the loads. (This may not apply if the shaft has its own support bearings and is flexibly coupled to the gear drive. This configuration will be discussed in our section on gear drives.)
- When the material cost per unit mass is high enough that hollow shafting will save enough money to offset the increased labor costs associated with hollow shafting. This will vary according to labor costs and the experience of the vendor making hollow shafts. As a rough guide, for major US vendors, it may be more economical to consider hollow shafts when the material of construction is 316SS and the required solid shaft diameter is 150 mm or larger. For higher alloys, such as Titanium, C-276 and the like, the transition may be less, possibly as low as 100 mm.
- When the weight of the wetted assembly would cause an unacceptably low bearing life in the shaft support bearing system (This is rare).

The use of hollow shafting can make a big difference in shaft critical speed. Usually, only part of the shaft is hollow, as it is difficult to use hollow shafts through seals and steady bearings. The solid shaft section is coupled to the hollow shaft by use of welded couplings resembling forged weld-neck flanges, though they are specially made and machined to assure concentricity and perpendicularity. We will illustrate this effect by converting the third span on our sample problem from solid to hollow, ignoring the coupling issue. This is shown in Figure 17.5

We replaced part of the shaft with a hollow section having an OD of 10.625" and an ID of 9.75", or a nominal wall thickness of 7/16". This may seem like a strange size; 10" Sch 80S pipe has an OD of 10.75 and an ID of 9.75. I have allowed for a 1/16" polishing material removal, which is admittedly rather severe, but serves to illustrate a possible design consideration. But the stresses are still OK, and the

Figure 17.6 Steady bearing dynamic analysis example output with hollow shaft

Carbon Steel
Stainless Steel
Titanium, Grade 2
Alloy C276

Top Entering
Side Entering

Calculate

		1st Critical Speed	2nd Critical Speed (RPM)
Customer Shaft Speed (RPM)	100	141.9 (70.5%)	533.3 (18.8%)
Temperature (F)	223		

Shaft Length (in)	Shaft O.D. (in)	Shaft I.D. (in)	Support Stiffness (lb/in)	Lumped Mass (lb)	Impeller Force (lb)	Impeller Power (HP)	Shaft Deflection (in)	Bending Moment (in-lbs)	Shear (lbs)	Tensile Stress (psi)	Shear Stress (psi)
1	7	0	1.00E+06	0.000	0.000	0.000	-4.642E-03	0.000E+00	-4.642E+03	3.584E+03	3.584E+03
19.7	7	0	1.00E+06	0.000	0.000	0.000	5.239E-03	-9.144E+04	5.978E+02	2.475E+03	3.833E+03
244.1	10.625	9.75	0.00E+00	726.000	458.000	81.000	4.124E-01	5.448E+04	1.398E+02	3.685E+03	2.890E+03
145.7	7	0	0.00E+00	726.000	458.000	81.000	4.456E-01	7.484E+04	-3.182E+02	3.459E+03	2.348E+03
145.7	7	0	0	741.000	1127.000	188.000	8.300E-02	2.847E+04	-1.445E+03	9.463E+02	5.235E+02
19.7	7	0	1.00E+05	0.000	0.000	33.000	1.445E-02	-1.470E-08	-7.974E-09	0.000E+00	2.183E-10
1	7	0	0	0.000	0.000	0.000	1.095E-02	-1.219E-08	-7.974E-09	0.000E+00	1.811E-10

The maximum tensile stress of 4.406E+03 psi occurs at 2.647E+02 inches.
The maximum shear stress of 3.833E+03 psi occurs at 2.070E+01 inches.

Figure 17.5 Steady bearing dynamic analysis example output with hollow shaft.

critical speed has increased from 127 rpm to 141.9, almost a 12% increase, changing N/N_C from 0.787 to 0.705. The total shaft mass has decreased by 788 kg. We could raise the critical speed more by increasing the portion of the shaft that is hollow. But it is often inconvenient to mount an impeller on the hollow portion, as we will see next. Moreover, the critical speed result is sufficient, even if we had added in coupling weights.

Hollow shafting has some drawbacks. Among them are:

- The larger diameter means the shaft couplings are larger also, which may affect the required agitator mounting flange size. This could reduce space available for other nozzles on the top head, unless a bottom entering agitator is used.
- Impeller hubs in most cases must be welded in place. Adjustment of impeller location is not possible if this is the case. With heavy wall pipe, clamped-on hubs may sometimes be possible if the hub location is machined accurately. In rare cases, a keyed design may be possible, but the wall thickness required for this may render the design uneconomical. External welded keys (welded to the shaft) have sometimes been used, with the hub slipped over the key.
- If impeller hubs are welded on, the required opening for installation is larger still. If installed through the agitator mounting flange, the flange diameter may need to be very large.
- Shaft couplings are more expensive than on solid shafts
- Large hollow shafts may need to be machined outside agitator vendor facilities due to lathe swing limitations, making quality control more difficult.
- A separately coupled solid stub shaft is generally required through the steady bearing, though this could be a plus for maintenance. The cost of the coupling can be offset by using it as the hub for disc-style lower radial turbines, whether Rushton-style or any of the concave styles.
- Except for small shafting or repetitive designs, FEA is almost always required to assure localized stresses in the hollow shaft are not too high. This is especially true if thin-wall pipe is used. Thin wall pipe may have a high enough section modulus for the bulk loads, but there can be high stress concentration factors where hubs are welded on. For example, a 20" OD pipe with a 1/8" thick wall is theoretically as strong as a 10.75" OD pipe with a 0.48" thick wall. But who would dare to attach anything to a 1/8" thick wall? The safest attachment is to use a cylindrical hub with stub blades integral to the hub. With heavy enough pipe wall thickness, it is possible to weld stub blades for axial turbines directly to the hollow shaft, but FEA must be used to check this, at considerable engineering expense. But if the stress concentrations are too high, one might witness (as I have) a phenomenon I call the can opener effect. (You can use your imagination.)
- Though not commonly a problem for fermenter designs, external pressure can be an issue for some hollow shaft applications, since the shaft is sealed and acts

as a pressure vessel. This is more likely to be an issue for hydrogenators than fermenters, as they run at much higher pressures. Still, some extremophiles may thrive at high enough pressures that this may need to be checked.

In short, hollow shafting can be a viable shaft design option for cost savings or critical speed considerations, but it is not without a few drawbacks.

Role of FEA in Overall Shaft Design-Simplified Discussion

Finite element analysis, or FEA, has been used in industry in general and agitator design in particular for decades. It can pinpoint design problems that manual calculations might not. Though the shaft design spreadsheet illustrated above is a sort of FEA program, most people associate FEA with color-coded stresses and deflections. Though I do not have a library of illustrations here, I make the below suggestions for where FEA should be used in agitator design to supplement manual methods.

- For most hollow shaft designs, as described above.
- For unusual impeller hub designs, or cases where the stub blade is less than 40% of the extension blade width.
- For deflection of concave blades, to make sure they do not “open up” too much under load, which would increase power draw and may even lead to catastrophic failure.
- For slots and bolt hole areas in disc turbine designs, especially those with adjustable diameter.
- To determine dynamic deflection and natural frequency of the impeller disc.
- To evaluate vessel head deflection and stress, as described in Chapter 16.

Agitator Gear Drive Selection Concepts

Though this topic could easily fill at least a whole chapter, I opted to keep all major mechanical design concepts in this chapter rather than create a separate chapter.

Except for the case of very small agitators, most agitator shafts and impellers do not operate at motor speeds. They need to be slowed down to allow large enough impellers to be used to control the full liquid volume, and to address such issues as shaft critical speed and seal life.

There are multiple ways of reducing speed and increasing torque. The most commonly used ones in agitator design are belt drive and gear drive. Though belt drives are simple and inexpensive, they are relatively inefficient under normal maintenance conditions (typically 92% per reduction stage), require belt tension adjustment frequently, and require belt replacement after a year or two; sometimes more frequently. Gear drives using helical or spiral bevel gears have a typical efficiency of 98.5% per reduction stage, and gear life is about 5–15 years in fermenter service, depending on design. Routine maintenance consists mainly of changing the oil and greasing the bearings.

Because belt drives are seldom used in production fermenters, we will discuss them no further, and concentrate on proper gear drive selection. We also will not discuss worm gear drives, as they are less efficient and wear out more quickly. We will confine our gear drive discussion to those using helical or spiral bevel gears. Planetary gearing would also be acceptable, but we are unaware of any agitator manufacturer offering this.

What we will do is go over the several ways a successful agitator drive system can be configured, offer opinions on what might be the best way, and then get into some specific details needed to specify a design.

But first, we will delve into a brief history of agitator design and construction, reliability problems encountered, the loads involved, and why agitator drives are a specialized product, distinctly different from a commercial gear drive.

Early History

The first manufacturers of agitators used whatever was commercially available to reduce speed and increase torque. But, because the magnitude of the loads was not well understood and the commercially available gear drives were not specifically designed for agitator service, most units had a short life span. Some failed catastrophically within days or weeks. Larger service factor helped, but people did not always know how to be sure of a selection that would be successful. The key to successful design turned out to be developing an understanding of the loads imposed on an agitator drive, and how to accommodate them. Specialized mechanical testing equipment was needed to develop load correlations, and most manufacturers did not have such equipment until the 1960s. Some still do not have them today. So, what are those loads and how do they differ from those a commercial gear drive is designed to handle?

Loads Imposed

Figure 17.6 depicts the loads a typical commercial gear drive is designed to handle. Such gear drives are generally suitable for close-coupled loads, such as a gear or chain-and-sprocket on the output shaft. These close-coupled loads create a small bending moment, well within the range of the gearbox output shaft

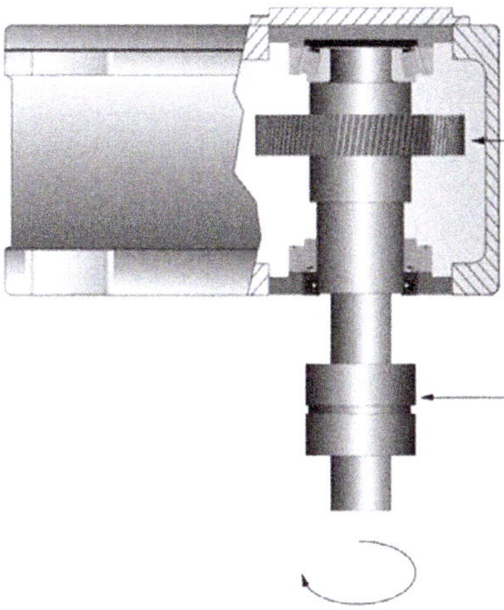

Figure 17.6 Commercial gear drive design load diagram.

bearing capacity. There is also a bending moment created by the gear reaction force. These two moments create a deflection of the output shaft, but it is small enough to create no problem for the gear alignment. Typically, the internal shafting is not much larger than that required to transmit torque. In fact, torque is the primary load a commercial gear drive is designed for. Axial thrust, or weight, imposed on the output shaft is limited to very small numbers compared to agitator loads.

An agitator experiences different operating conditions from most commercial gear drives. For example, it often runs 24 hours per day for at least 90% of the time in many plant applications, and the torque transmitted is relatively steady. An agitator also has a very different load diagram, as shown in Figure 17.7. In addition to torque, agitator service imposes such loads as the weight of the shaft and impellers, impeller thrust, impeller side loads and the attendant bending moment they create due to being far from the gear drive bearings, and pressure thrust if there is a seal and the vessel operates under pressure or vacuum. The bending moment not only affects bearing lives, but greatly increases gear tooth deflection due to driveshaft deflection. If not taken into account, bearing and gear failures can and will occur. That is what happened frequently in the early days of the agitator industry. There is an additional factor. The vast majority of commercial gearboxes are designed so that the output shaft bearings are a slide fit in the housing. This is fine for gear loads and side loads on the output shaft such as belt or sprocket loads but it is not so good if the gearbox must support an

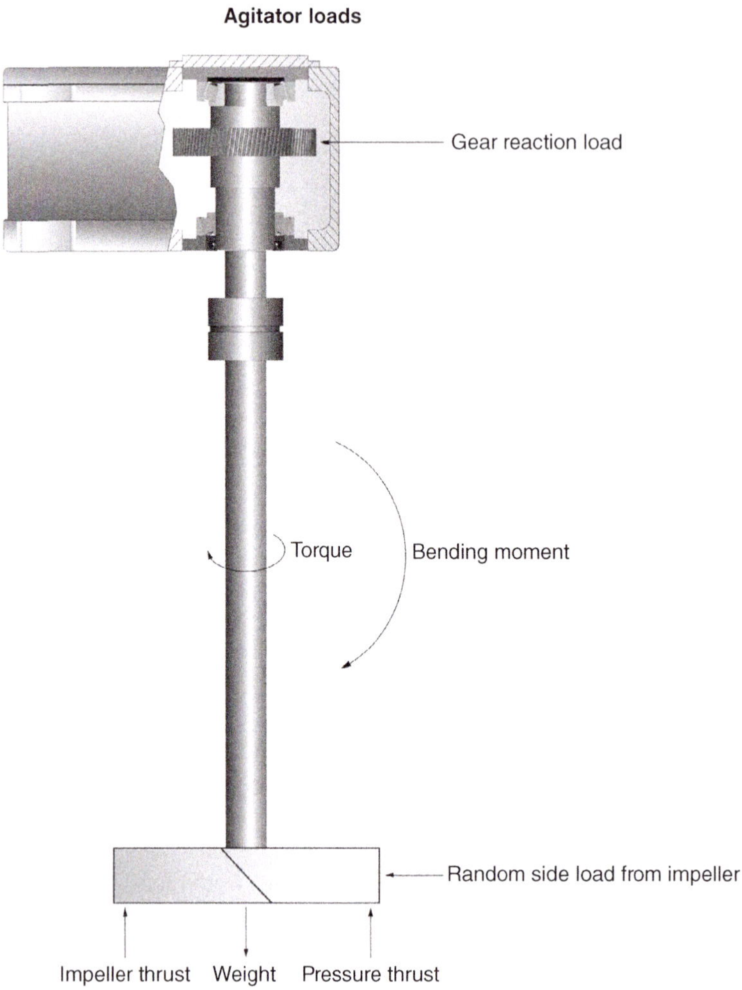

Figure 17.7 Agitator load diagram.

agitator shaft. The changing load directions can result in the bearing fretting in the bearing housing leading to a failure of the gearbox. If the bearings on the output shaft are very large for their duty then the rate at which the casing is damaged may not be enough to cause early failure. Also some smaller commercial gearboxes have bearing retention techniques that are not suited to a shaft which sees high levels of rotating bending moments. So how do we design to accommodate these loads?

Handle or Isolate Loads?

Broadly speaking, there are two major design approaches that can be successfully employed: designing the gear drive to handle the loads, or trying to isolate the loads from the gear drive by mounting the impeller driveshaft on a separate set of bearings. We will look at several approaches in each category.

Handle Loads Option 1: Oversized Commercial Gear Drive

It is sometimes possible to get satisfactory life in agitator service by using a larger commercial gear drive to get the necessary shaft diameter and robust bearings needed. Figure 17.8 conceptually compares a commercial gear drive with a service factor (ratio of AGMA rating/motor nameplate power) of 1.25 with one that has a service factor of 3.

Figure 17.8a shows the smaller gear drive; Figure 17.8b shows the larger one. If the smaller gear drive has a shaft diameter of d, the one with a 3.0 service factor will have a shaft diameter of about 1.3 d. It will also be taller and heavier, and its gears will theoretically last longer. However, due to the bending load, the gear deflection will be higher than what was intended for the product, so its life will be reduced somewhat. It is this author's experience that a service factor of 3.0 is pretty much the minimum required when applying a commercial gear drive to agitator service. We will get into more detail on service factors later.

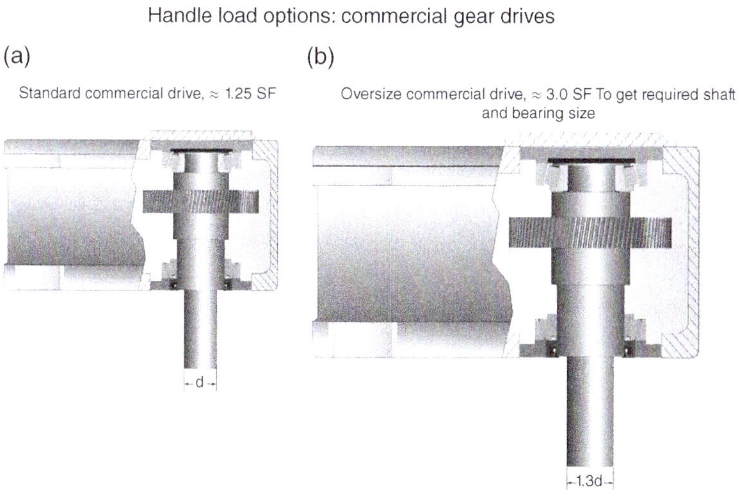

Figure 17.8 (a, b) Commercial gear drive size comparison.

The oversized gear drive approach has one advantage: it is usually the cheapest option. But it is still not usually as reliable as a purpose-built drive, and its longer internal shafting results in lower shaft critical speeds, reducing allowable shaft extensions, and its increased deflection will allow more wear on mechanical seals. It is also slightly less efficient, as there is more fixed loss in a larger gear drive.

Handle Loads Option 2: Purpose-Built Agitator Drive

The concept of this drive design, depicted in Figure 17.9, is to use an output shaft that is much larger than that of a commercial gear drive, combined with massive bearings that are mounted in such a way as to prevent rocking or fretting in the housing; a very rigid housing; and a short bearing span internal to the drive. The reason to use a short bearing span is to minimize deflection, both inside the gear drive and outside, at such locations as the mechanical seal and along the entire extension shaft. This logic can be understood by referring to Eq. (17.14), which can be derived from Ref. [6]:

$$\delta \propto S^3 / d^4, \text{ where } \delta \text{ is the deflection} \tag{17.14}$$

One can readily see that combination of a short bearing span and a big shaft diameter dramatically reduces deflection. This makes the gears last longer as well as improves seal life. It also raises the shaft critical speed, which can allow slightly longer shaft extensions. Compared to a commercial reducer, the shaft diameter is generally about 1.8 d. It is usually a solid shaft going through the gearbox.

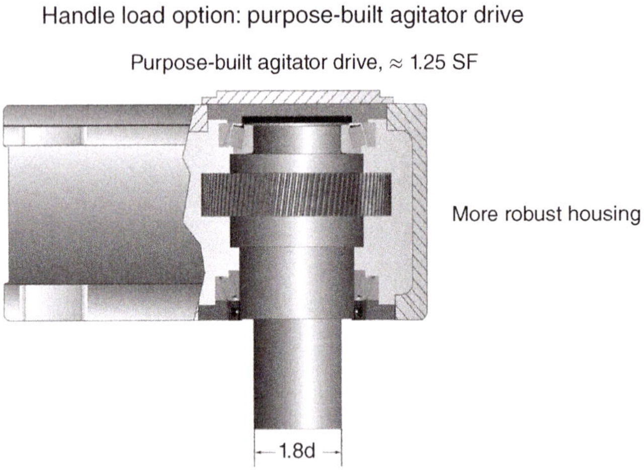

Figure 17.9 Purpose-built solid shaft agitator drive.

Such gear drives have been manufactured by major agitator manufacturers since the 1970s. In normal service, such a gear drive with a service factor of 1.25 will last for 15–25 years. It will have output bearing lives of several hundred thousand hours. For fermenter service, such drives should normally be applied with a service factor of 1.5 or preferably higher; 2.0 may be needed to achieve 15 year life.

Such drives have also been manufactured by companies that manufacture commercial gear drives. They are clearly advertised as agitator drives. In many services, they may give satisfactory results. However, in a study I did for a client, I found that they all had smaller output shafts and lower output bearing lives than those made by major agitator manufacturers; at least compared to those using tapered roller bearings on the output shaft.

The only disadvantage of the purpose-built drive is higher initial cost than even a commercial drive with a 3.0 service factor. However, lower downtime probably makes the total cost of ownership less.

A variation of this design used by some manufacturers uses a hollow output shaft, with the extension shaft rigidly locked into position. This is slightly less rigid, leading to lower shaft critical speeds and higher seal deflection and wear. If the hollow reducer shaft is made of steel and the extension shaft is made of an alloy, plant moisture has sometimes been known to cause galvanic corrosion between the two materials, making it very hard to remove the gear drive for servicing, or to change the mechanical seal in some designs.

Isolate Loads Option 1: Hollow Quill Integrated Drive with Flexibly Coupled Extension Shaft

This design concept has been in use since at least 1960 and is offered today by several agitator manufacturers as well as some gear drive manufacturers. A schematic of the concept may be found in Figure 17.10. The basic concept is that the output gear is mounted on a hollow quill shaft on its own set of bearings. The agitator driveshaft is concentric with the hollow quill, and it is mounted on its own set of bearings, attached to the same overall housing, above and below the quill shaft bearings. In principle, these bearings take all of the external loads described in the load diagram in Figure 17.7, whereas the quill shaft and its bearings take only the reaction loads from the low-speed gear set. The two shafts are connected by means of a flexible coupling located above the upper quill shaft bearing. It is claimed that such construction isolates the external loads from the gear shaft, reducing gear shaft deflection and promoting long life. This design has proven much more reliable than commercial gear drives for most applications. However, it does not actually work the way it is described in manufacturers' literature.

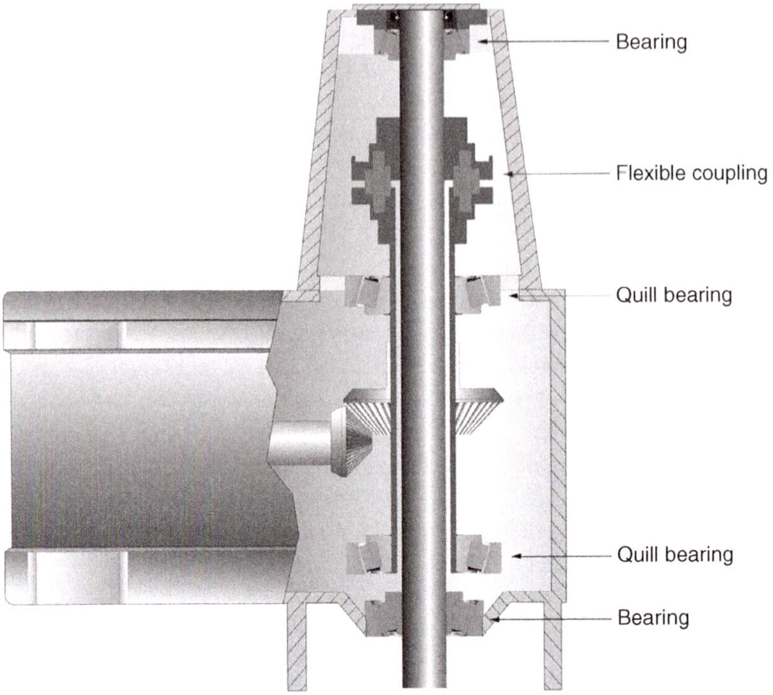

Figure 17.10 Hollow quill agitator drive.

To begin with, the degree of isolation of the extension shaft loads from the gear shaft is questionable. To understand this, one must understand a little about how flexible couplings are selected. Basically, such couplings have two selection criteria: the mechanical rating and the bore size. For a coupling to have long life and yet provide reasonable torsional and lateral flexibility, it should typically have a mechanical rating 125–150% of the applied torque. However, when the coupling must be sized to fit a very large shaft, such as the hollow quill shaft, it will be a much larger coupling than needed for the torque rating. In many cases, the actual coupling rating will be 5–15 times the applied torque. When applied at such a high service factor, the coupling will be almost rigid. It will provide no appreciable attenuation of torsional shock loads (although this is not a problem if the high-speed motor coupling is properly selected to do just that). It will also transmit much or most of the side loads directly to the quill shaft if the driveshaft starts to deflect. That means the quill shaft bearings must be designed to handle such loads. Fortunately, they usually are.

Secondly, we come to the matter of gear tooth deflection due to flexing of the quill shaft itself. In a Master's Thesis report (copy not available), done by R. Stephen Hill in the 1980s, it was determined that the common hollow quill designs of the day actually had greater gear tooth deflection from the gear reaction forces alone than the most common purpose-built agitator drives did from all loads combined. This is because the quill shaft was much less stiff (had a lower moment of inertia) than the massive solid shafts used in the purpose-built gear drives, and the hollow quill also had a longer bearing span.

Thirdly, we come to the matter of bearing lives. Most commercial gear drives as well as the solid-shaft agitator drive depicted in Figure 17.9 have two bearings supporting all of the output loads. The hollow quill design has four. Some might think this makes it more reliable, as the loads are spread out over more bearings. However, if any of the bearings fail, the agitator fails. And if all four bearings have the same life, the overall system reliability will be less. This may be easily derived from the common equation for mean time between failures (MTBF):

$$\mathrm{MTBF} = 1 \Big/ \left[\Sigma\left(1/\mathrm{L}\right)^2\right]^{0.5} \tag{17.15}$$

If the bearings have the same life, the above equation leads to an overall life of a four bearing design that is 70.7% as long as a two-bearing design with the same individual bearing lives. However, most hollow quill designs use ball bearings for the driveshaft support, so their life may well be even less.

Finally, we get to shaft deflection and critical speed. The quill shaft design has a substantially longer driveshaft than does the purpose-built solid shaft design, and its driveshaft is smaller in diameter as well. This leads to more shaft deflection and a lower critical speed. These two effects combine to produce lower seal life and a shorter allowable shaft extension.

The main positive points of this design are listed below.

- Design proven over at least 50 years of production.
- Much more reliable than commercial gear drives.

The main disadvantages are:

- Much more expensive than other purpose-built agitator drives as well as commercial gear drives.
- Rather tall. Overhead clearance could be an issue.
- Shorter maximum shaft extension, compared to solid shaft purpose-built drives.
- More deflection through the seal, reducing seal life.
- If built to the same gear rating and bearing life standards as solid-shaft purpose-built drives, they probably are less reliable overall.

Isolate Loads Option 2: Outboard Support Bearing Module

Shown in Figure 17.11, this concept can use either a commercial gear drive or a purpose-built agitator drive. However, there is little reason to use a purpose-built agitator drive as long as the commercial gear drive in question has ample gear and bearing life ratings.

The basic idea is that the extension shaft loads are supported in a bearing module independent of the gear drive. That module often also contains the mechanical seal as well. The shaft exiting the top of the bearing module is flexibly coupled to the gear drive above. The gear drive and the bearing module are normally connected together by means of a spacer pedestal. It is also possible to mount the gear drive independently, with a universal-joint shaft or similar shaft connecting the two, as shown in Chapter 16, Figure 16.8.

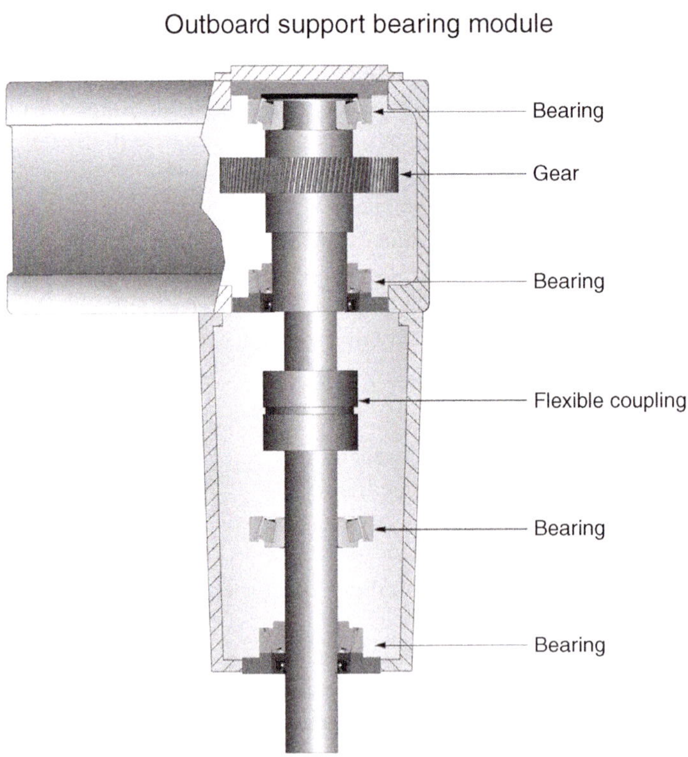

Figure 17.11 Outboard support module.

The bearing module approach allows a shorter bearing span for the shaft support than the hollow quill design. This allows standard seals to be used and permits longer shaft extensions. The bearing module shaft diameter can be as large as needed, independent of the gear drive selection. With this design concept, it is never necessary to oversize the gear drive to get the required shaft diameter. The shaft out of the top of the module is normally turned down so that a flexible coupling with a service factor of 1–1.5 can be used, better isolating the loads and possibly attenuating shock somewhat. Though this design approach is not common, it finds use when the size of the equipment is such that no standard agitator drive is available, and when vessel pressure creates thrust far beyond what standard agitator drives are designed for. Though high pressure is not normally a feature of fermentation processes, many of the world's largest fermenter agitators have been built this way, due to a lack of available purpose-built drives in the large sizes required. In most cases, for agitators of greater than 500 kW motor size, it is the only option available. Some pros and cons:

Pros:

- Cheaper than purpose-built agitator drives in large sizes.
- Available in larger sizes than purpose-built drives.
- Independent bearing module allows for high pressure or high bending applications without affecting gearbox life or selection.
- Can remove gearbox for service without having to support the shaft in the tank.

Cons:

- Even taller than quill design.
- Expensive in small sizes.
- More wearing parts; likely less reliable than a purpose-built agitator drive.

Bearing Life Considerations

To achieve a target gear reducer life of 15 years or more, the weakest bearing in the gearbox will usually need to have an L_{10} life (time to 10% failure) in the range of 30 000 hours minimum. It may need to be more than 50 000 hours if all the bearings have similar lives. Table 17.7 lists the requirements for bearing lives imposed by AGMA. In a competitive market, many commercial reducers just meet the bare minimum. One would logically conclude that we would want a minimum service factor of 1.75 when using a commercial gear drive. This would only apply if the agitator shaft loads were isolated from the gear drive, as in the bearing isolation module design. Much higher service factors would be needed to get the required shaft diameters and bearing sizes on the output shaft. Purpose-built gear drives, in general, have much higher bearing lives than AGMA requires, especially on the

Table 17.7 AGMA bearing life requirement.

Bearing life to 10% failing (L_{10})		
Gearbox service factor	**Ball bearing (hours)**	**Roller bearing (hours)**
0.8	2560	2376
0.9	3645	3519
1	5000	5000
1.25	9766	10520
1.5	16875	19317
1.75	26797	32292
2	40000	50397
2.25	56953	74630
2.5	78125	106032
3	135000	194704

For AGMA rated gearboxes the shortest calculated bearing lives should meet the above table.

output shaft bearings. Typical figures are a minimum L_{10} life in normal service of 30000 hours at a 1.25 service factor and more than 100000 hours on the output shaft. Such drives also have more rigid housings to accommodate the bending and thrust loads while keeping gear deflection well below AGMA allowances.

Based on the above, and a good bit of field history experience, I have compiled a rough guide to recommended gear drive service factors as a function of drive arrangement and service conditions. This is found in Table 17.8.

I emphasize that this is just a guide, and is intended to be conservative. For specific applications or designs, the vendor should calculate all loads and bearing lives for the particular design in question. The vendor should be required to provide a summary of gear ratings and bearing lives for the particular service. Nonetheless, if a vendor quote seems to have substantially lower service factors than recommended in Table 17.8, buyer beware!

To emphasize the comparison, I have included Table 17.9, which compares gear drive weight and output shaft diameter of a mid-range agitator drive, looking at a top-tier, purpose-built gear drive versus two standard commercial gear drives and a commercial gear drive manufacturer's "agitator" drive. This study was done for a client. Clearly, the top-tier product is more ruggedly built and can handle larger shaft loads.

Noise Considerations

Large agitator drives tend to be noisy. Using AGMA quality 10 or higher will reduce surface roughness and mitigate noise. At a given pitch line velocity,

Table 17.8 Recommended gear drive service factors.

	Gear drive type/arrangement				
Service	**Commercial**	**Purpose-built to handle loads by agitator manufacturer**	**Purpose-built by commercial gear drive manufacturer to handle loads**	**Hollow quill drive built by agitator manufacturer**	**Isolation module with commercial drive**
Waterlike, moderate agitation	3.0	1.25	1.5	1.25	2.0
Waterlike, intense agitation	4.0	1.5	1.75	1.5	2.5
Makedown of hard-to-wet solids	4.0	1.5	1.75	1.5	2.5
Viscous agitation	2.5	1.25	1.25	1.25	1.75
Gas-liquid contacting, e.g., fermenters	5.0	1.75	2.0	2.0	2.75

Table 17.9 Gear drive comparison.

Mid-sized gear drive, ~24 000 in-lb torque rating		
Description	**Weight, lb (Kg)**	**Shaft size, in. (mm)**
Top Tier Purpose Built	919 (418)	3.5 (89)
Commercial "A"	552 (251)	2.75 (70)
Commercial "B"	234 (106)	2.375 (60)
Commercial "Agitator" Drive	485 (220)	3 (76)

helical gearing is quieter than spiral bevel gearing. Therefore, for right-angle drives, it is preferable to make the first reduction stage helical rather than spiral bevel in a two or three reduction stage gear drive. That having been said, most agitators operating with a motor size of greater than 100 kW will require hearing protection in the area. Some will need area noise abatement measures.

Torsional Natural Frequency

Though torsional frequency analysis is beyond the scope of this book and the skill of this author, it can be an important consideration. Generally speaking, the torsional natural frequency is quite high compared to any oscillatory frequencies in the system. However, as the size of the equipment becomes greater, there is a greater likelihood that it can become a problem. I recommend that it be checked for all units with a motor size of 200 kW or greater. I have personally seen gear tooth destruction on a 300 kW unit. A potential resonance problem was detected in the design phase of multiple 750 kW units, and dealt with in the design phase.

What kinds of problems can torsional resonance cause? The main one is gear tooth chatter. If the torsional natural frequency is too close to the gear tooth meshing frequency of any of the gear sets, there can be a problem. Bearing roller frequencies are theoretically a possible problem area as well. For best attenuation, the torsional natural frequency should be well below any gear tooth meshing frequency. So, you might ask, How does one adjust the torsional natural frequency? Though the answer requires sophisticated finite element analysis, the concept is quite simple: select the correct torsional stiffness in the high-speed coupling, i.e., the motor coupling.

Most designers of high-power equipment prefer to use metal grid couplings, as they are very compact and relatively inexpensive for the torque transmitted. But they are also very stiff in both the torsional and lateral direction. To lower the torsional natural frequency, it is desirable to use a softer coupling. Some couplings use a tire-like rubber element. These may do the job, but they try to pull the motor shaft and gear drive input shaft together as the rubber element winds up, possibly reducing bearing life. A better design, in my opinion, uses a floating rubber or polyurethane element. Some designs use rubber blocks inside a "spider" metal housing, placing the rubber blocks under compression, but allowing them to float axially. With either kind of elastomeric element, there can be several choices of elastomer, resulting in different degrees of stiffness. Coupling service factor will also affect the stiffness. In short, coupling stiffness is the key to changing the torsional natural frequency, and there are many options available, covering a wide range of stiffnesses.

Most major agitator manufacturers should be able to calculate the torsional natural frequency. Alternatively, there are a number of independent mechanical engineering consultants who can do so.

Important or Useful Mechanical Design Features

- Dry well seal around low-speed shaft. This keeps oil in the gearbox in the event of a slow leak in the lip seal or grease seal at the lower output shaft bearing. In the case of pharmaceutical applications, it is needed to keep the area clean, and may be required by GMP standards. Overall, for top-entering units, this is a *must have*. For bottom entry units, it is not necessary, as the lower bearing cap is sealed. To keep the images simple, we did not show a dry well in our gear drive illustrations.

- Shaft coupling below gear drive. If it ever becomes necessary to remove the gear drive for servicing back at the maintenance shop, a rigid shaft coupling below the gear drive is a *must have*. Taper lock or compression fit shaft connections within a hollow gear drive have proven very difficult to remove, especially when the hollow gear driveshaft is carbon steel and the extension shaft is stainless or high alloy. Galvanic corrosion will almost certainly occur in a typically humid plant environment.
- Change gear feature. This can allow for future process changes, as well as allowing for some commonality of parts when the same gear drive size is used for several applications. A *nice-to-have* feature.
- Magnetic drain plug. Helps keep wear particles away from gears and bearings in between oil changes. *Nice to have*.
- Extended oil drain pipe to allow access away from the center of the vessel. Could prevent a dangerous slippery oil mess on the top head during oil change. *Nice to have*. In some locations, it might be a *must have*. This can easily be a field modification.
- Extended grease lube fitting pipes for ease of access. *Nice to have*.
- On units equipped with gear drive cooling, a filter and a temperature indication, as well as an interlock to prevent operating the agitator when no oil or coolant is flowing, and a temperature cooling water shutoff to keep the gear drive from getting so cold as to have internal condensation. A *must have*.

Summary of Chapter

We learned the basic considerations involved in agitator shaft design, as well as the most common options for shaft construction. Stress and natural frequency were both covered.

We also covered the various concepts used for gear drive selection and shaft support in agitator drives, including bearing life and gear rating considerations. We are unaware of this kind of analysis having been published previously.

List of Symbols

B	Distance of impeller from steady bearing
C	Distance of impeller from lower gearbox support bearing

d	Shaft diameter
D	Impeller diameter
F_D	Dimensionless hydraulic force
F_h	Peak value of hydraulic force
L	Shaft length, or impeller distance from support bearing
L'	Mean life of a piece of equipment or component
M	Bending moment
M_C	Combined bending moment and torque
MTBF	Mean time between failures
N	Shaft speed
N_C	Critical speed; AKA first natural lateral frequency
P	Impeller power
S	Bearing span
T_Q	Impeller torque. Shown as TQ in figures
T	Tank diameter
W	Impeller weight
W_e	Equivalent mass

Greek Letters

δ	Shaft deflection
σ_s	Shear stress
σ_t	Tensile stress

References

1 Hill, R. and Kime, D. (1976). How to specify drive trains for turbine agitators. *Chemical Engineering Magazine*: 46–51.

2 Young, W.C. and Budynas, R.G. (2002). *Roark's Formulas for Stress and Strain*, 7the. General reference. McGraw-Hill.

3 Glover, T.J. (1998). *Pocket Ref*, 2e (14th printing). Littleton, CO: Sequoia Publishing.

4 Ramsey, W. and Zoller, G. (1976). How the design of shafts, seals and impellers affects agitator performance. *Chemical Engineering Magazine* 30: 52–59.

5 Website, https://www.engineersedge.com/bearing/critical-speed-single-load.htm (accessed 11 October 2020).

6 Website, https://www.linearmotiontips.com/beam-deflection-equations-for-linear-systems (accessed 11 October 2020).

18

Sanitary Design

Introduction

This chapter is about agitator design and construction features related to sanitary operation. It is more or less divided into two sections: construction of wetted parts and surface treatment, i.e., polish. It does not cover sanitary design in the shaft seal area, as that topic was addressed in Chapter 15.

This chapter does not seek to answer the question of what degree of sanitary design is needed. There are fermenters operating with no sanitary design features at all. Most large fermenters making commodity products such as amino acids, antibiotics, or enzymes just have standard mill finish and bolted construction. Many smaller fermenters making high-value products, such as injectable drugs, as well as those with infection-sensitive organisms, plant cell, or mammalian cell culture, have very sanitary construction. How does one decide what is needed?

Years ago, when I was employed by an agitator manufacturer, while working on a fermenter project in China for a large international pharmaceutical company, I was present at a discussion between the company project engineer and the EPC (engineering, procurement and construction) company project manager. They were arguing about the degree of sanitary design needed, with the pharmaceutical company engineer lobbying for standard construction and the EPC firm engineer lobbying for sanitary design. Suddenly, they both turned to me and asked my opinion, as though they expected me to settle the argument. I replied that I cannot tell them what they need, but I can tell them what various options cost (and, sanitary design can add a lot to the cost).

The answer is still pretty much the same today. I cannot tell you what you need. But in this chapter, I can show you what some of the options are so that you can be in a better position to decide. I also hope to illuminate some of the issues and pitfalls in polished construction.

Agitator Design for Gas–Liquid Fermenters and Bioreactors, First Edition. Gregory T. Benz.

Definitions

Before we get into details, let us define some terms that are sometimes, albeit incorrectly, used interchangeably.

"Sterile" means having no living organisms present. Obviously, a fermenter cannot be sterile, or it would not ferment. However, the vessel and agitator will normally be sterile at the start of a batch or a production run. The growth medium should also be sterile until inoculated with the desired organism. The air supply should also be sterile. This is normally accomplished by filtering the air through filters that will not pass microorganisms.

"Aseptic" means having no undesirable organisms present. This is the condition that one strives for in a fermenter. If the desired organism is strong enough and present in large enough quantities, the fermentation may be completed before undesirable organisms gain a foothold.

These previous two terms describe process conditions, not the equipment.

The term "sanitary" describes the equipment design. Essentially, it means easy to clean and sterilize. In the next several sections, we will go over sanitary design principles and give examples of construction details intended to be as sanitary as possible.

Construction Principles

The basic idea of sanitary design is to avoid places where material can accumulate and hang-up or be shielded from steam or chemical cleaning solutions. That means that, to the extent possible, all surfaces should be sloped to permit good draining. There should be no cavities that can fill with broth and be protected from the tank environment, forming what a colleague dubbed "pus pockets." If there are holes or pits, they should be large enough to permit draining or cleaning. A 25-mm-diameter open hole is better than a 1-mm pit or crevice. Surface finish can also facilitate draining and cleaning. We will discuss how these principles are applied in subsequent sections.

Wetted Parts Construction Methods

Welded Construction

This is the most sanitary design method, provided all welds are ground smooth. It is still useful to have sloped surfaces, which are not automatically designed for just because the construction is all-welded. But welded construction completely seals off any potential "pus pockets" inside impellers, in-tank couplings, etc. It also is generally less expensive than other methods described herein. So, why are not all fermenter agitators made with all-welded construction? There are several reasons and several ways of dealing with the problems of all-welded construction. The main disadvantages follow.

- If welded at the agitator manufacturer, it may be hard to get into the vessel at the jobsite, as vessel openings must be large enough to pass the impellers, at least by rotating them through an opening. This requires some skill if the opening is smaller than the impeller diameter, and carries the risk of damaging the surface finish. With careful scheduling and coordination between the agitator manufacturer and the vessel fabricator, the agitator wetted parts can be shipped to the vessel fabricator before the vessel heads are welded on, so that the shaft and impellers can then be shipped inside the vessel. Careful design of shipping fixtures is needed to avoid damage during shipment. Also, shaft stresses due to shipping in a horizontal position must be checked, allowing for about 2 g of acceleration when the tractor-trailer hits a big pothole at highway speeds (this can actually cause more stress than the normal operating condition!).
- An option is to ship the wetted parts to the jobsite with some field welding required. Typically, hubs and coupling halves are welded at the factory; then, the blades are welded on and the couplings welded together inside the tank at the jobsite. This still requires a larger agitator mounting flange than standard. Since the impellers are often quite far off the tank bottom in a large fermenter, considerable rigging may be required. The welding crew must rigorously follow the manufacturer's welding specifications. Sometimes, the blades are bolted in place and then welded to the hubs; then, the bolts are removed and the bolt holes filled in with weld metal. A similar procedure can be used for coupling bolt holes. Both procedures involve a lot of weld metal. The advantage of using bolts first is that it assures proper alignment of all parts before welding. Some leave the bolts in place and just seal weld them. That is a structurally sound practice, but creates much extra labor if polished construction is required. Care must be taken not to weld in a way that distorts parts, especially the shaft.
- Once everything is welded and inside the tank, changes are quite difficult to make, as is repair of any broken impeller blades or shaft damage. A little extra conservatism in design will be helpful.
- If the shaft design requires a steady bearing (as most fermenter agitators do), having some crevices is unavoidable, so one may call into question whether there is any merit to an all-welded design.

In-Tank Couplings

Agitator shafts are rarely made in one piece on large units. Each individual section is joined to the next by means of rigid in-tank couplings. These can be welded to the shaft, with each section bolted to the next, or they can be removable, using a straight bore with shoulder fit or a taper bore fit. Figure 18.1 illustrates a simple welded in-tank coupling. Figure 18.2 illustrates a removable taper bore in-tank coupling.

From examining these cross-sectional views, one can readily see that the interior of the coupling will fill with broth, which will not readily be cleaned out between batches. Moreover, when the shaft size is large (say, >100 mm diameter), it will take a great deal of time for the interior of the coupling to reach sterilization temperature. Hollow shafts take even longer, as the air inside the shaft acts as an insulator. In short, in-tank couplings are probably the biggest pus pockets in the tank! For the best sanitary design, one should not have any in-tank couplings at all. However, there are a number of reasons why in-tank couplings may be unavoidable. A few of these are as follows:

- The stub shaft through the mechanical seal requires precision machining which may not be possible with a long piece of shaft.
- The above machining may require starting with an oversize shaft diameter for machining allowance, which might be expensive if done for the entire shaft length.

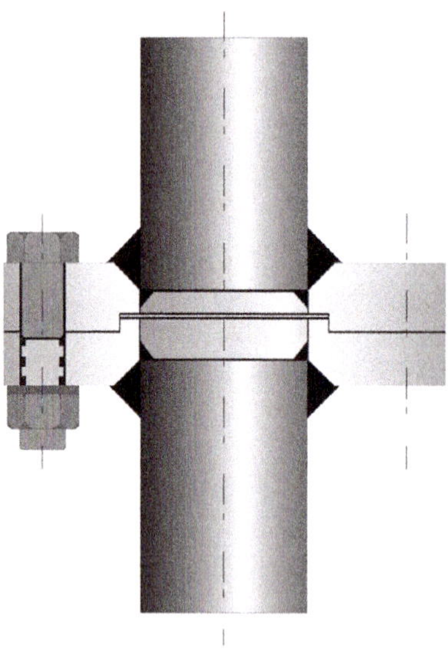

Figure 18.1 Welded in-tank coupling.

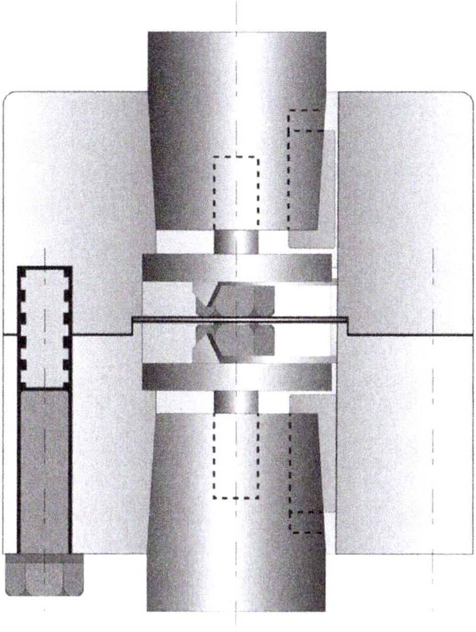

Figure 18.2 Removable in-tank coupling, taper bore style.

- The crane and rigging required to install a long one-piece shaft in the tank may be prohibitively expensive.
- If the shaft fails or the steady bearing wears through to the shaft, replacement cost, including rigging, may be very high or even almost impossible if the building is closed up.
- The agitator manufacturer may not have a lathe long enough to machine and straighten a one-piece shaft (typical shops have a lathe capacity of 10 m long or less).
- The required quality of round bar stock or hollow shaft material may not be available in sufficient length. Generally, large stainless bar stock, turned, ground, and polished, is limited to about 7.5 m maximum length.
- It may be desirable to have a removable stub shaft at the steady bearing to ease replacement if it becomes necessary.

So, in most cases, in-tank couplings are unavoidable. The question then becomes, How do we make them more sanitary or at least sterilizable? There are several ways, varying in cost and effectiveness. We will discuss a few here.

Over the years, I have heard many stories about how to deal with infections in a fermenter. Some had to do with in-tank couplings. Most fermenter agitators are equipped with mill finish, bolted construction, and welded in-tank couplings.

But that is not always satisfactory. One engineer I talked to found that he could greatly reduce infection and contamination problems by cleaning the metal parts thoroughly, then applying food-grade RTV silicone over the entire coupling, bolts, parting lines, and all. This is a rather cheap method, and it seems to work. Eventually, adhesion of the RTV silicone will fail, and the cleaning and sealing will have to be repeated. But it may be a way to improve productivity on an existing system that has no sanitary design features. I do not have an illustration of this.

A simple modification can be made to a welded in-tank coupling to improve its sterilizability. In essence, it involves sealing the interior of the coupling, just inside the bolt circle, with an O-ring or a gasket. What this does is allow sterilization to occur by getting the area from the O-ring outward sterilized; it is unnecessary to get the coupling interior up to sterilization temperature, as it is sealed off from the process. Field experience suggests that this can take hours off the sterilization time required. This modification is illustrated in Figure 18.3.

One drawback of this design is that there is no way to directly detect when the O-ring has taken a compression set and starts to leak. The main indication is that increased batch contamination will occur. Field experience suggests that the O-ring should be replaced at least every 5 years. This is not an easy task, as it may require rigging to support the shaft while the O-ring is replaced.

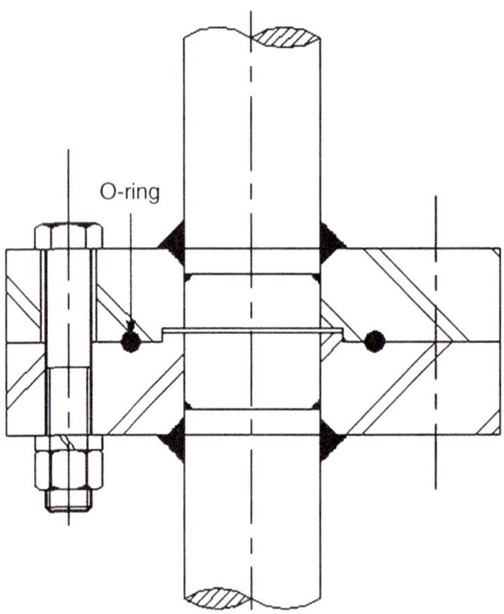

Figure 18.3 Sterilizable coupling.

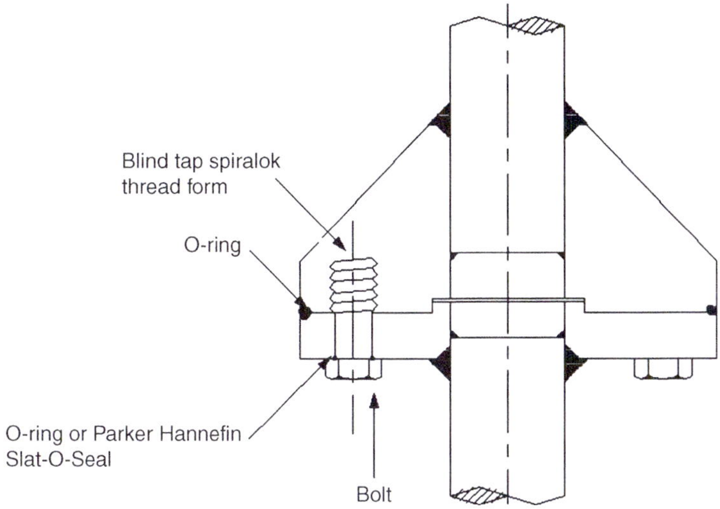

Figure 18.4 Sanitary in-tank coupling.

A more sanitary design can be found in Figure 18.4. This design has a sloped upper surface, is sealed at the OD between coupling faces, and has each bolt sealed by either an O-ring in a chamfered hole or a specialty seal such as the Parker Hannifin Stat-O-Seal. Thus, there are no exposed threads and no large pus pockets. As noted for the simpler coupling above, the O-rings can take a set and should be replaced at least every 5 years.

The shaft sections can also be joined using internal threads. Care must be taken to never operate the unit in a reverse rotation from the direction it was designed for, as the shaft will disassemble, leading to damage to the shaft and tank internals. But this may be the most sanitary design short of a fully welded joint. This is illustrated in Figure 18.5. Same note on the O-ring. If the threaded section must be larger in diameter than a straight shaft would allow, a thick sleeve or larger shaft can be welded on, with a sloped top surface to promote drainage. This option is not illustrated here.

Mounting Flange Area

We will not get into seals here, as they were covered in Chapter 15. However, we did not address some of the sanitary issues relative to mounting flanges there, so we will discuss those here. Figure 18.6 depicts the area where the agitator flange mounts to the vessel flange, with the shaft entering through the hole in the middle. With normal flange design, especially on large units, there

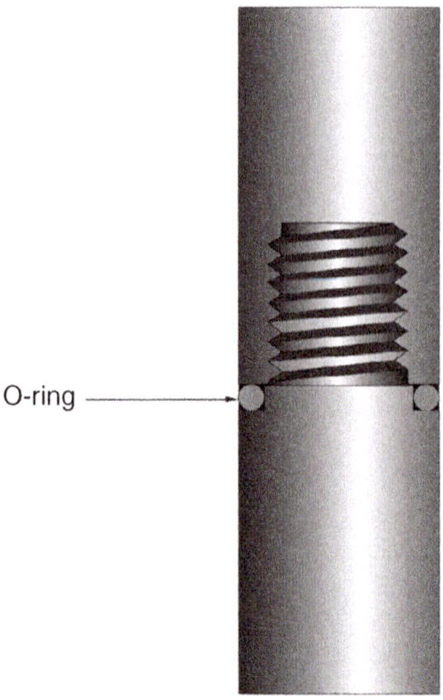

Figure 18.5 Internally threaded shaft coupling.

should be a spacer ring just inside the bolt circle to control gasket compression and support the agitator bending loads, essentially making a rigid mount (we do not want the drive to "wobble" due to gasket flexure). The gasket is placed just inside the spacer ring. Some gaskets may be rigid enough that they do not need the spacer, but we will assume it is needed here. The sanitary problem arises from the fact that the inside diameter (ID) of the gasket is typically larger than the ID of the vessel flange, as shown in Figure 18.6a. This creates a ledge where material can and will accumulate in a fermenter, as there is a mist throughout the vapor space, caused by the vigorous aeration of the broth. This is also in a difficult area to clean. The right side of Figure 18.6b shows a simple modification: adding a PTFE tape that fills in the ledge area. The same individual who told me about the RTV silicone treatment told me about this one. He reported that this also greatly reduced contamination. An alternative to this approach is to custom-order a gasket that has an ID matching the vessel flange ID or very slightly smaller. Though the drawing shows a relatively tight clearance around the shaft for compactness of the illustration, it is better to have a large opening to allow for spray ball streams to fully wash the area and to pass the top in-tank coupling. A steam port in the gland below the shaft seal (not shown here) can also aid in sterilizing the flange area.

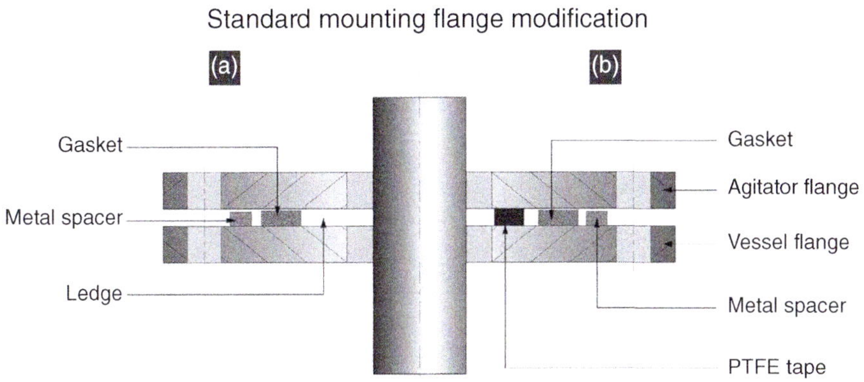

Figure 18.6 Standard mounting flange modification.

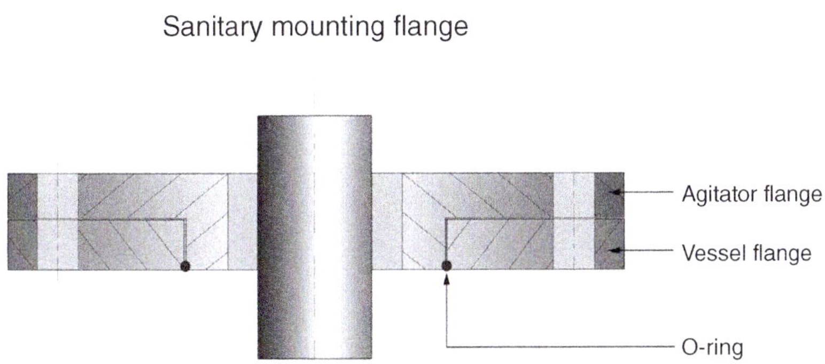

Figure 18.7 Sanitary mounting flange.

The above tweak is helpful for improving results for standard flange designs. It is also possible to design and build systems which are intrinsically more sanitary. One such design is shown in Figure 18.7. In the figure, the agitator flange is machined with a nosepiece that ends flush with the bottom edge of the vessel flange and is sealed by an O-ring as close to the bottom as possible. Note that the flanges touch metal-to-metal, so there is no problem supporting the agitator loads. The vessel fabricator must machine the flange exactly as instructed by the agitator vendor, or the agitator vendor can supply the vessel flange, to be welded in place by the vessel vendor. On small vessels, sometimes tri-clover fittings are used, but these are not suitable for large agitators.

Axial Impellers

As noted earlier, most production fermenters use bolted construction. Figures 18.8 and 18.9 illustrate two versions of bolted high-solidity impellers. While differing in construction details, both have many areas where it is difficult to clean, such as

- Bolts and bolt holes,
- V-shaped troughs formed between the extension blade and the stub-blades or along hub gussets. These do not readily drain.
- Small annular gap between the hub and shaft,
- Keyways,
- Locking device at the keyway.

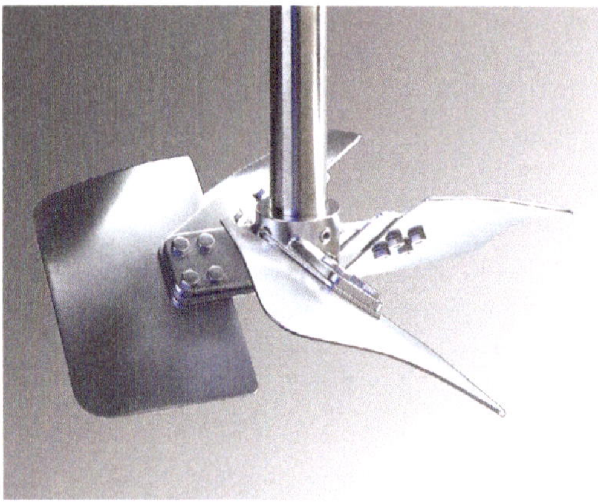

Figure 18.8 Lightnin A-315. *Source:* Courtesy of SPX FLOW, INC. © 2019 SPX FLOW, INC.

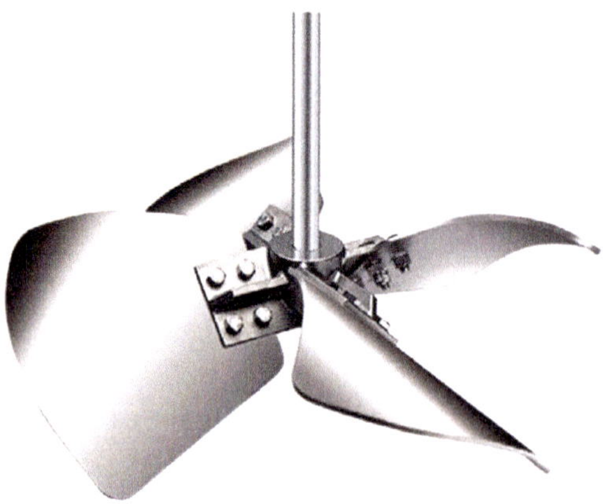

Figure 18.9 Chemineer Maxflo W-4. *Source:* Photo courtesy Chemineer, a brand of NOV. Permission granted by NOV.

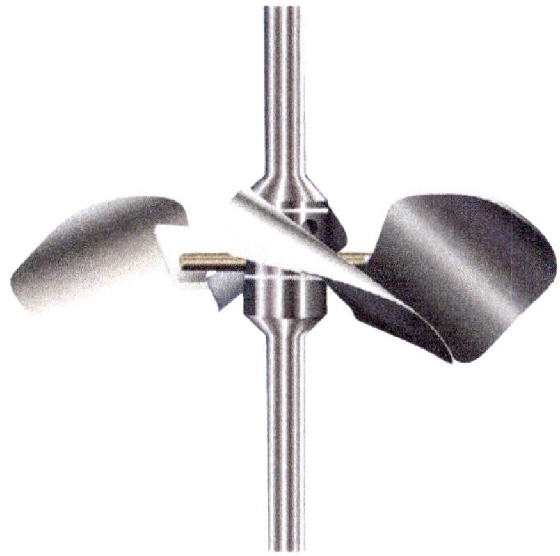

Figure 18.10 Smoothline impeller. *Source:* Photo courtesy Chemineer, a brand of NOV. Permission granted by NOV.

All of the above problems can be dealt with by using an all-welded construction, which may not be practical for some installations. It is also possible to weld the hub to the shaft at the factory and bolt the blades on or field-weld them. A welded hub may help somewhat. Generally, the agitator mounting flange must be large enough to pass the hub, which may require a non-standard flange size. Coating the entire hub area with food-grade RTV silicone is another option, as it was for in-tank couplings.

It is also possible to design an impeller that assembles in pieces and yet has no exposed fasteners. An example of such construction is the Smoothline™ impeller by Chemineer, a brand of NOV. It is illustrated in Figure 18.10. The sections are driven by internal threads and pins and are sealed at the OD by O-rings or a gasket. This construction is expensive, particularly when multiple impellers are needed. But such a design greatly simplifies installation and vessel design requirements. The opening required for installation needs to be just a little larger than the impeller blade width.

Radial Impellers

Though the process performance advantages of deep concave turbines are undeniable, as outlined in Chapter 5, they do have one disadvantage compared to a Rushton turbine: they are more difficult to clean. The deeply cupped blades make access from a spray ball stream problematic. They are also more difficult to polish.

Figure 18.11 shows a rendering of a Rushton turbine, courtesy of Fusion Fluid. One can see that the blades are fairly easy to clean. This rendering has a set screw-driven hub, which would only work on small units (typically up to 3.7 kW, at shaft speeds of 230 rpm or higher). Larger units would typically be welded or key driven.

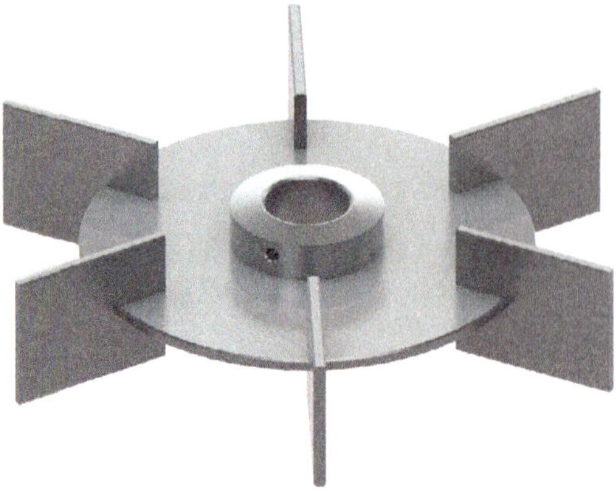

Figure 18.11 Rushton turbine. *Source:* Courtesy of Fusion Fluid.

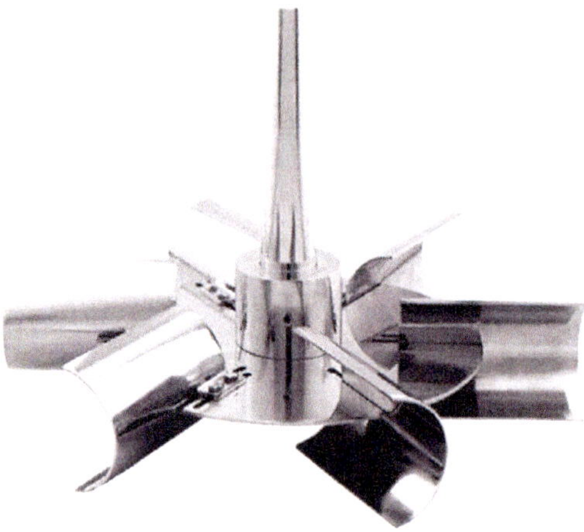

Figure 18.12 Chemineer CD-6. *Source:* Photo courtesy Chemineer, a brand of NOV. Permission granted by NOV.

Figure 18.12 depicts the Chemineer version of the Smith turbine, called the CD-6, with 180° of concavity. It is of polished construction but has bolted blades and adjustable slots for varying the impeller diameter. Thus, the polish is more aesthetic than sanitary, as the bolts, etc., provide many places where material can accumulate.

Figure 18.13 depicts a deep concave impeller, the Chemineer BT-6, with blades welded to the disc. Ideally, for the most sanitary design, the disc would be welded to the hub and the hub welded to the shaft. The polish would then add some value from a sanitary viewpoint.

One can readily see that as the blades become deeper, spray access is diminished. Also, there is probably no practical way to extend the Smoothline™ concept to a disc turbine, regardless of blade shape. Many who operate fermenters actually immerse the lower impeller in cleaning/disinfecting fluids rather than relying on spray balls. Steam is often used between batches, and it will at least kill whatever hangs up on the blades.

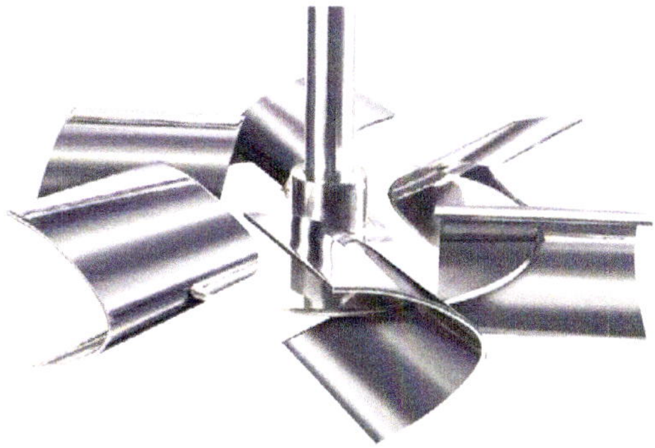

Figure 18.13 Chemineer BT-6.
Source: Photo courtesy Chemineer, a brand of NOV. Permission granted by NOV.

The presence of the disc also impedes spray access, unless there is also a spray ball below the disc. And a flat disc does not drain fully, potentially allowing material build-up. The Ekato Phasejet patent drawing shows a conical disc instead of a flat one, which may provide better draining. However, it is more difficult to manufacture, and the underside is difficult to reach with a spray ball.

Though rare, some have opted to use radial impellers with no disc. These are easier to clean. However, they must be about 5% larger in diameter to draw the same power. Some worry that gas may bypass the impeller without the disc. However, in an unpublished study that the author is aware of, the flooding point was unaffected by the absence of a disc as long as the power was equal to that of an impeller with a disc, and the gas was sparged using a ring 80–100% of the impeller diameter. Mass transfer was not investigated in this study. Some feel that an open impeller will result in a lower $k_l a$ value even at equal gassed power. Others say that only the gassed power matters. For now, it is an open question.

Bolts and Nuts

When bolted construction is used, it is important to make the design as freely draining as possible. This means that blind tapped holes should be avoided unless they can be sealed off at the bolt head. Any unnecessary crevices should be avoided. This should be obvious, but I remember a discussion with a client many years ago that tells me it might not always be so obvious. The client specified a highly polished construction with bolted construction. He further specified that the nuts should be castle nuts with cotter pins. (Ugh!) I asked, What was the point of polishing with such an unsanitary bolting method? He basically replied, "Just do it!"

One of the more egregious bolting mistakes is the use of acorn nuts, as shown in Figure 18.14. They look pretty, especially when polished. But an examination of their cross section reveals that they create a pus pocket that cannot be cleaned. It is better to just use standard bolts and nuts.

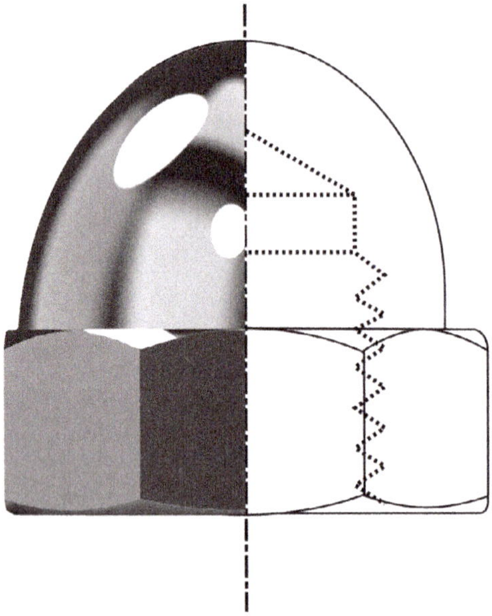

Figure 18.14 Acorn nut.

Steady Bearings

From a sanitary design standpoint, the best steady bearing is none at all. But in most large fermenters, they are unavoidable. Each agitator vendor has its own "standard" steady bearing design. Some have several. There is no industry-wide steady bearing design. But a common feature of all of them is that they are far from being sanitary. An example of a so-called "tripod" steady bearing is found in Figure 18.15. One can readily see that there are many cavities in this design, and the flat plate that supports the housing does not drain well.

A simple conceptual drawing of a sterilizable, sort of sanitary steady bearing can be found in Figure 18.16. To the extent possible, all surfaces are sloped for good draining. Exposed threads are minimized, but one is needed to keep the bushing from rotating or falling out. The bushing is shown as having a taper. This requires machining of the bushing. This may not be possible for all bushing materials. It works for PTFE and similar materials, as well as bronze. But it will not work for brittle materials such as graphite or metallized carbon, especially when placed under compression by a shrunk-on sleeve. Other designs are possible.

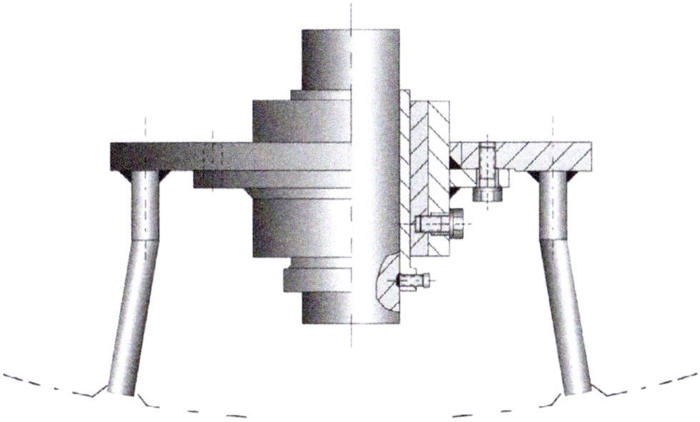

Figure 18.15 Tripod steady bearing.

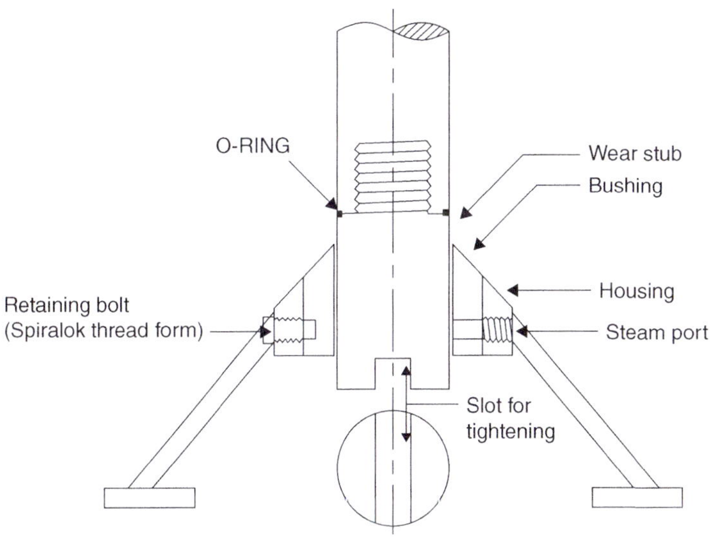

Figure 18.16 Sterilizable steady bearing.

Use of Castings, 3-D Printing

For economy, many manufacturers cast, rather than fabricate, their axial flow impeller hubs. They do this up to a certain size limit that varies by manufacturer, but a typical maximum shaft bore for a cast hub is about 150 mm. Large sizes that are made in small quantities are usually fabricated.

Castings vary in quality. Sand castings are rather rough and are generally not used for wetted parts in the agitator industry, but they may be used for items outside the tank such as pedestals. Most cast turbine hubs are made by investment casting. This produces a dimensionally accurate hub with relatively few imperfections. However, all castings can have some pits. If pits are found, they will not drain or clean out easily, as they tend to be small and may be larger beneath the surface, making a mini pus pocket. Each user must decide whether castings are acceptable. My own opinion is that if bolted construction is used, castings should be OK, as they are not any worse at creating crevices than bolt threads. However, if welded and polished construction is deemed necessary, it is probably best to avoid castings.

3-D printing is a relatively new technique and, to my knowledge, has not been applied to large agitator parts, such as hubs. But it has the potential to be used to make complex hub designs in any size. It is not clear to this author whether 3-D printing can produce a pit-free result.

Polishing Methods and Measures1: Polishing vs. Burnishing

The term "polishing" is used to refer to a lot of things that may have different purposes or techniques. For example, people refer to polishing their cars or polishing furniture. These kinds of polishing are for aesthetics only and actually involve adding material (typically some kind of wax or acrylic) to fill in pores or scratches and then buffing to a smooth, shiny surface. But for fermenter design, we are not referring to that kind of polish. We are interested in it for sanitary reasons, not aesthetics per se. That means we do not want material to hang up nor do we want crevices where organisms can hide.

This type of polishing involves removing material, not adding it. Basically, we scratch the surface of the metal deliberately to get a certain surface finish. There are several quantitative measures of polish and several qualitative measures that can also affect results. Different techniques affect these measures also. We will discuss these measures in turn. But first, we need to make a distinction between polishing and burnishing.

As stated above, polishing involves material removal, using abrasives. Strictly speaking, burnishing is done by rolling a hard, smooth surface over another so as to flatten any peaks, thereby making a smoother, shinier appearance without removing any metal. This is fine for decorative metal applications, but is not good for sanitary applications. It may not be obvious why I even bring this up, since the processes of polishing and burnishing seem so different that no one would get them confused. But burnishing can inadvertently occur when the polishing technician uses the same abrasive media too long. This results in dull edges on the media rather than sharp, cutting edges. These dull edges can bend peaks over and cause a burnished result, which may look good but is not sanitary. In Figures 18.17 and 18.18, we compare polishing with a sharp abrasive to inadvertent burnishing with dull abrasive, in an artist's conception of a metal cross section.

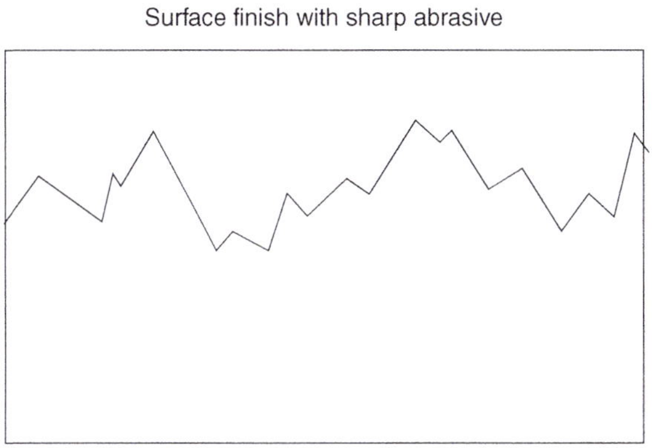

Figure 18.17 Polished surface.

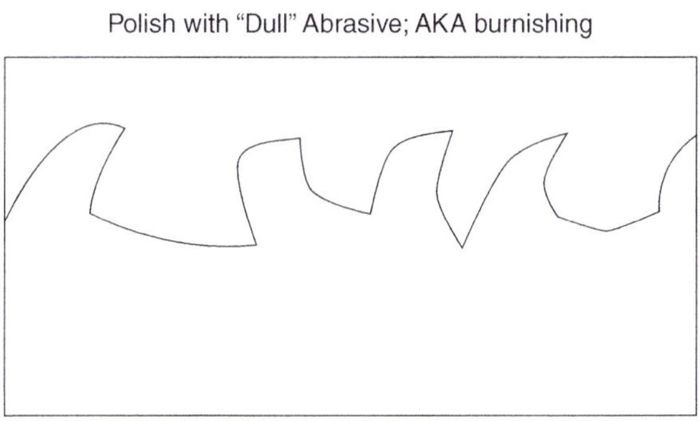

Figure 18.18 Burnished surface.

One can see that the valleys in the "polished" surface remain open, but in the dull abrasive, "burnished" metal, the peaks may be bent over enough to overhang the valleys. This can impede cleaning, trapping material and organisms there. So, polishing personnel must change abrasives frequently to achieve acceptable results.

Polishing Methods and Measures2: Lay

Since polishing is done by rubbing abrasives across the metal surface, this action creates a scratch pattern, called its "lay." There are a number of different possible patterns that may be created, either deliberately or just incidentally. Some are for decorative purposes. The subject of lay is discussed in more detail in Ref. [1]. For our purposes, there are two main lay patterns used for sanitary polishing: unidirectional and random lay. These patterns mainly depend on the method used to move the abrasive across the metal surface. For example, belt sanders produce a

Figure 18.19 Unidirectional lay.

unidirectional lay. This is usually how plate material used for impeller blades are polished, at least before bending. Also, flapper wheels such as used on a traversing fixture to polish a shaft produce a unidirectional lay. Hand sanding often approximates a unidirectional lay, with some curvature in the scratch lines. A unidirectional lay is illustrated in Figure 18.19.

Rotary sanders and orbital sanders, on the other hand, often approximate a random lay, with circular segments that cross each other again and again. Figure 18.20 illustrates this concept.

So which is better? In general, unidirectional lay is thought to yield better washability and drainability. But the direction of the lay matters also. For impeller blades, the lay should be as close to vertical as possible, meaning in the direction of the shaft axis when in an elevation view. This avoids microridges that might impede draining if the scratches were in a radial direction or normal to the shaft axis.

For radial impeller discs, the ideal pattern would be radial. But a flat disc does not drain well anyway, so the lay probably has little overall impact. In any case, it would be difficult to create a radial lay on a disc with any common polishing method.

Shaft polishing is often done by mounting an abrasive flapper wheel on a lathe and having it rotate, while the shaft is slowly rotating, and simultaneously, the flapper wheel unit traverses the length of the shaft. This method produces a very uniform finish. Unfortunately, the resulting unidirectional lay is circumferential. It would be better to make it longitudinal. However, a longitudinal lay would be more difficult to automate and thus may cost more.

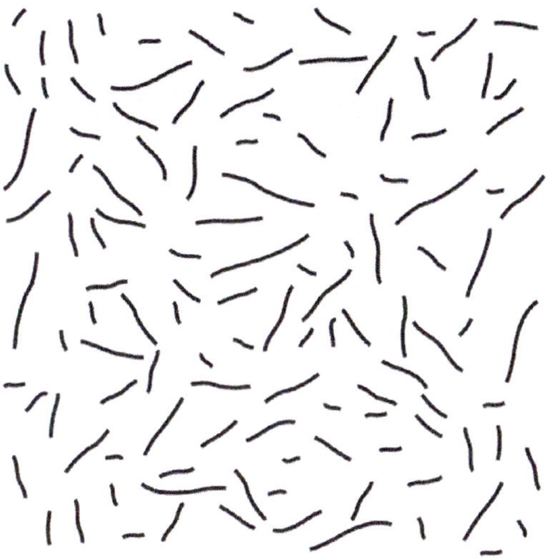

Figure 18.20 Random lay.

Many polishers use successively finer grits and change the direction of the lay with each grit size to make it easier to see when the coarser scratches are obliterated. But the final passes should be in the desired lay direction.

The project engineer, in conjunction with microbiologists, should decide the relative importance of the polishing lay and incorporate the decision into the equipment specifications. This is rarely done. Most specifications merely address the surface roughness, which is discussed in the next section.

Polishing Methods and Measures3: Roughness Average

One quantitative measure of a surface finish is its relative roughness. This concept is discussed in a fair amount of detail in Ref. [2]. We will give an abbreviated version here and discuss alternate ways of describing the finish.

Historically, surface finish has been described in a number of ways. There are numbered finishes, such as number 4, 2, and 2b. Some refer to a "dairy" finish. Others use the size of the abrasive, e.g., 120 grit finish and 180 grit finish. The problem with the numbered finishes is that they are a bit qualitative. The problem with using grit size to define the finish is that it actually just defines the tool used to make the finish, not the finish itself, which depends a lot on technique. We will illustrate that a little later. But first, we will define a quantitative means of measuring surface finish: the roughness average or RA.

RA is defined as the two-dimensional arithmetic mean value of the deviation of the peaks and valleys in the surface from the mean depth of those peaks and valleys. The two-dimensional measurement is taken across, or perpendicular to, the predominant lay. It is measured by a profilometer, which is a mechanical or optical instrument. Since most such instruments simply average the readings, RA is the most common measure.

Some argue that root mean square, or RMS, roughness figures may be more valid, as they give greater weight to the rougher readings. There are some instruments that will directly give such readings. Otherwise, the raw data could be numerically processed to give the RMS figure. However, with proper polishing technique, it has been found by experience that the RMS value is about 10% higher than the RA value. So, if an RMS value is specified, most vendors will measure RA and shoot for about a 10% lower value than the RMS value specified. Roughness can be stated either in microns or microinches. This author recommends the use of RA values over other means of describing surface finish. Table 18.1 lists some fairly common surface finishes with several descriptions in addition to the RA value. Note that the relative prices are for the polishing operations only, not the entire equipment cost.

Table 18.1 Surface finishes.

		Roughness values, RA		RMS	
ISO grade numbers	**Other description**	**Microns**	**Microinches**	**Microinches**	**Relative polish cost**
N10	Mill finish	12.7	500	550	0
N9	Mill finish	6.4	250	275	0.1
N8	Ground welds	3.2	125	137.5	0.3
N7	150 grit; No. 2 polish	1.6	63	69.3	0.5
N6	!80 grit; dairy finish	0.81	32	35.2	1
—	180–220 grit; No. 4 polish	0.64	25	27.5	1.1
N5	240 grit; No. 6 polish	0.41	16	17.6	1.3
N4	280–320 grit; No. 7 polish	0.20	8	8.8	1.7
N3	500+ grit	0.10	4	4.4	3
N2	Mirror finish	0.051	2	2.2	10
N1	True mirror finish	0.025	1	1.1	20

Figure 18.21 Fine vs. coarse abrasive with same RA.

So what finish is needed? I have no way of knowing. But the labor costs associated with polishing increase exponentially as the RA value goes down. Castings often cannot be polished down to an RA value of 0.5 μm or below. In my experience, most pharmaceutical applications get polished to an RA value in the 0.4–0.8 μm range or 16–32 micro inches. A few have been down around 0.1 μm or less. This can double the price of the equipment, especially if a hidden flaw shows up that cannot be polished out, thus requiring the part to be scrapped and fabrication to be started over.

All things considered, specifying RA values is a good quantitative way to characterize the finish. However, reality is a bit more complicated than that. It is possible to get the same measured RA value with different grit sizes by varying pressure and lubrication. A coarse grit with low polishing pressure and the use of heavy grease can cut shallow, but wide, scratches of the same depth as a finer abrasive used dry with more pressure. The grooves are narrower with the finer abrasive but may have the same average height. This is illustrated conceptually in Figure 18.21. I actually saw a demonstration of this many years ago, where a plate was polished with 80 grit to a 32 RA value (microinches) using a grease lube, and side by side was one polished with 180 grit (dry), also to 32 RA. The finer grit result looked better, but both were equally smooth to the touch. The 32RA finish made with the coarse abrasive was arguably more sanitary, as its wider grooves were more accessible to cleaning agents. It is not clear which is cheaper, as the use of the coarse abrasive requires the extra step of grease removal.

Electropolish

Just as mechanical polish involves material removal, electricity can be used to selectively remove material. This process is called electropolishing. In essence, the part to be polished is placed in a vat of electrolyte. The part is then connected to a positive DC current, making it the anode. A separate conductor or conductors are connected to a negative voltage, making them the cathodes. The passage of electric current causes the part's metal to dissolve and migrate toward the cathode. The rate of metal removal is directly proportional to the current applied. What makes this a polishing technique is that the rate of removal is not uniform. The current flux (current per unit area) is much higher on protruding surfaces, especially sharp ones. This means more metal is removed from peaks on the

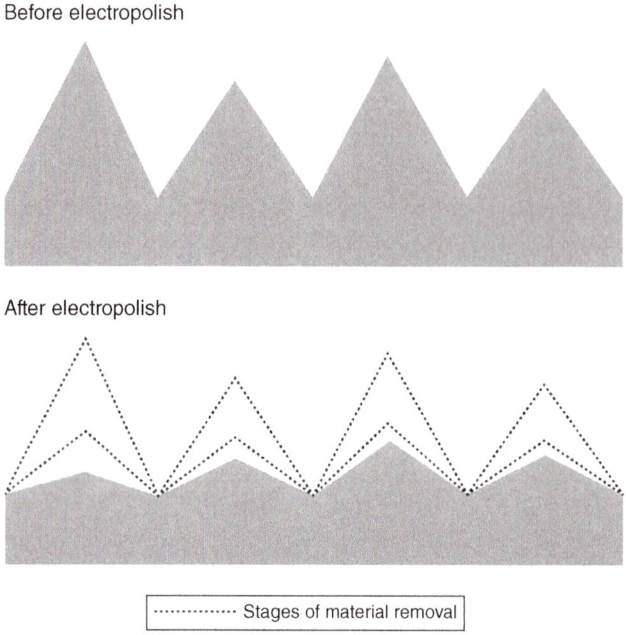

Figure 18.22 Electropolish process.

surface and less in the valleys. This both rounds over the peaks and in principle, reduces the RA value. It can potentially reduce the RA value by a lot. The surface also tends to be more reflective (shinier) and definitely easier to clean. The process is illustrated conceptually in Figure 18.22. The dotted lines indicate a couple of profiles as material removal progresses.

Generally speaking, electropolishing is cheaper for the degree of surface improvement gained than mechanical polish. However, for it to work well, the surface must first be properly mechanically polished, to make sure there are no burnished areas. Referring back to Figure 18.18, if the surface is burnished due to either rolling the surface or using a dull abrasive, electropolishing can open up the hidden valleys and result in an actual increase in the measured RA values. However, this is still an improvement in the sanitary function, as it is more cleanable. RA value can also increase if there is excessive turbulence in the electrolyte, which must be circulated to keep the dissolved salt concentration uniform. Whereas it is normal for electropolishing to improve the RA value by 2 or 3 ISO grade levels, I have personally seen cases where the starting RA value was 20–30 microinches and the RA value after electropolishing was more than 250. A combination of poor surface prep and too much turbulence was likely responsible. So, it is wise to enforce proper mechanical polishing techniques and choose a reliable electropolish vendor to get the desired result.

Passivating

While not a sanitary topic per se, some will pre-passivate stainless steel to improve its corrosion resistance. The process usually involves brushing or immersing the parts in an oxidizing acid, to help form the protective oxide film. It may be done at the manufacturers shop if safety and environmental concerns are addressed. However, it is this author's opinion that doing this is a waste of time and money if done anywhere other than in the tank the agitator is installed in. This is because the handling involved in shipping and installation will almost certainly result in damaging the finish in a few spots. These spots will need to be touched-up, and if passivating is actually needed, it will have to be done again at the jobsite, within the tank.

Effect on Mechanical Design

Since polishing involves material removal, an allowance for the amount of material removed must be figured into the mechanical design calculations. Typically, this could be up to 1 mm per side or 2 mm on a shaft diameter. From a strength viewpoint, the effect is greater for small parts, because the thickness of material removed is mostly independent of the size of the part for a given degree of polish. I can recall once where we had to polish a 150-mm-diameter concave turbine. We figured on a generous polishing allowance. However, by the time the polishing was done, the impeller had lost 1/3 of its weight. It was still strong enough, but the agitator was a small unit designed to run above first critical speed. The weight reduction caused the unit to be too close to critical speed, so we had to add a compensating mass above the impeller to fix the critical speed problem. Yes, we had to polish that, too!

On large fermenter units, the material loss normally does not cause problems, but still must be planned for. In places where precision parts fit-up is required, such as a shaft through a mechanical seal or through an impeller hub, polishing cannot be done. Instead, precision machining with a very fine cut is required.

Summary of Chapter

We have described the principles of sanitary design and provided several specific examples. We have discussed various degrees of polishing. The project engineer should use these principles to specify what is required and clearly communicate these to the vendors.

Additional Sources of Information

An entire book could surely be written on the issues raised in this chapter. But two additional resources that should be useful are publications by the International Society for Pharmaceutical Engineering (www.ispe.org) and the ASME (American Society of Mechanical Engineers) BPE (Bioprocessing Equipment) guides. An example link is https://www.asme.org/codes-standards/find-codes-standards/bpe-bioprocessing-equipment-(1). Published guides are updated periodically.

List of Symbols

$k_l a$ Overall mass transfer coefficient

References

1 https://en.wikipedia.org/wiki/Surface_finish (defines lay) (accessed 12 October 2020).
2 https://en.wikipedia.org/wiki/Surface_roughness (accessed 12 October 2020).

19

Aspect Ratio

This chapter presents an overview of what effects aspect ratio has on the power and capital cost of fermenters, and some comments on process performance and heat transfer capability. As noted in Chapter 2, aspect ratio is one aspect (pun intended) of the design choices that can be made, and it can potentially be part of the overall process optimization.

Acknowledgment

Much of the inspiration for this chapter, and some of the calculations and figures, came directly from Reference 1, an article I wrote for Chemical Engineering Progress (CEP). All figures attributed to CEP were used with blanket permission.

Definition and Illustration of Aspect Ratio

For the purposes of this article, aspect ratio is the ratio of the ungassed liquid level, Z, to the tank diameter, T, rather than the total height of the tank divided by its diameter. Generally, the lower limit of aspect ratio for an industrial fermenter is about 0.5, though it is possible to build a shallower geometry than this. Too shallow will require multiple agitators (this means the whole machine, not just the impellers). Though it is possible to build very tall fermenters, the tallest aspect ratio I have seen in a mechanically agitated fermenter is about 5, though bubble columns and airlift designs may be taller. Figure 19.1 illustrates aspect ratios of 0.5 and 5 with the same volume, drawn to scale, so that one may visualize how different these geometries are. The dotted horizontal line represents the ungassed liquid level.

Agitator Design for Gas–Liquid Fermenters and Bioreactors, First Edition. Gregory T. Benz.

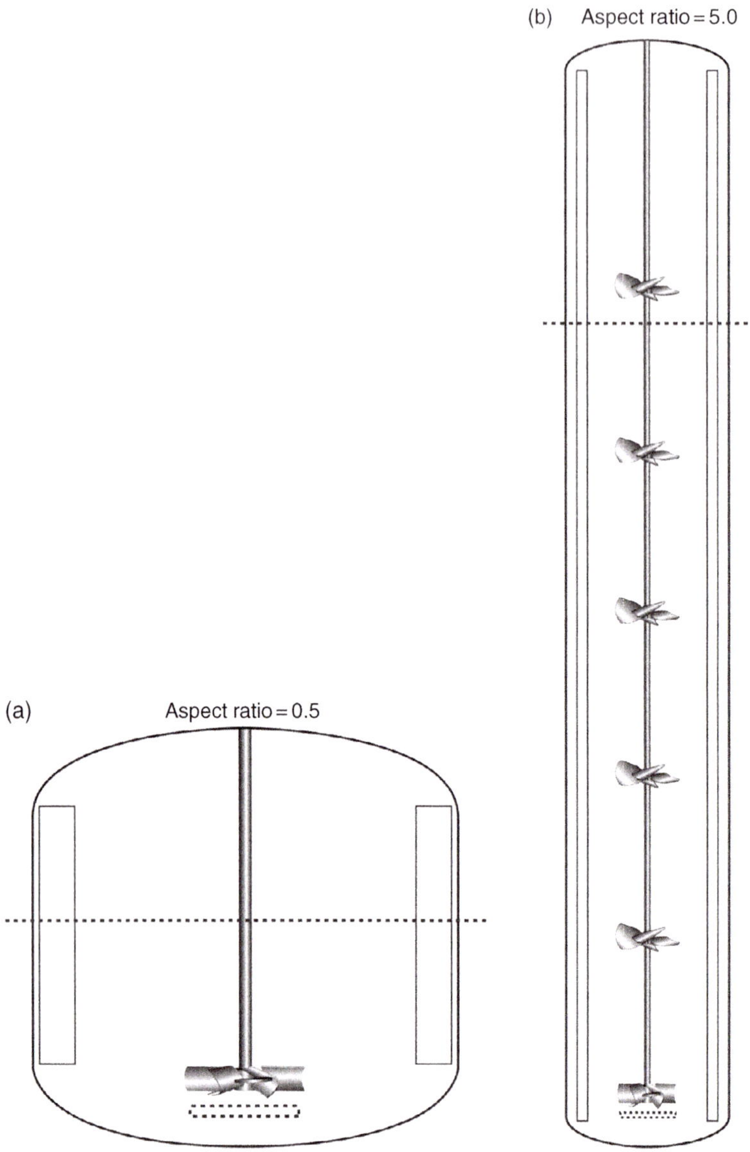

Figure 19.1 (a) and (b) Aspect ratio illustration

What Is the Optimum Aspect Ratio?

One question I have often been asked over the years is "What is the optimum aspect ratio?" The real answer is, "It depends." But what does it depend on? The optimum Z/T ratio depends on every detail of the design requirements, including, but not limited to: the mass transfer rate desired, dissolved gas uniformity required, cost of electric power, cost of the agitator, cost of the vessel, cost of real estate, heat transfer requirements, mass transfer correlation used, back pressure imposed on the tank, site restrictions, time horizon of the plant economic calculations, and probably a few others. The only way to find the optimum is to do detailed calculations over an array of Z/T ratios and then interpolate if needed. Some aspect ratios will be eliminated as infeasible in many cases. What this chapter seeks to do is identify the main effects of Z/T ratio and then illustrate the concept with example fermenter problems.

Effects of Z/T on Cost and Performance at a Given Working Volume

Vessel Cost

The effect of aspect ratio on vessel cost is not entirely straightforward. In general, the vessel becomes more expensive as Z/T increases. It has more surface area, and the greater height may increase installation cost. However, there may be a step change at the transition diameter between shop-fabricated and field-fabricated tanks. This will depend on the plant location and available logistics. In general, most locations that require truck shipment will be limited to a 14 ft (4.27 m) vessel diameter. Also, heat transfer surface requirements affect this trend. Tall, thin vessels may not need internals for heat transfer, so in some cases they may be less expensive.

Agitator Shaft Design Difficulty

Agitator shaft design becomes more difficult as Z/T increases at constant working volume. The longer the shaft extension required, the more difficult it is to keep the natural lateral frequency of the shaft at least 25% higher than the operating speed. Use of a bottom steady bearing is pretty much a given on almost all fermenters except maybe ones with very shallow aspect ratios. Especially tall designs may also require hollow shafts and sometimes even an intermediate steady bearing. This is not desirable from a maintenance standpoint, as scaffolding must be set up inside the tank to change out worn steady bearing components. An intermediate steady bearing is also less sanitary than just having one at the bottom. And partial batches can be a concern; the intermediate steady bearing must be submerged in liquid whenever the agitator is operating. Moreover, as Z/T increases, shaft speed tends to increase also, because smaller impellers must be used. This further increases the shaft design difficulty.

Power Required for Mass Transfer

Required agitator power for mass transfer goes down as Z/T increases. This is because of two reasons: The absolute pressure at the bottom of the tank becomes higher, increasing the mass transfer driving force by increasing the saturation concentration of the gas to be transferred, and the gas superficial gas velocity tends to increase due to less tank cross sectional area per volume, which will increase the overall mass transfer coefficient at a given power input, so agitator power input can be lessened.

Agitator Cost

Agitator capital cost may go up or down. The reduced agitator power associated with higher Z/T ratios will allow a smaller, less expensive motor, and the required gearbox size and cost decreases as well due to lower torque requirements (tall tanks tend to be forced into higher shaft speeds to avoid excessively large impellers; hence, less torque is required even if the power is the same. But this makes the shaft design problem even more difficult). However, more impellers and a longer shaft are required, tending to increase the wetted parts cost.

Airflow Requirements

Required airflow decreases as Z/T increases due to higher $k_l a$ at higher superficial gas velocities but there is a minimum needed based on stoichiometry. Also, for any given geometry, volume, oxygen transfer rate, back pressure, temperature, and mass transfer correlation, there is an optimum combination of agitator power and airflow that yields minimum total power. A procedure for determining this is described in Ref. [1] and our Chapter 8. In our example, we will be using the optimum for each aspect ratio.

Compressor Power

Compressor power may go through a minimum. Although airflow required decreases with increasing Z/T ratio, the liquid head increases, which will increase the compressor power required for a given airflow rate.

DO Uniformity

Dissolved oxygen gradient increases as Z/T increases. This is due both to a larger pressure gradient from top to bottom, increasing gas solubility at the bottom, and a longer blend time due to the greater distance each element of fluid must travel before blending to a certain degree of uniformity. Also, the fact that multiple impellers do not have perfect melding of their individual flow patterns increases blend time.

Heat Transfer Capability

Potential heat transfer area increases as Z/T increases. Sometimes, there may be a transition between external heat exchangers vs internal surfaces. In tall sizes, there may be a transition between a combination of jackets plus coils to jacket only or coils only. In other words, at low aspect ratios, it may be impossible to get the required heat transfer using a jacket and/or internal heat transfer surfaces. Thus, an external heat exchange loop may be needed. As the aspect ratio increases, an external exchanger may no longer be needed.

Years ago, I had a customer who required a fermenter of 5:1 aspect ratio for exactly that reason. Using such a tall vessel allowed the use of a jacket only for heat transfer, with no internal heat exchange surfaces. This was done for sanitary design reasons. Unfortunately, it then needed two steady bearings, which reduced the sanitary aspects somewhat and increased maintenance costs.

Real Estate/Land Usage Issues

In some plant locations, the greater footprint of a small aspect ratio design may lead to higher real estate cost. More likely, this can become an issue in an existing plant that is landlocked, so the only way to add capacity is vertically.

Building Codes; Noise

Though this may seem strange, some locales have height limits on buildings, which may narrow the choice of aspect ratio. Also, community noise regulations may require that noise sources be kept below a wall surrounding the fermenter, so tall units may not be allowed, at least with top-entering agitator drives.

Illustrative Problem Number 1

This problem is mainly taken from reference 1. Therefore, the capital costs are considered to be current for the calendar year 2019.

In the next several sections, we will look at various attributes of fermenters designed with aspect ratios of 0.5, 1, 1.5, 2, 2.5, 3, 4, and 5, on an ungassed basis. In each case, the airflow and power will be optimized for the conditions chosen, using the techniques described in Ref. [2] and in Chapter 8 of this book. The calculation details will not be shown here; just the results. The common conditions for all of these aspect ratios are enumerated below.

- Working volume of 40 000 gal or 151 cubic meters, ungassed basis (the gassed volume will be about 20% higher, depending on airflow and aspect ratio)
- Oxygen transfer rate (OTR) of 150 mmol/l-h
- Dissolved oxygen concentration (DO) of 1 mg/l at top of tank. DO at bottom will be higher, using the author's experience to estimate the value
- Saturation oxygen concentration at 21% oxygen, 1 atmosphere pressure is set at 6.8 mg/l
- Fluid density of 1020 kg/m^3
- Fluid viscosity of 50 cP
- Barometric pressure of 1 bar
- Back pressure of 0.69 bar is applied to the head space.
- The v'ant Riet ionic k_la correlation (Ref. [3]) applies: $k_la = 0.25(P/V)^{0.7}U_S^{0.2}$, where k_la is the overall mass transfer coefficient (1/s), P is the agitator power input under gassed conditions (w), V is the liquid mass (kg), and U_S is the superficial gas velocity, m/s
- It is assumed that 95% of the oxygen consumed is returned to the exhaust gas as CO_2
- The lower impeller is a deeply concave radial turbine (Figure 19.2). If more than one impeller is used, the upper impellers are up-pumping high solidity axial flow impellers (Figure 19.3)
- Agitator mechanical efficiency is deemed to be 95%. Compressor efficiency is assumed to be 70%
- Total pressure drop in the air handling system (piping, filters and sparge ring) is set at 1.72 bar.
- Additional relevant data are provided in each section below.

Vessel Dimensions

The vessel diameter, straight side, and ungassed liquid level are listed in Table 19.1 for each aspect ratio studied here. In most cases, the largest vessel

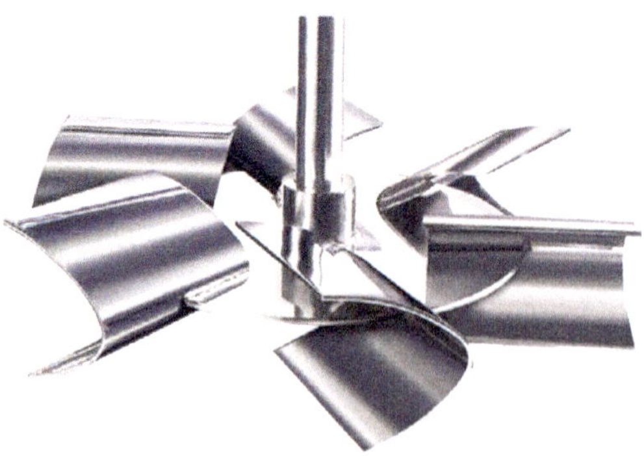

Figure 19.2 Deep concave turbine, Chemineer model BT-6. *Source:* Photo courtesy Chemineer, a brand of NOV. Permission granted by NOV.

diameter that can be shipped over the road is about 4.27 m. So, the vessels in this example problem with an aspect ratio of less than 2.5 would probably have to be field-fabricated.

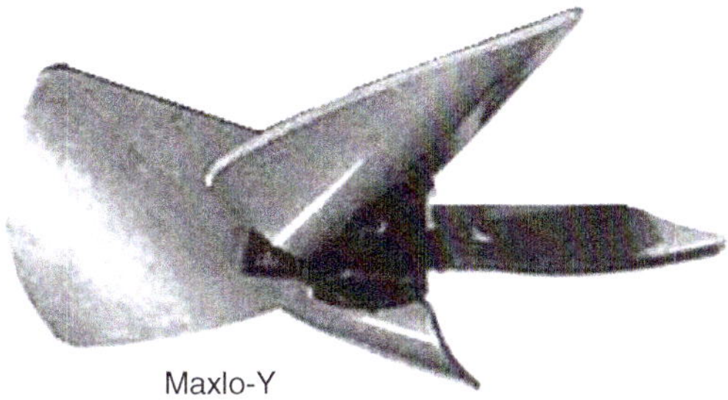

Figure 19.3 High solidity up-pumping impeller, Chemineer MY-4. Source: Photo courtesy Chemineer, a brand of NOV. Permission granted by NOV.

Table 19.1 Vessel dimensions.

Vessel Data (English units)			
Nominal ungassed aspect ratio	**Tank diameter, ft.**	**Ungassed Liquid level, ft.**	**Tank straight side, ft.**
0.5	25	12.5	14.5
1	19.5	19.2	23.5
1.5	17	24.7	31.5
2	15	31.3	39.5
2.5	14	35.7	45
3	13	41.2	52
4	12	48.1	61
5	11	57	72.5
Vessel Data (Metric units)			
Nominal ungassed aspect ratio	**Tank diameter, m**	**Ungassed Liquid level, m**	**Tank straight side, m**
0.5	7.62	3.81	4.42
1	5.94	5.85	7.16
1.5	5.18	7.53	9.60
2	4.57	9.54	12.04
2.5	4.27	10.88	13.72
3	3.96	12.56	15.85
4	3.66	14.66	18.59
5	3.35	17.37	22.10

Airflow and Power

Using the methods of Chapter 8, we chose the combination of airflow and agitator power that would yield the required mass transfer rate with the minimum sum of agitator plus compressor brake power (brake power as used here means power required at the output of the motor shaft, with efficiencies as described earlier in the problem statement).We varied the airflow in increments of 100 SCFM (0.042 std. m^3/s) for the optimization. Table 19.2 shows these results.

As expected, the optimum airflow requirements decrease somewhat with increasing aspect ratio, due to increasing liquid head. The compressor power goes through a minimum, however, as the increased head requires increased power for a given airflow, though the reduced airflow at higher aspect ratios allows the net

Table 19.2 Optimum power summary.

Optimum power (English units)				
Nominal ungassed aspect ratio	**Airflow, SCFM**	**Agitator BHP**	**Compressor BHP**	**Total BHP**
0.5	2900	406	424	830
1	2700	353	414	767
1.5	2600	322	413	735
2	2500	295	413	708
2.5	2400	287	407	694
3	2400	269	419	687
4	2300	257	416	673
5	2200	246	415	661
Optimum power (Metric units)				
Nominal ungassed aspect ratio	**Airflow, nm^3/s**	**Agitator Brake Power, kw**	**Compressor Brake Power, kw**	**Total Brake Power, kw**
0.5	1.21	303	316	619
1	1.13	263	309	572
1.5	1.09	240	308	548
2	1.04	220	308	528
2.5	1.00	214	303	518
3	1.00	201	312	512
4	0.96	192	310	502
5	0.92	183	309	493

power to decrease for modest increases in aspect ratio. The required agitator power drops consistently. The total power also drops consistently. So, energy cost favors higher aspect ratios. Because compressor power is in a fairly narrow band, we will assume that the capital cost will not vary, so a compressor capital cost will not be estimated for this book.

Heat Transfer Data and Assumptions

As aspect ratio increases, available heat transfer area increases. We will see what effect this has on design. Several assumptions apply:

- The metabolic heat load is 69 000 J/L-h or a total of 1.04E10 J/h (9.9E6 BTU/h). This corresponds to 460 000 Joules per gram mole of oxygen consumed, a fairly typical value. (1)
- The process temperature is 37 C (98.6 F).
- The cooling water inlet temperature is 15 C (59 F), which requires chilled water.
- The thermal properties of water apply.
- The fermenter is cooled via vertical tube bundles within the vessel and a surrounding cooling jacket (Figure 19.4). The flowrate of cooling water to the jacket and vertical tube bundles is 7570 L/min (2000 gpm) for each – i.e., 7570 L/min for the jacket and 7570 L/min for the entire set of tube bundles (we did not undertake the additional possible optimization step to see if there was a net heat transfer advantage by varying the flow split between the tube bundles and the jacket, though it is possible that a 50/50 split is not optimum).
- The process fluid has a density of 1020 kg/m^3, a viscosity of 50 cP, a thermal conductivity of 0.519 J/m-s-C (0.3 BTU/h-ft-F), and a heat capacity of 3977 J/kg-C (0.95 Btu/lb-F)
- The usable height of jackets and tube bundles is the ungassed liquid level. The tube bundles consist of six groups of three rows each. The tube bundles are made of 2.5-in. Schedule 20 pipes, 2.5-in. spacing, nominally up to a maximum of 20% of the tank diameter projecting inward from the tank wall (to allow for impeller clearance), which determines the number of tubes per row.
- Due to the many possible jacket configurations, a reasonable jacket-side heat transfer coefficient of 2271 J/m^2-s-C (400 Btu/h-ft.^{2}-F) was used for all cases. In some cases, this value might be a little low.
- Impeller sizes and shaft speeds are as listed in the capital cost section.

Below, we list some of the equations used to calculate the convective film coefficients, which are embedded in the Nusselt numbers. The fouling resistance, tube conductivity resistance, and the inside and outside convective coefficients are combined into an overall coefficient, or U value, and the U value is used to compute cooling duty by using the equation $Q = UA\Delta T$, where A is the cooling area

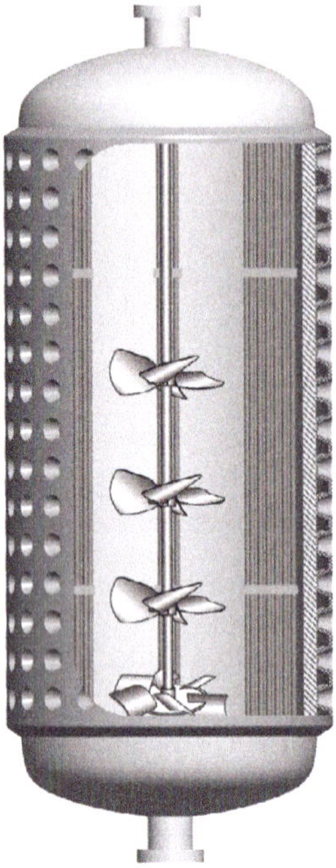

Figure 19.4 Jacketed tank with tube bundles.

and ΔT is the log mean temperature difference between vessel contents and the medium. Solving this is an iterative process, because the outlet cooling temperature of the medium varies with how much heat is removed.

The side wall process side coefficient is based on:

$$N_{Nu} = 0.4\left(N_{Re}\right)^{(2/3)}\left(N_{Pr}\right)^{(1/3)}\left(\mu/\mu_w\right)^{0.14}\left(T/Z\right)^{0.15} \tag{19.1}$$

where N_{Nu} is the Nusselt number, N_{Re} is the Reynolds number, N_{Pr} is the Prandtl number, μ is the bulk viscosity, μ_w is the wall viscosity, T is the tank diameter, and Z is the liquid height.

The bottom process side coefficient is based on:

$$N_{Nu} = 0.5\left(N_{Re}\right)^{(2/3)}\left(N_{Pr}\right)^{(1/3)}\left(\mu/\mu_w\right)^{0.14} \tag{19.2}$$

The tube outside coefficient is based on:

$$N_{Nu} = 0.09\left(N_{Re}\right)^{0.65}\left(N_{Pr}\right)^{0.3}\left(2/\#\text{of tube bundles}\right)^{0.2}\left(\mu/\mu_w\right)^{0.14} \tag{19.3}$$

An efficiency factor of 0.846 was applied for having 3 rows of tubes per bundle (we factor by 0.92 for each row beyond one, so, for three rows, the factor is $0.92^{(3-1)} = 0.846$).

The tube inside coefficient is based on:

$$N_{Nu} = 0.023\left(N_{Re}\right)^{0.8}\left(N_{Pr}\right)^{0.4} \tag{19.4}$$

Reasonable values for fouling factors were assumed.

Cooling duty includes agitator and compressor power input as well as metabolic heat. No credit is taken for evaporation of liquid into exhaust gas.

Heat Transfer Results

As expected, the total heat transfer area increases with aspect ratio (Table 19.3). So does the cooling capability. Because heat transfer correlations are approximately ±30% accurate, it is recommended that the calculated cooling capability be at least 30% higher than the cooling duty. The shaded cells do not meet that criterion. For an aspect ratio of 0.5, even with the use of both a jacket and vertical tube bundles, the heat transfer is insufficient. Likely, an external heat exchange loop would be needed. For aspect ratios from 1.0 to 4, the combination of jacket and tubes is sufficient. For an aspect ratio of 5, the tubes alone are sufficient; no jacket is needed. The capital cost figure in the next section does not include a jacket for an aspect ratio of 5, but includes both jacket and tube bundles for the rest. Note, however, that aspect ratios of 3 and 4 almost meet our 30% safety factor criterion when only tubes are used. Some design tweaks can be used to improve marginal heat transfer results. Among these are as follows:

- Use a higher cooling water flowrate
- Use a colder chilled water temperature
- Use smaller tubes, with more tubes per row. This must be balanced against the risk of fatigue failure of the tubes due to vibration caused by fluid forces imposed on the tube bundles by the agitator
- Use more tube bundles. This would have to be modeled for power draw effects; installations with 8–12 bundles show a marked decrease in actual power draw compared to standard baffling. Extreme cases have been known to cause 30–40% power reduction.
- Use 4 rows of tubes per bundle instead of 3. This would increase heat transfer from the tubes by about 15–20% at the same water flowrate.
- Chill the air coming from the compressor using an air-to-air exchanger. Since the compressor contributes a lot of power and heat, this could make a significant difference.

Table 19.3 Heat transfer results.

(English units)										
Nominal ungassed aspect ratio	Jacket area, ft^2	Number of tubes	Tube bundle area, ft^2	Jacket cooling, BTU/h	Tube bundle cooling, BTU/h	Total Cooling, BTU/h	Cooling duty, BTU/h	Ratio of cooling to cooling duty, jacket plus tube bundles	Ratio of jacket only cooling to cooling duty	Ratio of tube only cooling to cooling duty
0.5	1555	216	1767	6 234 024	8 328 607	14 562 631	12 012 350	1.21	0.52	0.69
1	1525	162	2035	6 127 773	10 532 408	16 660 181	11 852 015	1.41	0.52	0.89
1.5	1584	144	2328	6 343 310	12 145 860	18 489 170	11 770 575	1.57	0.54	1.03
2	1682	126	2581	6 680 469	13 308 199	19 988 668	11 701 860	1.71	0.57	1.14
2.5	1750	108	2523	6 931 823	13 503 203	20 435 026	11 666 230	1.75	0.59	1.16
3	1838	108	2912	7 235 075	14 902 610	22 137 685	11 648 415	1.90	0.62	1.28
4	1946	90	2833	7 604 141	14 870 654	22 474 796	11 612 785	1.94	0.65	1.28
5	2081	90	3357	8 072 973	16 638 467	24 711 440	11 582 245	2.13	0.70	1.44

(Metric units)										
Nominal ungassed aspect ratio	Jacket area, m^2	Number of tubes	Tube bundle area, m^2	Jacket cooling, kw	Tube bundle cooling, kw	Total Cooling, kw	Cooling duty, kw	Ratio of cooling to cooling duty, jacket plus tube bundles	Ratio of jacket only cooling to cooling duty	Ratio of tube only cooling to cooling duty
0.5	145	216	164	1827	2441	4268	3520	1.21	0.52	0.69
1	142	162	189	1796	3087	4883	3473	1.41	0.52	0.89
1.5	147	144	216	1859	3560	5419	3450	1.57	0.54	1.03
2	156	126	240	1958	3900	5858	3429	1.71	0.57	1.14
2.5	163	108	234	2032	3957	5989	3419	1.75	0.59	1.16
3	171	108	271	2120	4368	6488	3414	1.90	0.62	1.28
4	181	90	263	2229	4358	6587	3403	1.94	0.65	1.28
5	193	90	312	2366	4876	7242	3394	2.13	0.70	1.44

Shaded cells have insufficient heat transfer capability

For highly energetic fermentations (with an OTR greater than 200 mmol/L-h, for example), it may be necessary to use an external heat exchange loop.

Blend Time, DO Uniformity

Though we do not have a specific table of values for blend time or DO gradient for this example problem, the blend time always increases with increasing aspect ratio. This, combined with the steeper pressure gradient, causes a larger DO variation from top to bottom. Past experience can be used as a guide to determine the DO gradient. It is now possible to model DO gradient using computational fluid dynamics (CFD) techniques, as illustrated in Figure 19.5, though such models have not been validated by field measurements. The cost of such CFD models in 2020 is about $35 000 to set up the model and about $2000–$3000 per run. The print version of this book will show this figure in black and white; the digital version will be in color.

Figure 19.5 DO distribution by CFD. *Source:* Benz et al. presentation.

Capital Cost (Agitator Plus Vessel Only)

The agitator costs used in Table 19.4 were estimated by the author based on more than 44 years of experience in the agitation business. The vessel costs are based on a combination of author calculations, vendor data, and private communications between the author and users who have bought such vessels. As such, there may be some error in these numbers. However, they illustrate the fact that capital cost does not vary in a linear way with respect to aspect ratio. Interestingly, the total capital cost does not seem to vary by much between aspect ratios of 2 and 5. The majority of operating fermenters have aspect ratios between 2 and 3. The project engineer is encouraged to go through the detailed calculations of capital and operating costs to find the optimum solution for the particular case at hand.

Other Operating Costs

Though total power requirements decrease with increasing aspect ratio, some operating costs may increase. Clean-in-place (CIP) chemical, Steam-in-place (SIP) steam, and antifoam usage may all be greater in a tall vessel, due to more surface area to clean and sterilize and higher superficial gas velocity, which may lead to more foaming. In addition, in some cases, two steady bearings may be required. We have assumed this would be the case for the 5:1 aspect ratio

Table 19.4 Capital costs.

Capex (CY 2020 basis)						
Nominal ungassed aspect ratio	Tank diameter, ft	Ungassed Liquid level, ft	Tank straight side, ft	Agitator CAPEX, $K	Vessel CAPEX, $K	Total CAPEX, $K (vessel + agitator)
0.5	25	12.5	14.5	610	1060	1670
1	19.5	19.2	23.5	620	760	1380
1.5	17	24.7	31.5	630	715	1345
2	15	31.3	39.5	525	730	1255
2.5	14	35.7	45	535	670	1205
3	13	41.2	52	470	758	1228
4	12	48.1	61	495	720	1215
5	11	57	72.5	505	700	1205

Shaded cell has insufficient heat transfer capability.

design presented here. Such a steady bearing would require scaffolding to change and would probably take at least a day to service, compared to a few hours for a bottom steady bearing.

So What Is the Optimum Aspect Ratio for This Problem?

For our particular example problem, the total capital cost is relatively constant from an aspect ratio of 2 up to 5. Power cost continues to decrease as aspect ratio increases. So, based on the data and assumptions chosen, the optimum ratio may be close to 5. Since maintenance costs were not listed, the higher cost of maintaining an intermediate steady bearing, plus needing a taller building and a higher personal lift (which costs were not accounted for in this problem), the optimum might be more in the range of 3–4.

Illustrative Problem Number 2

For this problem, everything is the same as before, except we will use an air/water $k_l a$ correlation proposed by Bakker in reference 4. That correlation is:

$$k_l a = 0.946\left(P_g / V\right)^{0.6}\left(U_S\right)^{0.6}, \text{where V is expressed in kg; the rest is in watts, meters, seconds.} \quad (19.5)$$

We will not go into all the details here, but will summarize the results in tabular form.

Table 19.5 summarizes the agitator and compressor power requirements. Compared to using the V'ant Riet correlation, more airflow and more total power are required for the shortest aspect ratio. However, the total power and the agitator power fall off more rapidly as the aspect ratio increases, due to the higher sensitivity in the Bakker correlation to superficial gas velocity.

Since the agitator power decreases with aspect ratio, the impellers get smaller, reducing the heat transfer capability a little. But the agitator invests less power, reducing heat load by a little. Compressor power and heat input are somewhat higher than in the first problem. The results are shown in Table 19.6. The end result is that the ratio of cooling capacity to cooling duty drops a little, but the conclusions are still the same: for the 0.5 aspect ratio, we will need an external cooling loop. For aspect ratios of 1-4, we need a jacket and internal tube bundles. For an aspect ratio of 5, we can get by with tube bundles only; we do not need a jacket.

Because the agitator power is lower, in many cases a smaller gear drive can be used, and smaller impellers are required. Though the vessel costs remain the same

Table 19.5 Problem 2 optimum power requirements.

Problem 2 (English units)				
Nominal ungassed aspect ratio	**Airflow, SCFM**	**Agitator BHP**	**Compressor BHP**	**Total BHP**
0.5	4100	406	599	1006
1	3400	313	521	834
1.5	3100	266	492	759
2	2800	237	463	700
2.5	2700	214	457	672
3	2500	214	436	650
4	2400	184	434	618
5	2300	166	434	599
Problem 2 (Metric units)				
Nominal ungassed aspect ratio	**Airflow, nm^3/s**	**Agitator Brake Power, kw**	**Compressor Brake Power, kw**	**Total Brake Power, kw**
0.5	1.71	303	447	750
1	1.42	234	388	622
1.5	1.29	199	367	566
2	1.17	177	345	522
2.5	1.13	160	341	501
3	1.04	159	325	485
4	1.00	137	324	461
5	0.96	123	323	447

as in the first problem, the agitator cost is less. Therefore, the total Capex is less. Table 19.7 summarizes the capex.

We did not go out and get estimates on compressor cost, but it will likely go down with aspect ratio for this problem, as both the airflow and power decrease with aspect ratio increasing.

For this problem, the aspect ratio with the lowest CAPEX, excluding the compressor, is 2.5. However, depending on how present worth of power cost is figured, each kW saved may be worth $1000–3000 (this is not cost per year, but discounted power cost over the life of the facility or process). Thus, it may be worthwhile to consider a taller aspect ratio than 2.5 in this case. I would still avoid the aspect ratio of 5 due to the difficulty in maintaining the equipment.

Table 19.6 Problem 2 heat transfer results.

(English units)										
Nominal ungassed aspect ratio	Jacket area, ft^2	Number of tubes	Tube bundle area, ft^2	Jacket cooling, BTU/h	Tube bundle cooling, BTU/h	Total Cooling, BTU/h	Cooling duty, BTU/h	Ratio of cooling to cooling duty, jacket plus tube bundles	Ratio of jacket only cooling to cooling duty	Ratio of tube only cooling to cooling duty
0.5	1555	216	1767	6254406	8787141	14562631	12406062	1.17	0.50	0.71
1	1525	162	2035	6122513	10396126	16660181	11958523	1.39	0.51	0.87
1.5	1584	144	2328	6334009	11949567	18489170	11795261	1.57	0.54	1.01
2	1682	126	2581	6675612	13094334	19988668	11651342	1.72	0.57	1.12
2.5	1750	108	2523	6917351	13115625	20435026	11580464	1.76	0.60	1.13
3	1838	108	2912	7222769	14604261	22137685	11527019	1.92	0.63	1.27
4	1946	90	2833	7595133	14580777	22474796	11449396	1.96	0.66	1.27
5	2081	90	3357	8044830	16009882	24711440	11405877	2.17	0.71	1.40

(Metric units)										
Nominal ungassed aspect ratio	Jacket area, m^2	Number of tubes	Tube bundle area, m^2	Jacket cooling, kw	Tube bundle cooling, kw	Total Cooling, kw	Cooling duty, kw	Ratio of cooling to cooling duty, jacket plus tube bundles	Ratio of jacket only cooling to cooling duty	Ratio of tube only cooling to cooling duty
0.5	145	216	164	1333	2575	4268	3636	1.17	0.50	0.71
1	142	162	189	1794	3047	4883	3505	1.39	0.51	0.87
1.5	147	144	216	1856	3502	5419	3457	1.57	0.54	1.01
2	156	126	240	1956	3838	5858	3415	1.72	0.57	1.12
2.5	163	108	234	2027	3844	5989	3394	1.76	0.60	1.13
3	171	108	271	2117	4280	6488	3378	1.92	0.63	1.27
4	181	90	263	2226	4273	6587	3355	1.96	0.66	1.27
5	193	90	312	2358	4692	7242	3343	2.17	0.71	1.40

Shaded cells have insufficient heat transfer capability.

Table 19.7 Problem 2 Capex summary.

(CY 2020 basis)						
Nominal ungassed aspect ratio	**Tank diameter, ft**	**Ungassed liquid level, ft**	**Tank straight side, ft**	**Agitator CAPEX, $K**	**Vessel CAPEX, $K**	**Total CAPEX, $K (vessel + agitator)**
0.5	25	12.5	14.5	610	1060	1670
1	19.5	19.2	23.5	560	760	1320
1.5	17	24.7	31.5	505	715	1220
2	15	31.3	39.5	450	730	1180
2.5	14	35.7	45	410	670	1080
3	13	41.2	52	405	758	1163
4	12	48.1	61	425	720	1145
5	11	57	72.5	375	700	1075

Shaded cell has insufficient heat transfer capability.

Illustrative Problem Number 3

In the previous problem, we kept everything the same except the correlation. We could have changed many other things, such as the working volume, back pressure, OTR, changes in heat transfer surfaces, and/or cooling media. Rather than fill up the book with many combinations, we will just delve into one more. We will look at what happens if we cut the OTR in half to 75 mmol/l-h. All other conditions remain the same as in our first problem. We will increment airflow by half as much also.

Table 19.8 summarizes the results. One can see that the total power and airflow are drastically reduced. This also cuts the heat transfer requirements about 50%, because the metabolic heat generation is 50% of what it was before.

Since there is much less heat load, we may need less heat transfer area. Table 19.9 summarizes the heat transfer situation using the same heat transfer surfaces, media flow rates, and temperatures as before. The shaded areas are insufficient. But we can readily see that we do not need both the jacket and the tubes for any of these aspect ratios. The tube bundles alone will suffice for aspect ratios from 0.5 to 3. The jacket alone will suffice for ratios 4 and 5. In fact, that is exactly the reason why some users prefer tall vessels: so they have no internals to clean.

We could also investigate different tube bundle arrangements, media temperatures, and flow rates. It would be very helpful if we could use cooling tower water at 30C instead of chilled water at 15C. Table 19.10 summarizes the result of that set of calculations, with the same heat transfer surfaces and media flow rates. Only the 5:1 aspect ratio, using both a jacket and tube bundles, appears feasible. But using cooling tower water as opposed to chilled water saves both capital and energy costs.

The above heat transfer results have capital cost implications. Tube bundles are a very significant part of the vessel cost. Table 19.11 show capital cost based on using chilled water, with jacket only or tubes only as appropriate. Table 19.12 shows capital cost with both tubes and jacket, but only the 5:1 aspect ratio will work with cooling tower water.

Overall, the optimum for this scenario is likely to be a 4:1 aspect ratio using chilled water, though a case can be made for 5:1 with cooling tower water if the extra maintenance costs are acceptable.

Table 19.8 Optimum power at 75 OTR, problem 3.

Optimum power, problem 3 (English units)				
Nominal ungassed aspect ratio	**Airflow, SCFM**	**Agitator BHP**	**Compressor BHP**	**Total BHP**
0.5	1350	199	197	397
1	1300	167	199	367
1.5	1250	154	198	352
2	1200	141	198	340
2.5	1200	130	203	334
3	1150	130	201	330
4	1100	122	199	321
5	1100	111	207	319
Optimum power, problem 3 (Metric units)				
Nominal ungassed aspect ratio	**Airflow, nm^3/s**	**Agitator Brake Power, kw**	**Compressor Brake Power, kw**	**Total Brake Power, kw**
0.5	0.56	149	147	296
1	0.54	125	149	273
1.5	0.52	115	148	262
2	0.50	105	148	253
2.5	0.50	97	152	249
3	0.48	97	150	246
4	0.46	91	148	239
5	0.46	83	155	238

Table 19.9 Problem 3 heat transfer results.

(English units)										
Nominal ungassed aspect ratio	Jacket area, ft^2	Number of tubes	Tube bundle area, ft^2	Jacket cooling, BTU/h	Tube bundle cooling, BTU/h	Total Cooling, BTU/h	Cooling duty, BTU/h	Ratio of cooling to cooling duty, jacket plus tube bundles	Ratio of jacket only cooling to cooling duty	Ratio of tube only cooling to cooling duty
0.5	1555	216	1767	6 239 325	8 787 141	14 562 631	5 932 497	2.45	1.05	1.48
1	1525	162	2035	6 106 101	10 396 126	16 660 181	5 860 219	2.84	1.04	1.77
1.5	1584	144	2328	6 321 041	11 949 567	18 489 170	5 826 244	3.17	1.08	2.05
2	1682	126	2581	6 663 046	13 094 334	19 988 668	5 794 813	3.45	1.15	2.26
2.5	1750	108	2523	6 896 419	13 115 625	20 435 026	5 780 943	3.53	1.19	2.27
3	1838	108	2912	7 184 044	14 604 261	22 137 685	5 775 853	3.83	1.24	2.53
4	1946	90	2833	7 5705 32	14 580 777	22 474 796	5 751 421	3.91	1.32	2.54
5	2081	90	3357	8 024 621	16 009 882	24 711 440	5 745 185	4.30	1.40	2.79

(Metric units)										
Nominal ungassed aspect ratio	Jacket area, m^2	Number of tubes	Tube bundle area, m^2	Jacket cooling, kw	Tube bundle cooling, kw	Total Cooling, kw	Cooling duty, kw	Ratio of cooling to cooling duty, jacket plus tube bundles	Ratio of jacket only cooling to cooling duty	Ratio of tube only cooling to cooling duty
0.5	145	216	164	1829	2575	4268	1739	2.45	1.05	1.48
1	142	162	189	1790	3047	4883	1717	2.84	1.04	1.77
1.5	147	144	216	1853	3502	5419	1707	3.17	1.08	2.05
2	156	126	240	1953	3838	5858	1698	3.45	1.15	2.26
2.5	163	108	234	2021	3844	5989	1694	3.53	1.19	2.27
3	171	108	271	2105	4280	6488	1693	3.83	1.24	2.53
4	181	90	263	2219	4273	6587	1686	3.91	1.32	2.54
5	193	90	312	2352	4692	7242	1684	4.30	1.40	2.79

Shaded cells have insufficient heat transfer capability.

Table 19.10 Problem 3 alt 1 heat transfer results.

(English units), cooling tower water										
Nominal ungassed aspect ratio	Jacket area, ft^2	Number of tubes	Tube bundle area, ft^2	Jacket cooling, BTU/h	Tube bundle cooling, BTU/h	Total Cooling, BTU/h	Cooling duty, BTU/h	Ratio of cooling to cooling duty, jacket plus tube bundles	Ratio of jacket only cooling to cooling duty	Ratio of tube only cooling to cooling duty
0.5	1555	216	1767	1 982 764	2 703 552	4 686 315	5 932 497	0.79	0.33	0.46
1	1525	162	2035	1 942 045	3 209 262	5 151 306	5 860 219	0.88	0.33	0.55
1.5	1584	144	2328	2 012 246	3 724 476	5 736 722	5 826 244	0.98	0.35	0.64
2	1682	126	2581	2 121 931	4 081 996	6 203 927	5 794 813	1.07	0.37	0.70
2.5	1750	108	2523	2 193 526	4 026 307	6 219 833	5 780 943	1.08	0.38	0.70
3	1838	108	2912	2 284 907	4 481 388	6 766 296	5 775 853	1.17	0.40	0.78
4	1946	90	2833	2 409 427	4 465 385	6 874 813	5 751 421	1.20	0.42	0.78
5	2081	90	3357	2 553 122	5 015 081	7 568 203	5 745 185	1.32	0.44	0.87

(Metric units), cooling tower water										
Nominal ungassed aspect ratio	Jacket area, m^2	Number of tubes	Tube bundle area, m^2	Jacket cooling, kw	Tube bundle cooling, kw	Total Cooling, kw	Cooling duty, kw	Ratio of cooling to cooling duty, jacket plus tube bundles	Ratio of jacket only cooling to cooling duty	Ratio of tube only cooling to cooling duty
0.5	145	216	164	581	792	1373	1739	0.79	0.33	0.46
1	142	162	189	569	941	1510	1717	0.88	0.33	0.55
1.5	147	144	216	590	1092	1681	1707	0.98	0.35	0.64
2	156	126	240	622	1196	1818	1698	1.07	0.37	0.70
2.5	163	108	234	643	1180	1823	1694	1.08	0.38	0.70
3	171	108	271	670	1313	1983	1693	1.17	0.40	0.78
4	181	90	263	706	1309	2015	1686	1.20	0.42	0.78
5	193	90	312	748	1470	2218	1684	1.32	0.44	0.87

Shaded cells have insufficient heat transfer capability.

Table 19.11 Capex based on chilled water, problem 3.

Problem 3 Capex (CY 2020 basis), with chilled water							
Nominal ungassed aspect ratio	Tank diameter, ft	Ungassed Liquid level, ft	Tank straight side, ft	Agitator CAPEX, $K	Vessel CAPEX, $K	Total CAPEX, $K (vessel + agitator)	Comments
0.5	25	12.5	14.5	460	1000	1460	Tubes only
1	19.5	19.2	23.5	550	700	1250	Tubes only
1.5	17	24.7	31.5	560	650	1210	Tubes only
2	15	31.3	39.5	420	660	1080	Tubes only
2.5	14	35.7	45	365	600	965	Tubes only
3	13	41.2	52	350	700	1050	Tubes only
4	12	48.1	61	372	416	788	Jacket only
5	11	57	72.5	385	420	805	Jacket only

Table 19.12 Capex based on cooling tower water, problem 3.

Problem 3 Capex alt 1 (CY 2020 basis) cooling tower water							
Nominal ungassed aspect ratio	Tank diameter, ft	Ungassed liquid level, ft.	Tank straight side, ft.	Agitator CAPEX, $K	Vessel CAPEX, $K	Total CAPEX, $K (vessel + agitator)	Comments
0.5	25	12.5	14.5	460	1060	1520	Tubes + jacket
1	19.5	19.2	23.5	550	760	1310	Tubes + jacket
1.5	17	24.7	31.5	560	715	1275	Tubes + jacket
2	15	31.3	39.5	420	730	1150	Tubes + jacket
2.5	14	35.7	45	365	670	1035	Tubes + jacket
3	13	41.2	52	350	758	1108	Tubes + jacket
4	12	48.1	61	372	720	1092	Tubes + jacket
5	11	57	72.5	385	700	1085	Tubes + jacket

Shaded cells have insufficient heat transfer capability.

Summary of Chapter

We have shown that there is no simple answer to the question of "what is the optimum aspect ratio." There are many factors affecting the answer. Hopefully, we have identified most of them. The only way to determine the optimum is to go through the numbers for a range of aspect ratios. Though our sample problems all seemed to favor taller aspect ratios than typical that will not always be the case. And there are other factors besides energy and vessel costs that could not be fully addressed here. For example, the larger DO gradient associated with tall vessels may be problematic for some fermentations. The designer of the facility must take all relevant factors into account.

List of Symbols

$k_l a$	Overall mass transfer coefficient
N_{Nu}	Nusselt number for heat transfer, dimensionless
N_{Pr}	Prandtl number for heat transfer, dimensionless
N_{Re}	Impeller Reynolds number, dimensionless
OTR	Oxygen Transfer Rate
P	Power
P_g	Gassed power
T	Tank diameter
U_S	Superficial gas velocity
V	Liquid volume
Z	Liquid height
Z_g	Gassed liquid height
μ	Bulk viscosity
μ_w	Viscosity at wall

References

1 Benz, G. (2019). Optimize aspect ratio in industrial fermenters. *Chemical Engineering Progress* 115 (4): 26–31.

2 Benz, G. (2003). Optimize power consumption in aerobic fermenters. *Chemical Engineering Progress* 99 (5): 100–103.

3 v'ant Riet, K. and Tramper, J. (1991). *Basic Bioreactor Design*. New York: Marcel Dekker.

4 Bakker, A. et al. (2012). How to disperse gases in liquids. *Chemical Engineering Magazine*.

20

Vendor Evaluation

Up until now, we have focused on how fermenter agitators should be designed. We have covered major aspects of process design, as well as some fundamental issues in mechanical design. So, if all of the information from this book is used, along with any needed supplemental information such as a well-defined pilot program, one should be able to get a pretty good idea of what agitator design should be specified. However, unless the user plans to build the agitator in his company's maintenance or machine shop, he will need to buy it from an agitator vendor. This chapter focuses on how to select that vendor, keeping in mind that vendors change over the years as new people come into the company and others leave, especially when a new owner buys the company. So, buying from the vendor you always bought from before may not be the best decision, even if you have been satisfied in the past.

Both the company and the product should be considered. We will go over some factors for each.

Product Considerations

What does a buyer of agitation equipment value most? Hopefully, it is process performance. Poor process performance will result in poor production rate or yield, and loss of profits. Equally important are ruggedness and reliability. Unexpected downtime due to equipment failure is not what people expect when they buy a piece of process machinery. Indeed, the cost of downtime not only dwarfs the price difference s between different vendors, it dwarfs the entire cost of the equipment itself.

Let us look at some individual product features and discuss their importance.

Agitator Design for Gas–Liquid Fermenters and Bioreactors, First Edition. Gregory T. Benz.

Gear Drive Ruggedness

Unlike shaft seals and steady bearings, which must be replaced regularly and are therefore consumable items, the gear drive is expected to last for many years with nothing more than changing the oil and greasing the bearings. And, in fact, this is entirely possible if the right kind of gear drive is used and the service factors are ample, as described in Chapter 17. To achieve a life of 10 years or more, it is usually necessary for the gear drive to be purpose-built as an agitator drive, preferably made by the agitator manufacturer. Buyers should be careful and vigilant when considering a bid that uses a commercial gear drive. Service factors should be as described in Chapter 17, and each bearing should have an L_{10} life of at least 30 000 h; 50 000 would be better. It would be ideal if the *entire set* of bearings had a combined L_{10} life of 30 000 h or more. This means each individual bearing would need to have an L_{10} life substantially greater than 30 000 h.

Design Technology

This heading covers a variety of things. Mainly, it refers to the vendor's ability to design all parts of the agitator to be strong enough, avoid critical speed issues, and bearing failures. Finite element analysis (FEA) capability should preferably be in-house. FEA should be used in the development of all standard products and should be used on each unique design that requires non-standard components. Large impellers, hollow shafts, and higher than catalogued power are but a few examples of where this is needed. Dynamic shaft design should be used for critical speed determination. Power draw predictions should be based on extensive testing, ideally in the vendor's own well-equipped laboratory.

Impeller Selection

We have discussed impeller technology in this book, especially in Chapters 5 and 6. The buyer should bear in mind, however, that impeller technology is constantly changing. What has been recommended in this book is sound as of the year 2020, when it was written, but new impellers may supersede these recommendations. Still, the buyer should get the most up-to-date impeller technology, provided that such technology has been carefully vetted for stability and process performance, rather than just new for newness' sake. The buyer should not accept vendor claims on faith. Rather, the claims of impeller performance should be proven and able to be measured or witnessed by the buyer. Desirable features in the impeller selection process include the following:

- High gassing factor (P_g/P_u)
- High gas-handling ability
- Higher k_la at the same power input than other impellers, if possible (there are differing opinions on that)
- Ability to distribute gas evenly
- Mechanically stable; low vibration velocity
- Ability to fight foaming
- Impellers work as a system, with optimized power distribution
- Designed to reduce weight by blade/hub design (affects cost also)
- Good overall mixing; top-to-bottom blending
- Sanitary, if possible, and the budget permits

Shaft Design

Solid shafts are preferred. If hollow shafts are used, FEA is highly advised, especially for impeller-to-shaft attachment methods. Details of steady bearing design and in-tank coupling design are important for long life, maintainability, and cleanability.

Company Considerations

Not only is the company as important as the product, there are many different attributes or measures of a company that are important. We will go over many of them below. The list is not all inclusive, nor is it necessarily in any particular order of importance, as what is most important may vary according to circumstances. But hopefully, most of the important attributes are covered here.

Reputation with Customers

Ultimately, for a vendor to stay in business and maintain or increase market share, it must satisfy its customers. Therefore, what long-term users have to say about a vendor's quality, reliability, and service is very important. However, companies do change with time, turnover among key personnel, and ownership. A balance between long-term reputation and recent history is needed, especially if there has been a major management or ownership change.

Company Size

There can be good, conscientious companies of almost any size. The larger companies can support more R&D, a larger sales network, better field service capability, more integrated manufacturing, and more internal expertise. However, there is something to be said for the personal relationships one can develop with the owner or head of engineering of a small agitator manufacturer. Some such companies do very well with a few key customers and a few key industries, as they do not have the resources to take care of everyone. They often command considerable loyalty among their customers due to their focus on personal relationships and service.

The larger companies often have significant market share in many markets, have broad geographic sales coverage, and are able to service their equipment anywhere in the world. They have the financial strength to back up their product. So, in general, they are a safer choice than small companies. But if your facilities are in a location that can be well served by a smaller vendor, and that vendor has a good track record in your industry or in your company, that vendor very well might be the best choice for you.

Years in Business

It takes time for a vendor to build up the resources needed to be a top-notch supplier. Those resources are not just financial or physical, such as the manufacturing plant, its machines, and inventory. They include the intellectual property (IP) accumulated over the years. Such IP encompasses process design, mechanical design, and manufacturing technology, among other things. So, there are many advantages to older companies. On the other hand, some older companies grow moribund and are slow to adopt the latest technology. This will be evident when looking at their R&D program. The ideal is an older company with fresh, innovative ideas.

Years Under New Ownership

Most agitator manufacturers in existence today were founded by an individual or a small group of individuals. At some point, the original owners often decide to retire, and sell the company to another company or to the employees. Almost all new corporate owners come in with the attitude that the previous management allowed the company to grow fat with employees and that they can improve the efficiency of the operation by trimming the staff. Often, they do not really understand the business, treating it as a commodity rather than an engineered product enterprise. They often undervalue the importance of application engineering and R&D. And in many cases, they pay a hefty price to buy the company and are under financial pressure to improve the bottom line. One way they sometimes seek to save money is specifically to lay off senior people. Why senior people? Because they typically earn the highest wages and have the highest impact on health insurance premiums.

Unfortunately for the new owners, laying off senior people comes with a heavy cost that the new management may not even be aware of. They assume the business can continue to run as normal and that the knowledge to do so is retained in the records and structure of the business. However, the explicit knowledge retained in company written records and procedures is only a small fraction of the knowledge extant within any organization [1]. About 80–90% of an organization's knowledge exists in the form of tacit knowledge, i.e., in people's heads, and is not formally recorded anywhere. There tends to be more such knowledge in the possession of the most senior people. Therefore, when senior people are laid off, a lot of tacit knowledge walks out the door with them [2].

Based on my own observations, it takes an organization at least 20 years to replace such lost knowledge, if they ever do at all, by organic learning from within. In many cases, it takes 30 years or more. Therefore, when buying from a company that has had new owners for less than 20 years, extra due diligence is needed. The buyer should ask about layoffs that occurred within a year of purchase of the company and see whether a number of current employees are still there from before the ownership change. All of the below points should be examined with closer scrutiny as well.

Employee Turnover

A stable workforce is obviously a great benefit to the employer. But it is also a great benefit to the buyer. Aside from the loss of tacit knowledge that occurs when there is high turnover, the relationships between the buyer and individuals within the seller's organization may be severed. This costs time and money. Moreover, one has to ask why there is high turnover. It generally indicates an unhappy workforce. The buyer is generally wise to steer clear of companies that have a high turnover for an extended period of time.

Vertical Integration

Vendors vary significantly in the degree of vertical integration they have. Some basically buy all components outside, and just assemble and pack them for shipment. Most at least do fabrication and machining of shafts, impellers, pedestals, and seal housings in-house. Some even make their own mechanical seals. Oddly, very few manufacture their own gear drives, even though they are among the most expensive and critical individual items in an agitator assembly.

Vertical integration allows the vendor to control the quality of the product better than outsourcing the manufacturing, so vendors that are highly vertically integrated are preferred. Ironically, in the very largest sizes required for fermenters, many vendors outsource some components due to limitations of the machining equipment in their factory. Still, the experience they have gained from doing their own manufacturing enables them to better select sub-manufacturers and impose appropriate quality control measures on them.

For the very largest fermenters, it is common to outsource the gear drive, even if the manufacturer has the capability to make its own. The decision to make or buy is often based on competitive pressure. Due to low production volume, it is not unusual for a manufacturer's own gear drive to cost 2–3 times as much as a commercial gear drive of similar rating or 1.5–2 times as much as a commercial gear drive manufacturer's "agitator drive" with increased rating to compensate for weaker bearing lives. This makes the total agitator price to the consumer about 20–40% higher when using the vendor's own drive. That is hard to sell in a competitive market. However, if it can be demonstrated that the vendor's own drive is more reliable, it is money well spent. If I were in the business of selling agitators, I would quote both ways and discuss the options with the customer.

The main point of this section is that vertically integrated manufacturers are a safer choice in general. But there are other factors in play, as we shall see.

R&D Program and Publications

As in any other field of endeavor, agitation technology is constantly changing. This includes process technology, impeller technology, mechanical technology, and manufacturing technology. For a vendor to keep up with it, a significant portion of sales revenue should be devoted to R&D. My own opinion is that at least 2% of sales should be devoted to R&D. Some industry-leading vendors approach 10%. Since the absolute amount of money devoted to R&D is potentially greater in larger organizations, most new technology is developed by them. This can be a very strong advantage.

Ideally, the R&D should be organized by people who are recognized experts in their field. One way to evaluate the relative expertise of a company's R&D staff is to check for their publications. Most major agitator manufacturers publish frequently, in such publications as the AIChE Journal, Chemical Engineering Magazine, Chemical Engineering Progress, and many others, and they attend scientific and engineering conferences around the world, such as the North American Mixing Forum (NAMF), various AIChE meetings, and many other venues.

Smaller companies are less visible in their publications. Strangely, I know of one company that claims to be one of the largest and most technically capable in the world, with the largest staff of PhD engineers, but I have yet to see any articles published by them. They must be so great that they keep everything secret!

Depth of Application Engineering

Some agitator manufacturers allow their field sales people or representatives to do routine application engineering. This can be very efficient and lets the customer and field sales engineer have a direct dialog about the design. However, design of fermenters is almost always done in-house by application engineers. Smaller companies often have only 1 or 2 application engineers, who may also be the owner or VP of engineering. With a small application engineering staff, specialization is mostly impossible, unless the company has decided to serve only one market.

Larger companies can support more application engineers. The usual quantity for a large agitator company is 5–10 application engineers. With that many, though all can handle a variety of applications, they are often assigned primary market or technology areas, allowing some degree of specialization. Since fermenter design is complex, it is helpful to have an application engineer who specializes in this field.

The age distribution of the application engineers is important also. If they are all young, they will likely lack the vital tacit knowledge needed to tackle the toughest problems. If they are all 60+, they will be retiring in the near future, and their valuable tacit knowledge will go with them. The ideal is to have a spectrum of ages, so that the senior engineers can mentor the younger ones and transfer at least some of their tacit knowledge to them. I know that I personally benefitted greatly from the knowledge of my mentors early in my career. As Albert Einstein said, I have stood on the shoulders of giants.

As an aside, one major company has a very large application engineering group, and they do all the quotes in-house. This has some obvious advantages. But the drawback is that there are no local people who can intelligently answer questions, and communication can often break down.

So, although larger companies may have a larger application engineering staff than smaller ones, the real measure is if they have the right depth of knowledge for your industry. If not, see the comments later on the use of a consultant.

Testing Laboratory

Obviously, for a vendor to have a good R&D program, it must have a good testing laboratory (though some vendors farm out a lot of their R&D to universities and their laboratories). Such laboratories should be able to test gassed and ungassed power draw in a variety of vessel sizes, ranging from benchtop scale to at least several thousand liters. They should be able to measure stress and strain on impellers and shafts. They should have the ability to measure velocity profiles using such technologies as Laser Doppler Velocimetry (LDV), Particle Image Velocimetry (PIV) or similar technologies, and appropriate software to correlate the results.

The above are examples of capabilities the R&D laboratory should have and is not a comprehensive list. But equally important is the availability of the laboratory for customer-witnessed testing. Sometimes, this is done in a separate laboratory section to maintain confidentiality of the ongoing R&D work. But there are many things that may need to be done or available in a customer test laboratory. Examples of such things are listed below; the list is not exhaustive.

- Power draw tests in unusual tank geometries. This has often been done for large quantities of vertical tube bundles or panel coil baffles. The power draw can easily be 70% or less of the power draw using standard baffling.
- Gassed and ungassed blend time.
- Power draw and holdup at equal superficial velocity to proposed production equipment. This can involve really high gas flow rates when measured as VVM. Many tests I have done over the years required a VVM of 30 or more.
- k_la measurements. These can be used for some comparative results. However, broth-specific k_la correlations must be developed by the user with real fermenter broths, preferably during a "live" fermentation series. This is not something any agitator manufacturer I am aware of can do.
- Viscosity measurement. Simple viscometers are found in most mixing laboratories. A few have advanced rheometers, which are preferred for highly non-Newtonian liquids, such as soluble gums or polymers, as well as high-solid slurries.
- The ability to handle hazardous or flammable materials. Most vendors are very cautious about doing this, as they usually do not have explosion-proof laboratory equipment, and typical laboratory mixing equipment needed for a good mixing test is too large to fit into a standard laboratory fume hood. Some vendors have a custom-built, walk-in hood large enough to wheel in or out sufficiently large apparatus as needed.
- Having a bevy of almost all of the vendor's standard impeller styles in a variety of sizes to meet the test needs.
- Having the ability to make custom impellers as needed.
- Having a variety of vessel sizes available, with a variety of tank bottom styles.
- Most tests should be conducted in vessels made of clear materials so that mixing and gas dispersion can be observed visually.
- Proper disposal or return of materials subject to MSDS (material safety data sheet) requirements.
- Written safety protocols that are rigorously followed.

- The ability to record and edit laboratory tests for a condensed and relevant presentation of results. (Your management does not want to sit through an 8-hour long recording of a laboratory test!).
- Vendor may own the laboratory test data, or you may own it, depending on whether you pay for the test or not, and the details of any such lab test purchase orders.

The main takeaway of the above is that, if you need a lab test, work with a vendor that can develop the data you need and present it to you in a neat summary.

ISO Certification (Necessary vs Sufficient)

In the long run, for a company to deliver a good product and good service, it must do things consistently in a well-documented way. An ISO 9001 or similar certification is a good way to assure that. However, it should be considered a necessary condition, but not always sufficient. As some wags have said, it is possible under ISO 9001 to deliver a consistently bad product. So, more is needed than such certification to assure a good product is being produced.

Quality Control Program (Not Lot Sample; 100%)

Companies that make vast quantities of consumer goods can generally assure good quality by sample inspections or lot inspections. This is because they make repetitive products and need only check to see whether anything is drifting off the product specifications over time. The very nature of their manufacturing processes is geared toward repeatability.

Agitators are a different case. While there may be a number of standardized components, there are no truly standard agitators. Each one is custom designed for a customer's particular process, tank dimensions, and other specifications. In essence, it is an engineered product. As such, lot inspection is unacceptable. The agitator manufacturer should practice 100% inspection to all quality control standards applicable to each part of every agitator it produces, except for commodity items it purchases outside, such as motors, seals, fasteners, and bearings. For those purchased items, it must do vendor quality audits from time to time. A non-exhaustive list of inspections that should be carried out would include, at a minimum:

- Dimensional check on all finished parts
- Weld check by visual inspection, dye penetrant or X-ray as appropriate
- Shaft straightness/runout check

- Pressure test of mechanical seal assemblies
- Impeller balance check
- Gear drive noise check
- Gear tooth contact pattern check (if the agitator vendor makes its own gear drive)
- Material certifications as required
- Segregation of all raw materials by alloy designation
- Welding and machining of alloy components to be done in an area isolated from carbon steel fabrication and machining
- Paint job quality
- All bearings are greased
- Vapor phase rust inhibitor installed as needed
- Crating must be substantial to assure delivery undamaged. Special attention is needed for polished components.
- If wetted parts are polished, a profilometer test plus visual inspection are both required, and dye penetrant testing is mandatory to uncover cracks and pits.

Rep vs Direct Sales (a Good Rep Annoys the Manufacturer)

The vendor's selling system can be an important consideration.

Some vendors sell directly; the quote comes from company employees, who may be salespeople or application engineers. They may be located in a completely different country from where you are located. If all quotes are done by factory application engineers, they generally have a pretty good level of expertise about their product and technology. On the other hand, they are not necessarily familiar with you, your plant, or your process. Local sales people employed by the manufacturer can help bridge that gap. However, since they only sell agitators, the sales volume will be limited; there will usually not be that many direct sales offices. You may or may not have one near you. Whether or not they are knowledgeable sales people depends on the manufacturer. Some manufacturers train their salespeople almost as application engineers. Others use them only to meet customers, entertain them, and find inquiries. They may not be able to answer many technical questions.

Other manufacturers use manufacturer's representatives. These are independent businesses that sell on behalf of the manufacturer and get paid a commission. They generally sell a number of products besides agitators. That enables the manufacturer to have more of them, which increases geographic coverage. The fact that they have several products means they will usually be in your plant quite often, which enables them to know what is going on there, and find out about any problems that may need attention. This can be a very valuable advantage.

Manufacturers vary in terms of how much training they give their representatives. The better ones train their representatives enough so that they can do their own quotes for normal applications and answer most routine technical questions. Fermentation questions usually have to be answered by the factory application engineer.

From a manufacturer's perspective, the job of the representative is to represent the manufacturer to the customer and get orders. But in my experience, the best and most successful representatives almost take the attitude that their job is to represent the customer to the manufacturer, acting as an advocate. Though this often annoys the manufacturer, it is great for the customer. If you are lucky enough to have such a representative, he/she is well worth the commission being paid.

If the local manufacturer's rep is capable and conscientious, this is an advantage that should be considered in favor of the companies he represents. If none of the companies being considered have good reps or local sales people, you will have to do the best you can working with the manufacturers if possible.

Service Capability

Selling a piece of equipment is only part of the job of a vendor. Sooner or later, the equipment will need to be repaired. Hopefully, this will be a long time for such major items as the gear drive. However, seals and steady bearings will need more frequent replacement. A good vendor should have an adequate field service staff that can offer advice and training on proper installation and service procedures, such as aligning motors and steady bearings, changing seals, etc. Larger vendors can afford a larger, more geographically diverse field service staff. Some manufacturer's reps can offer aid in these areas also. Find out who has this capability and where they are located. The better manufacturers will have a 24 h per day emergency hotline that will be responded to within a short time of receiving the call or text.

Typical Delivery Times and Performance

All capital projects are planned based on a schedule of receipt of capital goods. It is easier to meet the planned schedule if the vendor has a short delivery cycle. However, the project manager should try to develop his plan based on the realistic deliveries typical of the vendors in a given category. It is unfortunate to have to choose a second-tier vendor just to meet a delivery schedule. That will be costly in the long run.

Perhaps even more important than a short delivery cycle is a predictable delivery cycle. Vendor performance and credibility on delivery promises are very important. A good vendor will deliver within +/– 20% of the promised delivery date, sometimes even better. But I have seen some vendors flat-out lie about deliveries. I have seen cases where the vendor sales people have claimed that the order was released to manufacturing and all materials ordered, when the drafting had not even started. I have seen cases when the drawings were not completed by the promised delivery date, and the actual delivery date was 6–9 months after the promised date, when the promised delivery time was 12 weeks. Once the order is placed, you are committed. Research the vendor's delivery credibility and be aware that it can change over the years, whether to changes in personnel, owners, new computerized business systems, etc.

It also helps if the customer project manager only releases the order to manufacturing after everything is checked and no change orders are anticipated. Every change order after release to manufacturing incurs delays and has potential cost impact. In my earlier career at an equipment vendor, I have seen the problems change orders cause. I have seen some customers issue 10–15 change orders during the manufacturing cycle, seemingly unaware of the headaches they cause. I have even seen some customers try to submit a change order after the equipment had shipped. As the song says, "It's too late, baby!" Of course, the vendor will charge for each of these change orders. But, in my experience, they almost never recoup the true cost of such changes. They just take the losses to maintain customer goodwill. You might not care about that. But you should care about what it will do to your project schedule.

Parts Availability

A good vendor will try to keep all normally wearing parts in stock. Custom parts, such as shafts and impellers, must be built-to-order. However, some manufacturers have a special rapid manufacturing program for custom-built parts that need to be replaced, either due to failure or customer-requested modifications, such as impeller retrofits or changes in impeller locations. Due to low-volume, large gear drives and their internals are not always kept in stock, as government taxes penalize companies for having a large parts inventory. This unfortunate tax structure discourages keeping low-volume parts on hand.

The location of the parts inventory is also important. If the manufacturer is located far from your plant, or even in a foreign country, do they have a local parts warehouse? If not, this may not disqualify a vendor, but may require the user to keep critical spares in stock. In my opinion, this should be done anyway. The cost of downtime dwarfs any possible savings by not keeping critical parts in inventory at the plant site.

Price (Least Important)

Many buyers of process equipment, when racking competitive bids, will focus on the price. I guess their company reward system encourages this illusory way of "saving money." And, indeed, if two or more vendors are completely equal in terms of process performance, mechanical reliability, and delivery, then a decision based on price makes sense. But are they ever really truly equal? In this book, we have tried to help the buyer get all vendors to an equal footing, but actual plant history can be a more powerful guide than what is presented in this book. Two or more vendors are rarely truly equal. If it can be proven, for example, that one vendor's offering is likely to have 5% less downtime than another's over, say, a 15-year period, then it would be better to buy that vendor's equipment at full list price or above than to accept the second vendor's equipment even if they gave it away for free. The difficulty is proving the differences. And, admittedly, it is a good idea to try alternate suppliers once in a while, as companies and products do change. But, at a minimum, if there are differences in gear drive quality and bearing lives, it is always a good idea to buy the best offering. Though purchasing agents may not like this idea, as they often get rewarded for "saving money," in many cases it would be best to have the key decision makers choose without even seeing the price, and then, after the price is revealed, negotiate with the chosen vendor to see whether a better deal can be attained. Getting all vendors to an "apples-to-apples" basis in the bidding cycle generally causes them to try to be as competitive as possible.

Willingness to Work with Consultants

Though most vendors are proud of their internal expertise, most are also willing to work with the customer's consultants, just as they would with a customer's own in-house mixing "guru." After all, the consultant acts as a member of the customer's team. The consultant not only aids in writing specifications and bid evaluation, he may need to inform the customer of installation problems he is aware of with a particular vendor and suggest that the customer check newer installations to see whether the problems have been solved.

Unfortunately, not all vendors are cooperative with consultants. I have personally been threatened with lawsuits for pointing out deviations in a vendor's bid from the specification, for analyzing process and mechanical performance problems in comparison with how I would do it, and for pointing out problem installations. (They even threatened to sue my customer in one case!) Yet these functions are legitimate and reasonable activities of an engineering consultant. If such suits were really actionable, Consumer Reports would be unable to stay in business. In my opinion, the real purpose of such lawsuit threats is intimidation. But who wants to work with such a vendor?

Vendor Audit Checklist

To aid in vendor evaluation, we have prepared a vendor audit checklist in Table 20.1.

This type of list will help take the measure of a company over the years. Interviews with customer references are very helpful if you can get them. Some of these list items may be difficult to ascertain by the end user. For example, many users of the equipment within the buyer's own industry will refuse to comment, as they may not wish to aid a competitor. In the end, you will have to make the best judgment call you can.

Table 20.1 Supplier Audit Checklist

Company name		Principal products	
Founding year		Number of employees	
Factory area, M^2		Factory employees	
Annual sales, USD		Administrative employees	
Annual parts sales, USD		Field Service Staff	
ISO certified?		Engineers, < 5years	
ASME?		Engineers, 5-10 years	
DIN?		Engineers, 11-20 years	
JIS?		Engineers, > 20 years	
CSA?		Code welders	
AGMA?		Skilled machinists	
Other qualifications (fill in)		CAD system used	
Top customers: Company name	Location	Contact person	e-mail or phone
Equipment history: Product type	First year sold	Cumulative quantity sold	Improvements made

Use of an Outside Consultant

Over the years, the methods used by CPI companies to specify, evaluate, and purchase process equipment have changed considerably. Since no one engineer can be expert in the design of all kinds or process equipment (though some think they are!), the problem has mainly been where to get the specialized knowledge that allows the best possible practice to be applied to the purchase of equipment. This applies to agitators as well as other engineered equipment. (Reference 3.)

In the 1960s and 1970s, many large CPI companies, and also a good many engineering firms, had staffs of specialists for a number of different kinds of process equipment. Examples include such items as heat exchangers, filters, evaporators, agitators, centrifuges, extractors, distillation columns, fractionation towers, dryers, and many other kinds of equipment.

In the 1980s, many CPI production companies began to regard such in-house expertise as a luxury they could not afford, so they reduced their staff size and even eliminated whole departments in some cases. Yet they still needed such expertise to function optimally. Some got it from engineering firms, though many engineering firms also cut specialized staff during the same time period. Most were forced to rely on the expertise of equipment suppliers. Though the equipment worked in most cases, it did so because the leading equipment manufacturers had adequate, highly specialized, and experienced staff. Moreover, the end user did not have the expertise to determine whether equipment was optimally sized, undersized, or oversized. In addition, energy versus capital cost considerations were not always explored. Frequently, the purchase decision was based on trust in the vendor or its representative, rather than any quantitative basis for comparing bids. Equipment vendors, being well aware of this, cultivated relationships and sought to avoid direct bidding on an equivalent basis to their competitors, so as to avoid price being the basis of the purchase decision. Also, equipment vendors have never had enough staff to get involved early in the process development stages where it may be useful to look at alternative approaches or to aid in the design of experiments to assure adequate data are taken for scale-up.

During the late 1990s through the present, equipment purchases by the CPI declined on an absolute basis, and the price of the equipment fell when adjusted for inflation. Manufacturing efficiency gains assuaged some of the effect on equipment profit margins, but there was still a net decline in profitability. Equipment manufacturers responded by reducing the size and experience level of their Application Engineering and R&D departments, in some cases suspending R&D altogether for a while. So, the level of expertise that end users had come to expect from their equipment suppliers was no longer there.

Today, the typical end user engineer is faced with having to make purchase decisions on process equipment without extensive personal knowledge of the equipment, without a corporate staff of specialists, without available specialists at engineering firms, and with reduced experience and expertise at the equipment vendors. How can an engineer improve this situation?

One way is to consider the use of consultants. Many of the highly experienced engineers who were laid off during the staff reductions at equipment manufacturers have gone into business as independent consultants. They can fulfill the same role that inside specialist staff once did, as well as experienced vendor staff.

When one chooses the right consultant for the job, many benefits accrue. Examples include the following:

- Guidance of R&D efforts to assure proper data are gathered for scale-up
- Assurance that process results are as expected from the equipment
- Assurance of mechanical integrity standards
- Objective bid specification to get all vendors on an equivalent basis ("apples-to-apples")
- Typically lower bid prices based on the above kind of specification, (vendors must bid on an equivalent basis, rather than a different basis of their own which can obfuscate the decision process) saving much more than the consultant's fee
- Objective bid evaluation to written criteria
- Liaison between end user's needs and terminology and that of the equipment manufacturers
- Much deeper involvement in the project than equipment vendor engineers have time for
- Aid in looking at optimization or modification of equipment with little or no capital expenditure. Process equipment manufacturer's engineers have little interest in this.

Choosing the right consultant is, of course, a key ingredient in the success of a project. The right consultant is a part of the project team, not just a source of expertise.

Here are a few of the attributes that make for a good consultant to put on your team:

- Extensive experience with the kind of process equipment being purchased (typically 20 years or more), especially with manufacturers of that kind of equipment
- Specific experience in the end user's industry or application area is a strong plus, though not always necessary
- Flexibility of scheduling, to assure timely responses
- Willingness to travel to jobsites or meeting sites as needed, especially on international projects
- Adaptability to different corporate or national cultures
- Personal reputation in the industry

- Desire to optimize and save the customer money
- Willingness to work on both small and large projects as needed

A typical scope of activity for a consultant, specific to fermenter design, may include some or all of the following:

- Develop an experimental protocol aimed at creating a broth-specific k_la correlation and fitting the data to such a correlation
- If no such experimentation is planned, use published correlations to develop potential production designs on a feasibility basis
- Scale-up and optimize the production design, taking into account any process constraints
- Develop specifications for bid
- Evaluate vendor bids
- Train the customer staff on relevant design issues
- Evaluate alternative vessel geometries and mass transfer rates versus volume required
- Coordinate CFD simulations when required.

A typical range of fees for a USA-based consultant, in 2020 dollars, would be about $5000-50,000, depending on the detailed scope of activity and deliverables.

The main difficulty in finding a consultant specifically for fluid agitation, especially for fermenter design, is that there are very few of us, and we are getting old. There are probably no more than three with such qualifications in the United States as of this writing. This book is written partially to let the user make as informed a decision as possible without guidance from a consultant.

Summary of Chapter

The choice of agitator supplier is an important part of overall process optimization. Equipment integrity, vendor service, and knowledge all pay a role in the decision process. An experienced consultant can be of help as well.

List of Symbols

k_la	Overall mass transfer coefficient
L_{10}	Life at 10% failure rate
P_g	Gassed power draw
P_u	Ungassed power draw
VVM	Volume of gas per Volume of liquid per Minute

References

1 O. Ivona, "The Importance of Tacit Knowledge Within the Organization", publisher unknown
2 Penya, A. (2013). Institutional Knowledge: When employees leave, what do we lose? *HigherEdJobs blog, November* 18.
3 Benz, G. (2006). Team up with a consultant for equipment selection. *Chemical Engineering Progress* 102 (8): 50–51.

A. Appendix to Chapter 20

Below is a partial list of agitator suppliers known to the author to have supplied agitators for fermenters and bioreactors. It is not an exhaustive list; I am sure there are many others. It is not in any particular order. The inclusion of a vendor on this list is not an endorsement. Neither is an exclusion intended to disparage any vendor; it is just a function of the author's limited awareness of agitator vendors who have supplied to this field, as probably 90% of fermenter agitators have been supplied by the top 4 companies globally.

Name	Website	City, State or Province	Country
Fusion Fluid Equipment	http://www.fusionfluid.com/index.html	Michigan	USA
Fluidmix	https://www.agitadoresfluidmix.com/en/	Madrid	Spain
Mixrite	http://www.jetmixing.net/	Mumbai	India
Proquip	https://proquipinc.com/	Ohio	USA
Cleveland Mixer	https://clevelandmixer.com/	Connecticut	USA
Ekato	https://www.ekato.com/en/		Germany
Chemineer/NOV	https://www.chemineer.com/	Ohio	USA
Lightnin/SPX	https://www.spxflow.com/lightnin/	New York	USA
Mixel	https://www.mixel.fr/en/		France
Philadelphia Mixing Solutions	https://www.philamixers.com/	Pennsylvania	USA
Hayward Gordon	https://haywardgordon.com/		Canada
GMM Pfaudler/ Mixion	http://www.gmmpfaudler.com		India
MGT	https://www.mgt.co.il/		Israel

21

International Practices

Introduction

If you have read this far, you have waded through quite a bit of technical information. Here, in the last chapter of this book, you will get a break from that. There will not be a single equation in this chapter! Therefore, you can relax and learn about some differences between the design, supply, and buying practices around the world, which might be useful if you decide to locate a plant in a different country from your own, or procure equipment or services from overseas.

This book has been written largely from a North American point of view, as the author has spent his career mostly in the USA. However, he has traveled a fair amount, and these notes reflect his international experiences. Keep in mind, however, that countries and their practices can and do change, so the comments herein are not to be viewed as dogmatic descriptions.

The comments are organized by region. South America, Australia, and a few other places were not described, basically because I have not conducted enough business there to know anything about their practices. I don't know much about agitator design practices in Antarctica, either.

North America

Vendors

Though there are literally dozens of agitator manufacturers in North America, probably 90% of the agitators are manufactured by the top five vendors. When it comes to agitators used for fermenters, probably 90% of sales come from the top three vendors, all of which are located in the USA. There are several reasons for this. One is that their products have a proven history of reliability. Another is that they traditionally have strong application engineering groups, with specialists in

Agitator Design for Gas–Liquid Fermenters and Bioreactors, First Edition. Gregory T. Benz.

specific applications such as fermentation. A third is their ability to support the products wherever the plant may be located.

Staff reductions in the major agitator manufacturers over the last 20 years or so have resulted in a considerable loss of tacit knowledge in key industry segments, including fermentation. The newer crop of engineers brings their own skills, such as high computer proficiency, but lack the tacit knowledge that is generally found after many years of experience, and mentorship has been cut off. This reduces their competitive advantage over the smaller suppliers. Over the last several years, I have seen several minor suppliers get orders for fermenter agitators. One of them supplied a 500+ kW bottom entry fermenter agitator even though they had been in business for less than 10 years. Time will tell how well they did on the design and manufacturing.

To this author's knowledge, none of the North American manufacturers presently has staff with a high degree of knowledge specific to fermentation.

In terms of market share, it has been estimated by internal studies that the top three US-based vendors account for about 90% of agitator sales in North America. For fermenter agitators, it is probably closer to 90% for the top two vendors.

Design Practices

Most North American agitator manufacturers do their own designs in-house. They do not normally design or supply the vessel, though some vessel fabricators build their own agitators. Agitators typically mount to a mounting nozzle supplied by the vessel manufacturer, but the agitator assembly includes a mounting flange attached to the pedestal (the transition piece between the flange and the gear drive). This flange is machined to accurately support and align the seal components, especially for mechanical seals. This practice is different from what was sometimes done in China in the latter part of the twentieth century, as we will see later.

Most North American vendors use a right-angle gear drive, with a horizontal motor. The belief is that such a design is easier when it comes to mounting and aligning the motor, particularly for large motors. Also, it requires less headroom above the vessel. However, auxiliary motor support is often required (never needed for vertical motors), and less space is available on the vessel head for nozzles. The use of vertical motors is increasing in North America, partly because the gear drive is less expensive.

Most vendors have extensive CAD (computer-aided design) capability as well as CAM (computer-aided manufacturing). Some of the smaller vendors may not.

Selling/Buying Practices

In general, North American agitator suppliers sell through a network of representatives or direct sales offices. Their contracts with those sales offices are exclusive; a buyer cannot bypass the office assigned to his territory. Sometimes the assigned territory is purely geographical. Sometimes it is sorted by industry segment. Sometimes it is both. The representatives are generally able to quote standard designs for routine applications. Special designs and special applications (which include fermenters) will usually be prepared by factory Application Engineers, but the final quote preparation, follow-up, and some of the service will be handled by the representative. A good representative is worth every penny of commission he earns.

Buyers of equipment in North America are accustomed to knowledgeable representatives and application engineers, so their specifications are often not very detailed. They describe what the agitator must do, but usually leave details such as power, shaft speed, and impeller system up to the manufacturer, as well as details such as bolt sizes, blade thicknesses, etc. For large fermenter projects, they may hire an engineering firm or a consultant to put more detail into the specification. That is how this consultant makes his living.

Most North American customers are willing to consider buying products made anywhere in the world, as long as they are of high quality and have adequate field support. That may be changing with current international tensions, however.

Degree of Vertical Integration

The largest suppliers in the USA are highly vertically integrated. In addition to doing their own fabrication and machining work on shafts, impellers, etc., they even manufacture their own gear drive, except in very large sizes. This gives them excellent control over the product. The smaller manufacturers will buy the gear drives from a well-known gear drive manufacturer, but usually, they fabricate everything else. A few outsource all manufacturing and just assemble and ship the units. This is not common for those who build fermenter agitators, however.

None of the major North American manufacturers manufacture their own motors or mechanical seals, though these items are part of their scope of supply in most cases.

Role of Design Firms

North American EPC firms are often involved in the bidding of large projects (though the largest fermentation installation in the world did not use any such firm.). They are typically hired to perform the following tasks.

- Prepare general mechanical specifications (this is often "boilerplate," or repetitive standard material)
- Define a list of acceptable bidders

- Prepare an agitator data sheet, defining what the agitator must do, as well as relevant details, sometimes including power, speed, impeller system, etc. If they do not have in-house expertise, they may hire an outside consultant to help them
- Assemble and compare bids
- Sometimes, they select the vendor, depending on what type of contract they have with their customer
- They sometimes provide installation, start-up, commissioning, and validation services.

R&D

The larger North American agitator companies, based in the USA, are active in mixing R&D. They regularly attend conferences devoted to mixing, and publish papers frequently. The amount of technical information they publish is surprising to some, as it seems to reveal proprietary information. The smaller companies tend to publish little and seem to rely on works published by academia for their technical knowledge. The universities publish a great deal, but some of their work is strange, such as detailed studies of solids suspension with a Rushton turbine, when no competent agitation engineer would use a Rushton turbine for solids suspension. Most of the agitation research from university students is concentrated in about half a dozen institutions, which have professors interested in the subject. Several large private companies have their own staff devoted to fluid agitation, and they contribute to the body of knowledge.

Culture

Though no one can describe a country or region's culture fully in a sentence or two, some characteristics stand out which affect business transactions. In North America, Canada, and the USA, people tend to have mid-way directness in communication. By that I mean they are direct about some things but less likely to be direct about negative comments toward individuals. They tend to be fairly egalitarian, which means they do not always accept authority and require managers to prove their case. They tend to prize individual achievement and are competitive. They are comfortable with new ideas. They are also less disciplined and more self-indulgent than most other regions. Of course, these stereotypes do not apply to all, but are intended to indicate average cultural behavior. More cultural resource links are available at the end of the chapter.

EU

Vendors

As in the USA, there are dozens of agitator manufacturers. Some are part of larger organizations that supply many other kinds of equipment. However, all but one are much smaller than the top three North American manufacturers. The one that is not is located in Germany, and it may very well be the largest agitator manufacturer in the world. It is the only one that has a significant presence outside the EU. Interestingly, it is the leading supplier to the Chinese market, in terms of economic value if not the number of units sold.

To my knowledge, none of the EU vendors have staff with high expertise in fermentation. The large German firm had such an individual until about a decade ago, but he passed away.

Design Practices

Little is known by this author about the design practices of most EU suppliers. None make their own gear drive, and several seem more comfortable with hollow shafts than their American counterparts. There seems to be a proliferation of specialized impeller types there. This is sold as an advantage. However, it takes a great deal of research to fully understand the process and mechanical attributes of any new impeller, so it is possible that some of these impellers are not well understood. It may be better to deeply understand a smaller number of impeller styles.

EU vendors are more apt to include elaborate lifting devices for mechanical seal change as standard. American vendors offer them as an option. Most North American customers seem content to devise their own lifting devices.

Selling/Buying Practices

EU vendors use both direct sales and representatives. The smaller ones use direct sales and have limited geographic coverage. None of the small ones have a significant presence in North America or Asia, for example. The large German firm has representatives all around the world. However, these representatives are, in general, not technically trained in fluid agitation and do not have sizing/pricing tools or sales data books. All proposals are prepared by application engineers in Germany, and all technical questions must be answered by them. The representatives basically find inquiries and take customers out to eat. In my opinion, this impedes communication, though it can be argued that the answers are more likely to be correct when they come from application engineers rather than sales reps.

In my experience, customers in the EU are somewhat more educated about agitation, including for fermentation, than their American counterparts. This may be due in part to the lack of information provided by the number 1 vendor. Or it may be because a major design institute doing research in agitation, BHR, is located in the UK. Or it may be because of the amount of R&D done by major universities in the field of agitation, especially gas–liquid contacting. Delft University of Technology, among others, has contributed greatly to our understanding of this topic. I have referenced the work of some of their alumni throughout this book.

Degree of Vertical Integration

The EU vendors are about as vertically integrated as their smaller North American counterparts. However, none are as vertically integrated as the top three United States-based vendors, in that they do not make their own gear drive, which is one of the most important parts of an agitator.

The top vendor in Germany, however, makes its own mechanical seals. This is highly unusual; it may even be unique. Most vendors buy seals from the major seal manufacturers. Some make their own seal housings; others buy the entire seal cartridge from the seal manufacturer.

Role of Design Firms

Mostly, their role is similar to that in North America. However, they are more apt to specify hardware details, including motor power, shaft speed, and impeller system. Many vendors take exception to this and quote their own design.

R&D

Based on publications in technical journals, the EU does a great deal of agitation research. Most of the published work is done at various universities. The main vendor publishes little, apparently preferring to keep its information confidential.

The problem with many university studies is that the researchers are not experienced in the field, and do a lot of re-inventing the wheel by publishing papers on Rushton turbines without adding much new information, and applying impellers in a way that is not useful in real applications (e.g., Rushton turbines used for solids suspension.). Oddly, few papers study concave radial impellers, even though the work that led to their development was mainly conducted at Delft Technical University and the University of Surrey. However, the sheer volume of papers coming out of EU universities is such that quite a bit of useful information is still being published. One just has to wade through a lot of papers to get at it.

Culture

The EU is composed of many countries, and each one has its own culture. On average, the EU is even more suspicious of authority than the USA. They respect authority if it has the expertise to back it up. They are highly individualistic but restrained and disciplined. They are generally quite direct in their communication. The term "brutally frank" comes to mind.

Japan

I was a little reluctant to write much about Japan, as my information is both old and rather limited. Therefore, take this as a simple guide with a grain of your favorite alkali metal halide.

Vendors

I am aware of an old market study that says there are no heavily dominant agitator manufacturers in Japan. In this study, it was estimated that the top five vendors combined accounted for about 60% of agitator sales in Japan. And, of course, that would make all of them smaller than the top US and German vendors. Only one seems to have any presence at all in the USA, and it is a very small presence in the area of portable agitators.

Design Practices

I have little information here. Based on inquiries from Japanese processing firms for plants in the USA, the end user often specifies details about impeller system, speed, and power. I have even seen them specify such details as blade thicknesses, bolt sizes, etc. This would seem to indicate a lack of technical leadership on the part of the agitator vendors.

Selling/Buying Practices

No information is available to the author, other than the seeming practice of end users to dictate a lot of design details. Also, the Japanese domestic buying practices almost never result in buying from a non-Japanese vendor. In my 25 years working for a major American agitator supplier, I cannot recall any units sold into Japan.

Degree of Vertical Integration

Based on limited observation, the vertical integration is low in most cases; similar to that of smaller vendors in the EU and North America.

Role of Design Firms

The role of engineering firms in Japan is similar to those in North America and the EU, with a tendency to specify things in even more detail. In contrast to the agitator vendors in Japan, which are small, several of the Japanese engineering firms are among the largest in the world. They bid on and execute projects all over the planet. They sometimes even own and license their own processes, and provide complete design, equipment, and commissioning services. At the risk of being repetitively redundant, here is a partial list of services they may offer, which mostly copies the North American list:

- Prepare general mechanical specifications (this is often "boilerplate," or repetitive standard material)
- Prepare detailed process and mechanical specifications where needed or expected by the client
- Define a list of acceptable bidders
- Prepare an agitator data sheet, defining what the agitator must do, as well as relevant details, usually including power, speed, impeller system, etc. If they do not have in-house expertise, they probably won't admit it, but may hire a consultant from within Japan
- Assemble and compare bids
- Usually, they select the vendor, depending on what type of contract they have with their customer
- They usually provide installation, start-up, commissioning, and validation services
- Sometimes, they own and license the process

R&D

Based on publications, relatively little agitation R&D comes out of Japan, even though much of the foundational work was done there in the 1940s by Professor Shinji Nagata. Most of the published R&D comes from various universities and engineering firms. The agitator vendors seem to publish very little.

Culture

Paradoxically, Japan is both extremely competitive and yet not very individualistic. They prize achievement more for the group they are a part of rather than individual achievement. They are moderately accepting of authority. They are pragmatic toward new ideas. They have a strong tendency to avoid risk and are restrained in their personal behavior. Like some other Asian cultures, they strive to avoid "losing face." Most Westerners do not understand this concept; it is not simply a matter of embarrassment. The shame felt after losing face can cripple a person's ability to function, and the individual may be shunned or not trusted for years.

China

As a country that has a very long history, China has seen many changes. These include changes in national boundaries, ruling classes, political, and economic systems, etc. Over the last 5000 years, China has had long periods of relative stability punctuated by periods of rapid change, often accompanied by violence and chaos. The twentieth century and the beginning of the twenty-first century have been one of those periods of rapid change. Since the 1940s, we have seen China go from an empire resembling a feudal state to a Communist regime to a solidly Capitalist economy ruled by a one-party system that calls itself Communist in spite of the Capitalist economic system.

On a smaller scale, industrial practices have changed in China as well and are continuing to change. Thus, what I describe here is approximately the way things have been from about 1990 to 2020, but things are continuing to change. I will try to so note where applicable.

Vendors

Most of China's vendors are small, and there are many of them. Some are basically machine and fabrication shops owned by the end user. Others may also be vessel fabricators. The largest vendor appears to be one that is an outgrowth of a reducer manufacturing company, named after China's number one tourist attraction.

Two of the top three USA agitator vendors have set up operations in China, including manufacturing. Their presence has affected the design practices and selling/buying practices, as we will see later.

The top German manufacturer also has facilities in China, which is their number one market. Small items are made in China; large equipment is made in Germany. The German company tends to focus on polymers and polymer intermediates rather than fermentation.

Design Practices

When I started my many sojourns into China in the 1990s, agitator manufacturers in China had very little in-house technology. The normal practice was for state-owned design institutes (similar to engineering firms) to specify everything, including detailed manufacturing drawings of every part. In essence, the agitator companies were really build-to-print fabrication shops. Many of them bought the gear drives from what is now the number one domestic agitator manufacturer in China, which is still the only one making its own gear drives. Now that company has a division which is at least partially USA-owned.

After China opened to doing business with the west, western agitator manufacturers started doing business there. The practice of doing their own designs in-house and having application expertise gave them a competitive advantage over the Chinese manufacturers, in spite of the higher cost. Chinese manufacturers slowly adapted, and most of them now have in-house design and application expertise. However, many of them limit their sales to very small industry segments. The large Western manufacturers generally serve all markets.

In the early 1990s, some agitators in China had the mechanical seal flange welded directly into the vessel head, with the pedestal connecting to the gear drive bolted directly to it (Figure 21.1a). The western practice has long been for the agitator supplier to include a mounting flange, machined for the seal mount, in its scope of supply, designed to be mounted to a nozzle or pad supplied by the vessel manufacturer (Figure 21.1b).

Probably due to Western influence, most Chinese installations today have the mounting flange supplied as part of the agitator rather than part of the vessel.

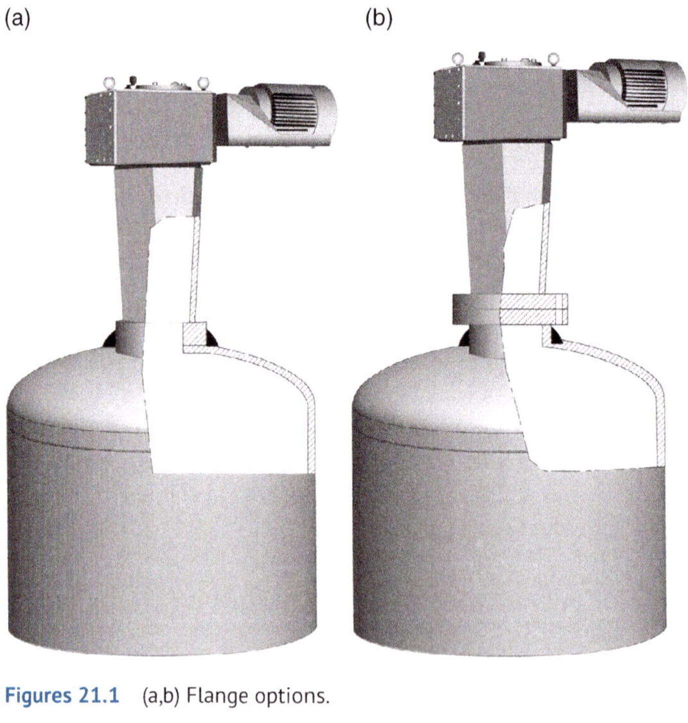

Figures 21.1 (a,b) Flange options.

However, the older Chinese style may save some capital cost, and it is more sanitary, as it eliminates a gasketed connection. The main problem is splitting the manufacturing up of a critical component between different vendors. If a problem occurs, there is a lot of finger pointing that will happen. Most agree it is better to have unit responsibility for this part and the seal assembly.

Due to lower labor costs in China, it is more common to find construction methods that save material but use more labor than Western designs. Examples include hollow shafts when not needed for critical speed, and alloy-clad carbon steel.

Up until the last decade or so, many Chinese-built agitators were basically direct copies of Western designs. I have seen several Chinese-made copies of the Chemineer BT-6 and similar impellers, for example. I have also seen cases of Chinese companies copying images from Western company literature and presenting it as their own product. It is becoming more common these days for Chinese companies to create their own designs.

Selling/Buying Practices

The smaller vendors that focus on specific industries are often local to their customers, and use direct selling. The larger companies use a system of representatives and direct sales offices. The two American companies doing business there have local application engineers as well as trained sales engineers. The large German company still does all of their quoting from Germany, but has a few local people with knowledge of agitation technology.

Traditionally, project engineers and owners in China have preferred to buy foreign brands if their budget would permit. This is because they perceived the foreign brands as having better technology and quality. Today they are somewhat more likely to buy local products. The global major manufacturers have responded by having a portion of their manufacturing done in China, so the price gap has lessened.

Not all agitator buying decisions are made by Chinese companies. Many foreign based companies who build plants in China, whether as a joint venture or wholly foreign-owned, prefer to choose their own vendors. Some choose the vendors that have served them well in their native countries. Others try to be a good corporate citizen in China by trying to buy from Chinese vendors when possible.

Copying designs or outright theft of IP is still a problem. As an example, I once designed fermenter agitators for a Chinese company when I was working for Chemineer. They planned to buy 10 eventually, so we wanted to be competitive. After negotiating the price down, they informed us they would buy only two. When pressed, they said they intended to copy our design after they received the unit and have a local fabricator build the other eight, and that is what they did. (They could not copy our gear drive, of course, and they had a lot of trouble with the gear drives they bought). Of course, we could have turned down the order. But we figured that it would be better to take this order at a small margin than to hand an installation to our major US-based competitor.

Degree of Vertical Integration

None of the domestic vendors other than the one owned by a reducer manufacturer is highly vertically integrated. They basically fabricate shafts, impellers, and pedestals, and buy their motors, gear drives, and seals. Of course, this degree of vertical integration is pretty much the norm globally.

Role of Design Firms

This is an area that has changed greatly in the last 20 years. Whereas once they basically designed agitators in great detail, now they mostly fulfill the same roles as engineering firms across the world. In that regard, they are similar to North American engineering firms. Mostly, they do work in China for Chinese customers, especially those involved in fermentation. The exception is engineering firms that focus on the petroleum and petrochemical industries. China has several that do business all around the world. Most of these are state-owned enterprises. Also, major engineering firms from all around the world have set up offices in China. Partly, it is to serve Chinese customers. Often, it is to serve companies that have headquarters in another country but have built or are building plants in China.

R&D

Judging once again by published articles, the agitator vendors in China do not seem to be doing much R&D work. Many are still copying the designs and technologies of others when they can. The largest Chinese domestic vendor has been touted as having a massive R&D program, with many PhD engineers. That may be so, but it is not evident from publications in international journals. I have yet to see any.

The visible agitation R&D in China has been carried out mostly by university students. There has been quite a lot of it, and most of it is of high quality. There is still too much published on Rushton turbines in ways that do not conform to good industrial practice. There is also quite a bit published by Chinese students studying at overseas universities, and many bring the product of their R&D back to China, though some stay overseas in their new host country. I have personally worked with a number of Chinese graduate students in the field of fermentation while they were interns, and later as they became full-time employees of multinational companies. In general, they have learned the technology quite well and are contributing to the body of knowledge.

Culture

China is a rather authoritarian society. People accept that there are those who have power over them, whether in government or industry. They tend to be collectivist in their thinking. They are fairly competitive and will sacrifice a lot personally to get ahead, though they care for the group they are part of. They tend to exercise personal self-control. And this stereotype is one of the things which seem to be changing due to Western influence. In fact, it seems that one of the goals of traditional Chinese leaders is to stop the invasion of Western ideas, leading to censorship of the Internet, among other things. Communication is less direct than in most Western cultures, though I have met some Chinese people who are very direct. Losing face is still an important consideration, but it has become less important over the years.

Summary of Chapter

This chapter is far from a complete treatise on how agitators are designed, bought, and sold in different regions of the world. It is hoped that this serves as a brief introduction to such differences. If I could list one main takeaway from this chapter, it would be that there are real differences, and those who seek to do business internationally need to educate themselves about those differences to avoid operational and cultural pitfalls. And, since cultures and practices change, one has to keep on top of the changes over the long haul.

Cultural Resources

Those who wish to study cultural differences along six social dimensions may find these sites interesting:

https://www.hofstede-insights.com/country-comparison
https://hbr.org/2014/04/a-tool-that-maps-out-cultural-differences

Afterword

This book began as a suggestion by the editors of Chemical Engineering Progress magazine, due to the many articles I have published there. My first reaction was that I did not have time for such a large undertaking, so I turned down the idea. I also was concerned that there might not be much of an audience for another book on agitation, albeit a specialized one. But after about 10 years of occasionally thinking about it, I decided in the spring of 2019 to see if any publisher would be interested. I outlined a draft proposal and sent it off to Wiley Publishing. Lo and behold, they accepted it! My first reaction was, Oh #$@*, now I actually have to write it! They allowed me two years to write it, and I completed the draft in one year, much to my surprise. Now that is it done, I hope you find it useful and worthwhile. Thank you for reading it!

Agitator Design for Gas–Liquid Fermenters and Bioreactors, First Edition. Gregory T. Benz.

Index

a

b

Agitator Design for Gas–Liquid Fermenters and Bioreactors, First Edition. Gregory T. Benz.

Printed and bound by CPI Group (UK) Ltd, Croydon, CR0 4YY

17/09/2026

14964320-0001